Lecture Notes in Business Information Processing 554

LNBIP reports state-of-the-art results in areas related to business information systems and industrial application software development – timely, at a high level, and in both printed and electronic form.

The type of material published includes

- Proceedings (published in time for the respective event)
- Postproceedings (consisting of thoroughly revised and/or extended final papers)
- Other edited monographs (such as, for example, project reports or invited volumes)
- Tutorials (coherently integrated collections of lectures given at advanced courses, seminars, schools, etc.)
- Award-winning or exceptional theses

LNBIP is abstracted/indexed in DBLP, EI and Scopus. LNBIP volumes are also submitted for the inclusion in ISI Proceedings.

Krzysztof Węcel
Editor

Business Information Systems

25th International Conference, BIS 2025
Poznań, Poland, June 25–27, 2025
Proceedings

Springer

Editor
Krzysztof Węcel
Poznań University of Economics and Business
Poznań, Poland

ISSN 1865-1348 ISSN 1865-1356 (electronic)
Lecture Notes in Business Information Processing
ISBN 978-3-031-94192-4 ISBN 978-3-031-94193-1 (eBook)
https://doi.org/10.1007/978-3-031-94193-1

This Springer imprint is published by the registered company Springer Nature Switzerland AG
The registered company address is: Gewerbestrasse 11, 6330 Cham, Switzerland

Preface

The 25th edition of the International Conference on Business Information Systems was held in Poznań, Poland. Since the first BIS edition in 1997, it has served as a platform for researchers and business practitioners to discuss up-to-date issues that include the development, implementation, and application of business information systems.

This year's theme of the conference was "AI-driven business transformation: challenges and opportunities." Since the last time the conference took place, we have observed an explosion in the development of artificial intelligence (AI), both in image and text processing, with a particular focus on generative AI. Artificial intelligence has rapidly transitioned into business applications and now underpins many services, disrupting traditional industries and challenging established business models. While the development of core AI technologies is largely driven by big tech companies, researchers continue to explore new applications, products, and research methods that leverage AI.

The Program Committee consisted of 57 members, and papers were reviewed by a total of 65 scientists. Based on their feedback, a set of 19 long and four short papers were selected from a total of 54 submissions, further grouped into seven sessions. In alignment with this year's conference theme, a significant number of contributions were dedicated to the field of artificial intelligence. This reflects the increase in research activity in the area, as well as the growing anticipation of its transformative impact on business practices. The presented works emphasize a strong orientation toward real-world applicability, with researchers offering innovative solutions capable of influencing practice.

The proceedings opened with a session on *Artificial Intelligence in Management*, which explored various frameworks for human-AI collaboration and examined how AI technologies are being integrated into established business functions, such as recruitment and decision-making processes. The second session, *Artificial Intelligence for Visualization*, highlighted the emerging potential of generative AI to automate the creation of business data visualizations, an area expected to shape the future of business intelligence and user interaction with data. The third session returned to one of the foundational themes of the BIS conference series: natural language processing (NLP) with AI. Here, large language models are leveraged across diverse applications including digital consultancy, sentiment analysis, the construction of patent claims, and the analysis of political narratives. The fourth session revisited business processes, a domain historically prominent at BIS, but with a new incarnation as *Process Mining*. The presented papers capture renewed interest in automation through innovations in process efficiency, process re-engineering, and increasing adoption through trust. This session can be a signal towards deeper integration of AI into core business processes, laying the groundwork for excellence in digital transformation. Addressing continued interest in fintech, the fifth session was devoted to *Finance and Privacy*. While cryptocurrencies have received considerable attention in past research, key challenges such as security and compliance remain underexplored. The session also touched on the transparency and explainability of machine learning models, a topic of increasing relevance in regulated environments. The sixth session, *Data Platforms and Deployment*, covered advances in infrastructure and implementation. Contributions here discuss architectures such as data lakehouses, continuous deployment methodologies, and the evolving role of agile IT departments in enabling digital transformation. Lastly, the seventh session, *Sustainable Operations and Smart Automation*, reflected the growing concern of our community with environmental responsibility. The presented works addressed pressing issues related to sustainability, such as greenhouse gas emissions and water resource management, demonstrating how digital technologies can support greener business operations.

We are confident that the proceedings offer valuable insights for both researchers and practitioners, and we hope they inspire further exploration, collaboration, and innovation at the intersection of business and information systems.

June 2025 Krzysztof Węcel

Organization

Program Committee Chair

Węcel, Krzysztof	Poznań University of Economics and Business, Poland

Program Committee Members

Apostolou, Dimitris	University of Piraeus, Greece
Arndt, Timothy	Cleveland State University, USA
Avila, Oscar	UniAndes, Colombia
Babar, Abdul	Western Sydney University, Australia
Bajaj, Akhilesh	University of Tulsa, USA
Benyoucef, Morad	University of Ottawa, Canada
Celebi, Remzi	Maastricht University, The Netherlands
Charoy, François	Université de Lorraine, FranceCorchuelo, RafaelUniversity of Seville, Spain
Dascalu, Sergiu	University of Nevada, Reno, USA
Debruyne, Christophe	Université de Liège, Belgium
Embury, Suzanne	University of Manchester, UK
Filipowska, Agata	Poznań University of Economics and Business, Poland
Fiore, Ugo	University of Naples Federico II, Italy
Florea, Adrian	Lucian Blaga University of Sibiu, Romania
Fodor, Paul	Stony Brook University, USA
Govindarajan, Vijay	Expedia Group Inc, USA
Hartig, Olaf	Linköping University, Sweden
Hernes, Marcin	Wrocław University of Economics and Business, Poland
Jelonek, Dorota	Częstochowa University of Technology, Poland
Klischewski, Ralf	German University in Cairo, Egypt
Koschmider, Agnes	University of Bayreuth, Germany
Kriksciuniene, Dalia	Vilnius University, Lithuania
Lei, Jeff	University of Texas at Arlington, USA
Lewańska, Elżbieta	Poznań University of Economics and Business, Poland

Lewoniewski, Włodzimierz	Poznań University of Economics and Business, Poland
Li, Julia	University of Massachusetts Dartmouth, USA
Marrella, Andrea	Sapienza University of Rome, Italy
Marx, Edgard	Leipzig University of Applied Sciences, Germany
Matulevičius, Raimundas	University of Tartu, Estonia
Oberweis, Andreas	Karlsruhe Institute of Technology, Germany
Oliveira, Toacy	Instituto Piaget, Insight Research Center, Portugal
Pastuszak, Zbigniew	Maria Curie-Skłodowska University, Poland
Pegoraro, Marco	RWTH Aachen University, Germany
Pröll, Birgit	Johannes Kepler University Linz, Austria
Pulvermüller, Elke	University of Osnabrück, Germany
Romero, David	Tecnológico de Monterrey, Mexico
Sakalauskas, Virgilijus	Vilnius University, Lithuania
Sandkuhl, Kurt	Rostock University, Germany
Smirnov, Alexander	SPIIRAS, Russia
Smolnik, Stefan	University of Hagen, Germany
Stróżyna, Milena	Poznań University of Economics and Business, Poland
Verjus, Herve	Université Savoie Mont Blanc, France
Weigand, Hans	Tilburg University, The Netherlands
Ziemba, Ewa Wanda	University of Economics in Katowice, Poland

Additional Reviewers

Acitelli, Giacomo
Casciani, Angelo
Fonger, Frederik
Golo, Marcos
Le, Vinh
Wang, Yi
Xiao, Lu
Zhang, Lin
Zifla, Ermira
do Carmo, Paulo

Organizing Committe

Lewańska, Elżbieta	Poznań University of Economics and Business, Poland
Lewoniewski, Włodzimierz	Poznań University of Economics and Business, Poland
Stróżyna, Milena	Poznań University of Economics and Business, Poland

Contents

Data Platforms and Deployment

Sustainable Operations and Smart Automation

Artificial Intelligence in Management

Challenges in Managing the Relationship Between Agentic AI Systems and Humans in Organizations

Christopher Wissuchek[1(✉)] and Patrick Zschech[2]

[1] Universität Leipzig, Grimmaische Str. 12, 04109 Leipzig, Germany
christopher.wissuchek@uni-leipzig.de
[2] TU Dresden, Helmholtzstr. 10, 01069 Dresden, Germany
patrich.zschech@tu-dresden.de

Abstract. Advancements in AI, particularly generative models, have facilitated the rise of agentic AI systems, which operate with increasing autonomy, make independent decisions, adapt to dynamic contexts, and proactively influence outcomes. These systems bring distinct considerations and challenges to organizational contexts, distinguishing them from traditional, passive AI systems. This paper examines these unique challenges using a three-pronged qualitative methodology comprising a systematic literature review, expert interviews, and an analysis of professional publications. Through qualitative coding, we delve into the dimensions of agentic AI challenges, focusing on the complexities of their integration into organizations. Our findings reveal challenges in the core human-agentic AI relationship as well as exogenous and internal organizational factors. Our findings offer valuable insights into these complexities and establish a foundation for actionable strategies in future research.

Keywords: AI Agents · Agentic AI · Generative AI

1 Introduction

Recent technological advancements, such as in computing power, storage, and algorithmic capabilities, alongside the exponential growth in data generation, have propelled AI into a general-purpose technology. AI systems harness advanced computational, analytical, and quantitative capabilities, supporting humans with tasks formerly primarily associated with requiring human intelligence [1, 2]. AI's conceptual scope continues to evolve in response to ongoing technological advancements. AI systems have been primarily understood as analytical tools optimized for data-driven decision-making and tailored to discrete and structured application scenarios [3].

Generative AI has recently marked a significant paradigm shift, expanding AI's capabilities [4]. In contrast to AI systems employing discriminative models, which learn the decision boundary between predefined classes, generative AI systems leverage multiple modeling principles, such as generative modeling or reinforcement learning, to create new content, simulate scenarios, and adapt dynamically across varied contexts [3].

K. Węcel (Ed.): BIS 2025, LNBIP 554, pp. 3–17, 2025.
https://doi.org/10.1007/978-3-031-94193-1_1

This evolution broadens AI's applicability, particularly in knowledge-intensive domains where tasks require high adaptability, creativity, or contextual understanding. In particular, large language models have become dominant within generative AI, employing vast datasets to perform complex language processing tasks [4, 5].

As an extension of these recent developments, there has been a shift from AI functioning predominantly as a passive tool to taking on more agentic roles [6]. Traditionally, AI systems required human initiation for clearly defined tasks, providing outputs in response to explicit prompts. In contrast, agentic AI can act with less human intervention, for example, initiating actions, adapting dynamically, and setting and pursuing cascading goals [5, 7]. These advancements have created new opportunities, solidifying AI as a strategic priority and ubiquitizing it on a broader scale [3, 4, 8]. This is marked by a significant market transition, with the release of various agentic AI systems, their integration into existing IT ecosystems, or even enabling users to build their own "no-code" agents [e.g., 9, 10]. Additionally, foundation models trained on extensive datasets and applicable across different use cases are now widely accessible, open-sourced, or commercially, allowing organizations to extend them to address unique scenarios [8].

However, implementing and managing AI systems introduces numerous challenges that organizations must carefully navigate to effectively integrate them into business processes and unlock their full potential [11]. Existing research examines AI challenges across various perspectives, such as technical (e.g., data quality, model complexities), organizational (e.g., management support, resource allocation), and environmental (e.g., privacy). However, these studies largely remain general [e.g., 12, 13], not discussing the unique complexities introduced by agentic AI systems – particularly their autonomy and evolving relationship dynamics with human users. While related research has begun exploring agentic AI-specific challenges [e.g., 5, 7, 14, 15], the critical aspect of bi-directional interactions between humans and agentic AI systems remains underexplored. We argue, that recognizing and addressing these novel tensions is essential for organizations that leverage agentic AI effectively. Therefore, we pose the following research question to guide our study: *What challenges impede the relationship between agentic AI systems and humans in organizations?*

We employed a three-pronged qualitative approach to address the research question, synthesizing insights from academic literature, expert interviews, and professional literature. The data were analyzed through structured coding, revealing the challenges of employing agentic AI systems within organizational contexts. This paper is organized as follows: In Sect. 2, we establish the theoretical foundations of AI, agency, and delegation; in Sect. 3, we detail the research method; in Sect. 4, we present the findings of the qualitative study; and finally, in Sect. 5, we discuss limitations and outline directions for future research.

2 Background

2.1 AI, Agency, and Delegation

We clarify our understanding of AI as a general-purpose technology that tackles existing tasks and enables the exploration of new avenues by taking over activities formerly associated primarily with human intelligence, such as decision-making, problem-solving, reasoning, acting, generating, perceiving, or even exhibiting creativity [1]. In this sense, AI systems have the purpose of performing tasks that were limited to humans in the past [2]. For this, an AI system integrates multiple components, such as an underlying machine learning model and infrastructural components. It is potentially nested in a broader IT ecosystem with user-facing interfaces to achieve its intended purpose by addressing stakeholders' needs and task requirements across organizational structures [16, 17]. Against this backdrop, we follow [18] to establish the term "human" as the individual interacting with an AI system already deployed in a productive state instead of individuals involved in its creation or configuration. Nevertheless, human experts conceptualize and develop agentic AI systems, meaning their design and behavior originate from human decisions.

Given that AI systems are endowed with intelligent capabilities, they can potentially assume the human attribute of agency [6, 19], which is the capacity to make independent choices, take actions, and exert control over the environment. Agency encompasses the ability to set objectives, make decisions aligned with one's intentions, and act on these to influence outcomes [20]. It implies autonomy and intentionality, positioning the individual as an active participant rather than a passive recipient of external forces [21]. Contrary to the traditional view of IS as passive tools operated by humans [18], AI systems can occupy various points along the spectrum of agency. For instance, analytic AI remains passive, revealing information to human users only upon request, leaving them to decide on subsequent actions [22]. In contrast, GitHub Copilot exemplifies exhibiting agency, as it generates code suggestions in real-time by observing human inputs and changes in the development environment [23].

The emergence of agency in AI systems signifies a change in the underlying human-IT relationship, necessitating a re-evaluation of the superiority traditionally assigned to human agency [3, 18]. In this context, task delegation offers a theoretical lens for examining the relationship between humans and agentic AI systems. Delegation, defined as "transferring rights and responsibilities for task execution and outcomes to another," [18] typically arises when one party lacks the resources, capabilities, or knowledge to achieve their goals independently or wants to offload tasks to another to focus on higher value activities. By delegating tasks appropriately, organizations and individuals can benefit from enhanced efficiency and effectiveness in task execution [24]. Although this work focuses specifically on single-agent interactions, it is important to recognize collaborative multi-agent systems – interactions involving multiple human and artificial agents – as another developing area of research [15].

In AI, research on delegation has focused predominantly on human-to-AI delegation, where humans are regarded as superior, retaining primary control over decision-making [14]. However, the discussion has recently shifted to include AI-to-human delegation, where AI systems can assume a superordinate role, initiating actions autonomously and assigning tasks to human counterparts [14, 25]. These delegation trajectories are typically viewed mostly as one-directional. However, as noted in the introduction, the advent of generative AI has been a critical driver in the proliferation of agentic AI systems, which adopt a more bi-directional, continuous, and even co-creative approach [4].

2.2 Conceptualizing Agentic AI Systems

To our knowledge, there is currently no agreed-upon conceptualization of what constitutes an agentic AI system. The terminology surrounding this concept remains ambiguously defined. From an IS-theoretical perspective, we conceptualize agentic AI systems as socio-technical artifacts designed by humans and endowed with autonomous capabilities, challenging traditional assumptions of passive technology operated solely by humans [18]. Our ontological stance emphasizes that these systems embody varying degrees of agency, actively participating in organizational processes, shaping interactions, and redefining the human-technology relationship. The design and use of these autonomous capabilities are fundamentally driven by human choices, values, and goals, linking human agency to AI agency.

For example, [18] conceptualizes agentic IS artifacts along a continuum of decision-making latitude, from elementary tasks (i.e., narrow agency) to full task completion with autonomy (i.e., expansive agency). With a focus on generative AI, [4] highlights the mode of interaction. AI can be employed for discrete or simple tasks, where users initiate the AI system to generate an output, referred to as transactional use (i.e., non-agentic). Alternatively, AI can increasingly assume a continuous, proactive, or even autonomous role, engaging in ongoing collaboration and potentially initiating interactions to work toward expansive, not necessarily predefined, objectives adapting to dynamic contexts (i.e., agentic). [5] define agentic AI as "the degree to which a system can adaptably achieve complex goals in complex environments with limited direct supervision." Rather than conceptualizing a spectrum, they identify attributes: the pursuit of complex goals (versus simple), operation within dynamic environments (versus static), adaptability to unexpected circumstances (versus predefined), and independent execution with minimal human intervention (versus explicit human initiation). Similarly, [7] describe dimensions of agentic AI by highlighting underspecification, which refers to the extent to which an AI system can achieve a given goal – set by its designers – without detailed instructions on how the goal should be accomplished. Another dimension they propose is the directness of impact, described as the degree to which the system's actions affect the environment without human mediation or intervention. Finally, they identify goal-directedness, which captures the system's capacity for long-term planning, indicating its ability to make cascading decisions to achieve objectives over extended time horizons.

In summary, for the purpose of the research objective, these perspectives contextualize agency in AI, guiding our research in uncovering challenges in organizational settings. Not all of these factors must be present to consider AI agentic, but they add to its overall expansiveness or level of agency.

3 Research Method

We adopted a three-pronged qualitative approach to investigate challenges between agentic AI systems and humans in organizational contexts, integrating data from academic research, expert interviews, and professional literature. This triangulation aimed to enhance the practical relevance of our findings and ensure alignment with the latest advancements in the field. The data set comprises 47 academic articles, six expert interviews, and six professional or applied research publications.

To establish the academic literature corpus, we followed [26] for SLR, using the search string: *("artificial intelligence" OR "AI") AND ("delegat*" OR "agency" OR "agentic")*. The first part of this query captures the broad scope of AI, while the second narrows the focus to key AI agency and delegation topics. During the validation process, we excluded the term "agent" due to its ambiguity and frequent use as an umbrella term that does not necessarily address AI's agentic nature. Also, we decided not to include the term "human" to encompass research focusing on the inherent challenges stemming from the nature of agentic AI systems. We further refined our search through a title filter to ensure alignment with our core focus. Searches were conducted across the AISeL, IEEE Xplore, Scopus, and Web of Science databases, encompassing perspectives from IS, computer science, and a broader interdisciplinary scope. This process yielded 157 articles. After eliminating duplicates (–32) and filtering based on title (–30), abstract (–33), and full-text analysis (–23), with additional articles identified through citation chaining (+8), we arrived at a final set of 47 relevant papers. We excluded articles that did not explore challenges or conflicts within agentic AI systems or examine delegation mechanisms linking AI and humans in organizational contexts.

Table 1. Overview of the conducted expert interviews

ID	Role/title	Expertise	Organization	Duration
E1	Data Engineer	Early adopter and experienced in using an agentic AI for coding and organizing knowledge	Multinational retailer	48 min
E2	Product Owner	Leading prototyping of an agentic AI system for organizational use	Enterprise software	47 min
E3	Account Manager	Managing executive discussions and business cases; AI community lead	Technology	72 min
E4	Technical Project Lead	An early adopter and experienced in using an agentic AI to organize knowledge in large-scale software engineering projects	IT Consulting	42 min
E5	AI Product Owner	Research and practical experience with AI/ML operations and platforms	IT Consulting	55 min
E6	Researcher	AI research with a focus on managerial and organizational factors	University	74 min

As a second data source, we applied qualitative semi-structured interviews with experts, enabling focused yet adaptable discussions that allowed participants to reflect on their experiences [27]. The objective was to assemble a balanced panel of experts representing a range of perspectives, including early adopters, researchers, and professionals actively involved in advancements in agentic AI (see Table 1). We identified the participants through our professional network and LinkedIn. The process began with four initial interviews, during which respondent saturation and sample adequacy were assessed. This initial sample size enabled a focused and targeted exploration of the research objective. Subsequently, two additional interviews were conducted to ensure sufficient depth and comprehensiveness in the insights gathered [28]. The interviews were structured in three parts: (1) participants shared their previous experiences to establish a mutual understanding of key concepts such as AI, agency, and task delegation. (2) we used an open-question format to investigate the challenges of collaborating with agentic AI systems in their unique organizational contexts. (3) For the third phase, experts were invited to reflect and collaboratively generate an accurate summary containing first-order data elements for subsequent coding. An AI-based transcription and summarization tool was employed to facilitate this. In collaboration with the interviewees, the AI tool was queried to produce a summary, recap discussion points, and identify any potential omissions. Once the experts confirmed the accuracy and completeness of the summary, it was incorporated as first-order data elements.

Finally, we added professional literature and publications as a third data source to enhance practical relevance and timeliness. Following established guidelines for including grey literature in qualitative research, we assessed each source based on factors such as the producer's authority, objectivity, or outlet type [29], deciding to include it in the sample. Our selection included documents from reputable sources, such as consulting firms (e.g., PwC) and providers of agentic AI systems (e.g., OpenAI), as well as articles from their applied research arms (e.g., Microsoft Research).

To structure the data, we followed established guidelines for qualitative analysis [30] to ensure rigor in our inductive exploration of challenges in the relationship between humans and agentic AI systems. This approach is particularly suitable for systematically processing unstructured qualitative data to uncover relevant categories and relationships. We used a three-step coding process to examine our phenomenon across different levels of abstraction [26, 30]. First, we applied open coding to the academic articles, professional literature, and interview summaries, extracting critical information that aligned with our research objectives of revealing challenges, totaling 290 first-order data elements. In the second step, axial coding, we identified relationships within these data elements, consolidating them into second-order challenges. Finally, we organized these into five overarching aggregate groups in the selective coding phase. With this final data structure, we developed a framework to illustrate the interrelationships among these groups. Results were repeatedly re-evaluated throughout the process by continuously linking and comparing data elements with their underlying statements. This iterative approach allowed us to consolidate findings, eliminate or reframe challenges, and refine nomenclature as needed. It involved moving back and forth between second-order challenges, aggregate groups, and their interrelationships.

4 Results

Here, we introduce the resulting framework with five aggregate groups, beginning with challenges inherent to humans and agentic AI systems. These challenges create core relational tensions, particularly in task delegation, resulting in suboptimal outcomes that fail to meet organizational expectations and reduce potential business value. Beyond relational challenges, organizations struggle to position agentic AI systems. Exogenous challenges – factors outside the organization's control – must be addressed in managing this evolving relationship. The resulting five aggregate groups and their relationships are visualized and summarized in Fig. 1.

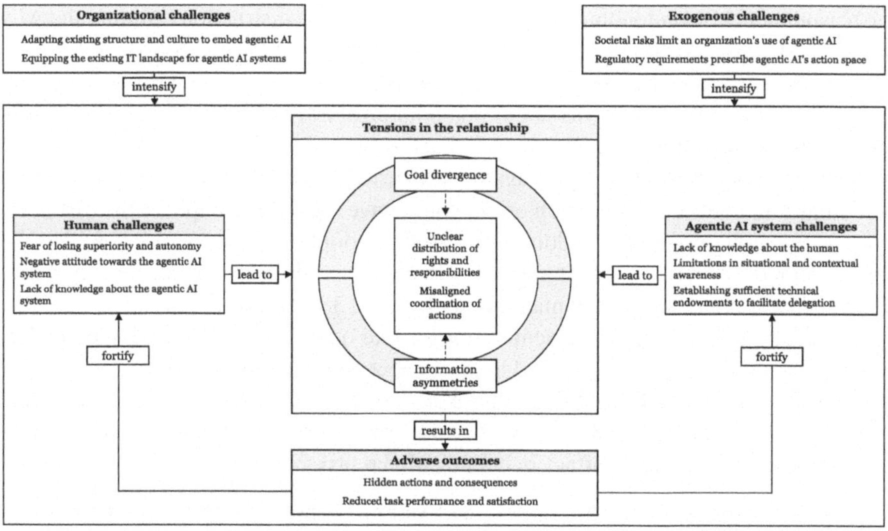

Fig. 1. Overview of challenges in the relationship between humans and agentic AI systems

4.1 Human Challenges

Fear of Losing Superiority and Autonomy. The results of the analysis reveal a potential fear among human users regarding perceived losses of superiority and autonomy when interacting with agentic AI. Such concerns are particularly significant given the shift from passive AI tools to agentic systems capable of autonomous decision-making and action. These apprehensions can stem from various factors: the potential erosion of individuals' self-perception as experts, a perceived reduction in professional purpose, the displacement of tasks they find fulfilling, and a general reluctance to defer to non-human entities in decision-making processes [31, 32].

These perceived threats to human agency may result in resistance behaviors, posing obstacles to effective AI integration. For example, users may obstruct implementation processes, withhold expert knowledge necessary for agent design due to fears of redundancy [E2], or even attempt to undermine or manipulate the agentic AI system [E1]. In light of these dynamics, it may be challenging for organizations to implement mechanisms that allow human agents to override or adjust the actions of agentic AI systems, which may be critical for fostering acceptance and minimizing the resistance of humans [E3, 14] by supporting them to maintain a sense of control and agency.

Negative Attitude Towards the Agentic AI System. Beyond the inherent or intrinsic fear of losing control, humans may also harbor negative attitudes toward agentic AI systems, which can significantly impact their willingness to collaborate. The perceived opaqueness of AI decision-making often exacerbates this reluctance, as humans may struggle to understand the underlying rationale behind autonomous actions. The lack of transparency in AI's decision logic and autonomous instructions can generate mistrust and hesitation, hindering cooperation [E5, 33–36]. Giving users greater visibility into the reasoning behind autonomous actions and access to the instructions guiding the AI's decisions [E1] is critical to fostering trust and encouraging collaboration.

However, transparency alone is insufficient when we move from passive to agentic AI. Addressing human concerns about potential errors in autonomous actions is challenging for organizations. Humans may feel uneasy about the potential consequences of erroneous actions performed by an agentic AI system without direct human oversight, particularly in contexts where such errors could have significant implications [E2, 37]. Privacy concerns represent an additional barrier to adoption, as users may fear that the data collected by these systems to operate effectively could compromise their personal information or organizational confidentiality [E4, 36, 38]. Lastly, humans may perceive agentic AI systems as lacking essential human-like qualities, such as empathy, understanding, or intuition, which can lead to disconnect and reluctance to fully engage with or rely on these systems [33]. Without these perceived human attributes, users may view the AI system as fundamentally alien to their way of working, reinforcing that the system is an outsider rather than a partner in organizational tasks.

Lack of Knowledge about the Agentic AI System. A potential challenge for humans is a lack of foundational knowledge about agentic AI systems, even with a willingness to cooperate. Humans must understand how the agentic AI system functions and how to operate it while maintaining realistic expectations regarding its limitations, as these systems are not infallible [E2, E3, E5]. A lack of this understanding can lead to inflated expectations, with users, as E1 noted, perceiving the agentic AI system as a "magic solution" rather than approaching it with a balanced skepticism. This knowledge gap can hinder effective interaction with these systems in various ways; for instance, users may struggle to engineer appropriate prompts or understand how to effectively incorporate agentic AI into their individual use case [E6, 31, 37]. Additionally, humans must discern when and how to delegate tasks to the AI system, as it may not be suitable for all types of work [5]. This requires humans to critically assess the AI's capabilities and their own, recognizing when human intervention is preferable [E3, 39].

4.2 Challenges of the Agentic AI System

Lack of Knowledge about the Human. While much of the research has traditionally focused on the human perspective, the opposite viewpoint – the agentic AI system's understanding of the human– remains comparatively underexplored [14]. We argue that this reciprocal perception is essential, particularly in bi-directional delegation relationships where the AI system must determine when to delegate tasks back to the human and accurately infer the human's intentions upon receiving instructions. While collaboration between AI systems and human agents can enhance task performance through complementary strengths, this synergy does not universally apply. Sometimes, the AI system or the human performs better independently, making joint efforts inefficient or counterproductive [39, 40].

In such situations, agentic AI systems must have the capacity to assess when collaboration is beneficial versus when solo performance is optimal. Agentic AI systems require access to rich information about human preferences, strengths, weaknesses, and specific abilities to make these assessments [14]. This knowledge enables agentic AI systems to make more informed decisions, predict likely outcomes, and leverage complementarities effectively. For example, an agentic AI system could analyze patterns in a human's past decisions to gauge areas where the human is skilled or where they might require additional support. The AI system can dynamically adjust its role by refining its understanding of human behavior and balancing autonomous actions with delegated responsibilities [E4, 14]. Ultimately, for bi-directional delegation to function effectively, agentic AI systems must be able to "learn" and adapt based on the unique characteristics and preferences of the humans they support.

Limitations in Situational and Contextual Awareness. Given the shift from passive to agentic AI, these systems are increasingly expected to operate in dynamic, unstructured environments with cascading goals, evolving objectives, and frequent changes. Here, agentic AI systems may struggle to attain the situational and contextual awareness necessary to understand the current state, changes, or actions occurring in their environment. Unlike humans, who possess innate intuition and creativity, AI systems face limitations in handling non-routine tasks under conditions of uncertainty [E3, 41, 42].

Furthermore, agentic AI systems may face challenges in accurately identifying implicit adverse or harmful actions, particularly those deemed acceptable in other contexts. This limitation is especially pronounced when goals are unclear, increasing the potential for unintended consequences due to a lack of situational awareness [43]. While advancements continue, it remains uncertain when AI will fully bridge this capability gap. Until then, organizations must solve the challenge of effectively involving humans in the loop through control and monitoring mechanisms, essential for responding to ambiguities and unexpected changes [E1, E5, 43]. Additionally, maintaining a connection to the decision environment – whether indirectly through human agents serving as information brokers or directly – is crucial for enabling agentic AI systems to adapt dynamically and learn effectively from experience [E6, 44].

Establishing Sufficient Technical Endowments to Facilitate Delegation. Employing AI applications is generally a technologically complex endeavor. However, within the scope of our research objective, an additional factor comes into play: fostering an effective relationship with the human user. The technical capabilities of agentic AI systems to enable effective delegation and collaboration are essential for fostering productive partnerships with human users. Much of the research focuses on enhancing these systems' transparency, aiming to reduce the opacity of AI-generated outputs [34].

However, this challenge intensifies as agentic AI systems take on more autonomous roles, acting as the delegator rather than solely the executor. In these cases, agentic AI must be capable of clarifying task instructions and articulating the rationale behind delegated actions, along with the associated benefits or risks [E1, E6]. Also, enabling users to adjust or even build their own agentic AI systems for unique application cases could be a hurdle for organizations as the industry moves in this direction [9], which should be achievable in a low-code or no-code manner with an intuitive user interface [36].

Moreover, effective collaboration demands that agentic AI systems interpret and understand input from their human counterparts, ensuring alignment and mutual comprehension. As agentic AI takes on more responsibility, these systems must also self-validate their outputs in line with task objectives before executing actions, especially when human oversight is not immediately available [E6, 5].

4.3 Potential Tensions in the Relationship and Adverse Outcomes

When organizations do not adequately address both agents' inherent challenges before adoption, our findings mount up to two fundamental clashes in task delegation. First, this can lead to **information asymmetries** between the human and agentic AI systems [14, 44], stemming from a lack of knowledge about each other, their environment, or themselves. Within a similar context, [5] refer to this as legibility, when either the human or agentic AI system is unaware or cannot ascertain crucial information of each other's "thought processes" or available information. Secondly, they can lead to **goal divergence**, where objectives may not align, which can be intentional or unintentional. For example, a human might manipulate the agentic AI system due to negative perceptions by withholding or distorting information [E1]. Unintentionally, goal divergence may occur due to conflicting interests, skills, or objectives. For instance, while an agentic AI system might focus on task completion efficiency, the human agent could prioritize other implicit factors [E1, E3, 14, 45]. Both agents must be synchronized on information, objectives, and restrictions for a productive delegation relationship [E6].

If not, this misalignment can adversely affect the **coordination of actions** and the **distribution of rights and responsibilities** as the core delegation mechanisms [18]. Ultimately, disruptions in both can result in reduced task performance and hidden actions and consequences, further fortifying the innate challenges of the human and Agentic AI systems [14, 40]. In sum, the primary goals should be to enable continuous collaboration between both agents [E3] while improving task performance [34], not complicating procedures, and substantiating the importance of solving these challenges.

4.4 Organizational Challenges

Adapting Existing Structure and Culture to Embed Agentic AI. Adopting agentic AI systems requires a strategic and cultural transformation within the organization, supported by a clear, sustainable, long-term vision and roadmap [38]. This transformation must address cultural and people-oriented dimensions as well as structural and governance mechanisms. The significance of these factors becomes even more pronounced with the introduction of agentic AI systems, which amplify the complexity and associated costs from these perspectives [38]. As agentic AI systems act as de facto "new team members" [E2], existing processes and routines must be adapted to accommodate them effectively, leading to challenges for organizations to establish novel governance mechanisms, including a clear definition of use cases, roles, and responsibilities, alongside the operational management of these systems [E2, E5, 46, 47]. Additionally, it may be challenging to establish control mechanisms to ensure that agentic AI systems perform as intended and adhere to organizational objectives [E2, E3, E4, 5, 38].

Beyond structural changes, introducing AI-based team members necessitates a cultural and mindset shift across the organization, leading to potential tensions that are challenging to overcome [E3, 25]. Transparency about the forthcoming changes is critical to minimize resistance from the workforce. As [E2] highlighted, organizations should maintain honesty with employees about the substantial impact on work routines. Moreover, investing in upskilling and employee education is paramount, given the wide-ranging influence of agentic AI systems [38]. While these efforts are inherently complex and time-intensive, they are essential to realize the full potential of agentic AI systems. Crucially, these systems must align closely with the overarching IT strategy [E2, 38], more so than passive systems. Such alignment ensures seamless integration with existing organizational structures and enhances the effectiveness of agentic AI systems in achieving strategic objectives.

Equipping the Existing IT Landscape for Agentic AI Systems. In addition to these factors, integrating an agentic AI system into an organization's broader IT infrastructure presents significant challenges. While data integration is often discussed [E1, E2, E4, 38], we argue that agentic AI systems require additional conditions compared to more passive systems. We have mentioned the need for a connection with the decision environment. Further, we assert that they must also be seamlessly amalgamated into the existing IT system landscape, enabling them to act across the organization and, in some cases, across organizational boundaries [E4], which has not been considered in the literature. Also, E4 raised the challenge of enabling agentic AI systems to access physical systems to act directly in the decision environment [42]. In this context, a critical challenge is the establishment of effective boundaries and setting constraints on the action space of agentic AI within the existing IT landscape [5]. This approach ensures that agentic AI systems are granted access only to the data, resources, and applications essential for executing their defined tasks [43]. Such restrictions are pivotal in mitigating the risk of unintended or adverse behaviors.

4.5 Exogenous Challenges

Societal Risks Limit an Organization's Use of Agentic AI. The results underscore the importance of addressing broader societal implications when organizations implement agentic AI systems, including risks. These systems may influence internal employee dynamics and external perceptions. A recurring concern and challenge is the external pressure to constrain agentic AI capabilities due to fears of human labor displacement and economic disruption [E1, E6, 5, 43, 48]. While this is a common apprehension with technological advancements, societal risks become more pronounced with agentic

AI, which [7] describes as "collective disempowerment." This phenomenon involves shifting decision-making authority from humans to agentic AI systems or concentrating power within a small group of organizations that oversee or control their underlying infrastructure. Such shifts could lead to ethical dilemmas and systemic harm, eroding human-centric values and shifting a significant spectrum of human decision-making power to agentic AI systems [E2, E6, 38, 49, 50]. The resulting impacts might be indirect or delayed, further complicating their identification and mitigation, which is challenging for organizations to anticipate [5, 7].

Regulatory Requirements Prescribe Agentic AI's Action Space. Given the challenges discussed, it is evident that regular oversight of agentic AI operating within organizations is essential, particularly in high-stakes environments, to establish boundaries on their autonomy [5, 7]. Organizations must also carefully navigate potential regulatory requirements before implementing agentic AI systems. Directives offer guidance, such as the right to explain automated decisions under GDPR [E3, 50]. Establishing the EU AI Office in 2024 and drafting the AI Act mark the first steps toward a comprehensive legal framework [51]. This act, approved by the European Council, introduces significant regulatory challenges for organizations. A key issue is the ambiguity surrounding accountability for harmful actions or damage caused by agentic AI systems, posing an obstacle to compliance and implementation efforts [E1, E5, 52].

5 Discussion and Concluding Remarks

Agentic AI presents significant potential for organizations to enhance efficiency and improve the work routines of information and knowledge workers. This paper integrates practical insights and research perspectives to highlight the unique challenges of agentic AI systems within organizational contexts.

Specifically, we identify five critical groups of challenges: human challenges, including fears around autonomy loss, negative attitudes, and limited understanding; agentic AI challenges, such as insufficient understanding of human counterparts, limited situational awareness, and technical limitations; relational tensions from information asymmetries and goal misalignment; organizational challenges encompassing structural, cultural, and IT integration complexities; and exogenous challenges stemming from societal risks and regulatory requirements. Proactively addressing these challenges will enable organizations to leverage agentic AI systems effectively. Ultimately, recognizing humans' pivotal role is critical in understanding and navigating these challenges, as it grounds the apparent autonomy of agentic AI within human responsibility and intentionality.

Our results suggest initial approaches organizations might adopt, such as knowledge-building initiatives, enhancing transparency in AI decision-making, clearly defining roles and responsibilities, establishing robust governance structures, fostering a cultural mindset conducive to collaboration, and implementing effective tracking mechanisms to proactively identify and manage tensions or adverse outcomes. However, detailed strategies, best practices, and pathways to comprehensively overcome these challenges remain crucial for future research.

As with any research, limitations and threats to validity exist. The nascent stage of agentic AI presents challenges in accessing robust data, identifying experts with tangible experience, and establishing a strong research foundation. Thus, our study was potentially constrained by limited data. Further, qualitative research inherently risks subjectivity and bias from the authors and the interviewees. Our interviewees were male, potentially influencing data interpretation and contributing to gender bias. Therefore, this paper serves as an initial exploration of this emerging field. Future research should build on these findings and progress toward identifying effective organizational strategies to address these challenges.

References

1. Rai, A., Constantinides, P., Sarker, S.: Next-generation digital platforms: toward human–AI hybrids. MIS Q. **43**, iii–x (2019)
2. Deng, L.: Artificial intelligence in the rising wave of deep learning: the historical path and future outlook [Perspectives]. IEEE Signal Process. Mag. **35**, 180–177 (2018)
3. Feuerriegel, S., Hartmann, J., Janiesch, C., Zschech, P.: Generative AI. Bus. Inf. Syst. Eng. **66**, 111–126 (2024)
4. Riemer, K., Peter, S.: Conceptualizing generative AI as style engines: application archetypes and implications. Int. J. Inf. Manag. **79**, 102824 (2024)
5. Shavit, Y., Agarwal, S., Brundage, M.: Practices for Governing Agentic AI Systems (2023). https://cdn.openai.com/papers/practices-for-governing-agentic-ai-systems.pdf
6. Dattathrani, S., Dé, R.: The concept of agency in the era of artificial intelligence: dimensions and degrees. Inf. Syst. Front. **25**, 29–54 (2023)
7. Chan, A., Salganik, R., Markelius, A., et al.: Harms from increasingly agentic algorithmic systems. In: 2023 ACM Conference on Fairness, Accountability, and Transparency. pp. 651–666. ACM, Chicago (2023)
8. Schneider, J., Meske, C., Kuss, P.: Foundation models: a new paradigm for artificial intelligence. Bus. Inf. Syst. Eng. **66**, 221–231 (2024)
9. Microsoft: Microsoft AI - Copilot, https://www.microsoft.com/en-us/microsoft-copilot. Accessed 25 Apr 2024
10. OpenAI: Introducing the GPT Store. https://openai.com/blog/introducing-the-gpt-store. Accessed 12 Dec 2024
11. Cubric, M.: Drivers, barriers and social considerations for AI adoption in business and management: a tertiary study. Technol. Soc. **62**, 101257 (2020)
12. Kar, S., Kar, A.K., Gupta, M.P.: Modeling drivers and barriers of artificial intelligence adoption: insights from a strategic management perspective. Intell. Syst. Acc. Fin Mgmt. **28**, 217–238 (2021)
13. Dwivedi, Y.K., Hughes, L., Ismagilova, E., et al.: Artificial Intelligence (AI): multidisciplinary perspectives on emerging challenges, opportunities, and agenda for research, practice and policy. Int. J. Inf. Manag. **57**, 101994 (2021)
14. Guggenberger, T., Lämmermann, L., Urbach, N., Walter, A.M., Hofmann, P.: Task delegation from AI to humans: a principal-agent perspective. In: ICIS 2023 Proceedings (2023)
15. Acharya, D.B., Kuppan, K., Divya, B.: Agentic AI: autonomous intelligence for complex goals—a comprehensive survey. IEEE Access **13**, 18912–18936 (2025)
16. Borges, A.F.S., Laurindo, F.J.B., Spínola, M.M., Gonçalves, R.F., Mattos, C.A.: The strategic use of artificial intelligence in the digital era: systematic literature review and future research directions. Int. J. Inf. Manag. **57**, 102225 (2021)
17. Haenlein, M., Kaplan, A.: A brief history of artificial intelligence: on the past, present, and future of artificial intelligence. Calif. Manag. Rev. **61**, 5–14 (2019)

18. Baird, A., Maruping, L.M.: The next generation of research on is use: a theoretical framework of delegation to and from agentic is artifacts. MISQ **45**, 315–341 (2021)
19. Ågerfalk, P.J.: Artificial intelligence as digital agency. Eur. J. Inf. Syst. **29**, 1–8 (2020)
20. Bandura, A.: Social cognitive theory: an agentic perspective. Asian J. Soc. Psycho. **2**, 21–41 (1999)
21. Emirbayer, M., Mische, A.: What is agency? Am. J. Sociolog. **103**, 962–1023 (1998)
22. Wissuchek, C., Zschech, P.: Prescriptive analytics systems revised: a systematic literature review from an information systems perspective. Inf. Syst. E-Bus. Manag. (2024)
23. Cui, K.Z., Demirer, M., Jaffe, S., et al.: The Productivity Effects of Generative AI: Evidence from a Field Experiment with GitHub Copilot. An MIT Exploration of Generative AI (2024)
24. Leana, C.R.: Predictors and consequences of delegation. Acad. Manag. J. **29**, 754–774 (1986)
25. Adam, M., Diebel, C., Goutier, M., Benlian, A.: Navigating autonomy and control in human-AI delegation: user responses to technology- versus user-invoked task allocation. Decis. Support. Syst. **180**, 114193 (2024)
26. Wolfswinkel, J.F., Furtmueller, E., Wilderom, C.P.M.: Using grounded theory as a method for rigorously reviewing literature. Eur. J. Inf. Syst. **22**, 45–55 (2013)
27. Myers, M.D., Newman, M.: The qualitative interview in IS research: examining the craft. Inf. Organ. **17**, 2–26 (2007)
28. Malterud, K., Siersma, V.D., Guassora, A.D.: Sample size in qualitative interview studies: guided by information power. Qual. Health Res. **26**, 1753–1760 (2016)
29. Garousi, V., Felderer, M., Mäntylä, M.V.: Guidelines for including grey literature and conducting multivocal literature reviews in software engineering. Inf. Softw. Technol. **106**, 101–121 (2019)
30. Gioia, D.A., Corley, K.G., Hamilton, A.L.: Seeking qualitative rigor in inductive research: notes on the gioia methodology. Organ. Res. Methods **16**, 15–31 (2013)
31. Chong, T., Yu, T., Keeling, D.I., De Ruyter, K.: AI-chatbots on the services frontline addressing the challenges and opportunities of agency. J. Retail. Consum. Serv. **63**, 102735 (2021)
32. Galsgaard, A., Doorschodt, T., Holten, A.-L., et al.: Artificial intelligence and multidisciplinary team meetings; a communication challenge for radiologists' sense of agency and position as spider in a web? Eur. J. Radiol. **155**, 110231 (2022)
33. Liu, B.: In ai we trust? Effects of agency locus and transparency on uncertainty reduction in human–AI interaction. J. Comput.-Mediat. Commun. **26**, 384–402 (2021)
34. Vössing, M., Kühl, N., Lind, M., Satzger, G.: Designing transparency for effective human-AI collaboration. Inf. Syst. Front. **24**, 877–895 (2022)
35. Wu, Q., Bansal, G., Zhang, J., et al.: AutoGen: Enabling Next-Gen LLM Applications via Multi-Agent Conversation (2023). https://arxiv.org/abs/2308.08155
36. artificiality: Agentic AI (aka AI Agents) - The future of AI is agentic. https://www.artificiality.world/content/files/2024/05/Agentic-AI.pdf. Accessed 12 Dec 2024
37. Maurya, H., Agrahari, A., Kumar, A.: Human- AI collaboration: cognitive challenges in interacting with generative AI agents. In: SIGHCI 2023 Proceedings. (2024)
38. pwc: Agentic AI - the new frontier in GenAI. https://www.pwc.com/m1/en/publications/documents/2024/agentic-ai-the-new-frontier-in-genai-an-executive-playbook.pdf. Accessed 12 Dec 2024
39. Fügener, A., Grahl, J., Gupta, A., Ketter, W.: Cognitive challenges in human-artificial intelligence collaboration: investigating the path toward productive delegation. Inf. Syst. Res. **33**, 678–696 (2022)
40. Hemmer, P., Westphal, M., Schemmer, M., et al.: Human-AI collaboration: the effect of ai delegation on human task performance and task satisfaction. In: Proceedings of the 28th International Conference on Intelligent User Interfaces (2023)

41. Song, J., Lin, H.: Exploring the effect of artificial intelligence intellect on consumer decision delegation: the role of trust, task objectivity, and anthropomorphism. J. Consum. Behav. **23**, 727–747 (2024)
42. Huang, Q., Wake, N., Sarkar, B., et al.: Position Paper: Agent AI Towards a Holistic Intelligence (2024). https://arxiv.org/abs/2403.00833
43. Fourney, A., Bansal, G., Mozannar, H., et al.: Magentic-One: A Generalist Multi-Agent System for Solving Complex Tasks (2024). https://arxiv.org/abs/2411.04468
44. Sidorova, A., Rafiee, D.: AI agency risks and their mitigation through business process management: a conceptual framework. In: HICSS 2019 Proceedings (2019)
45. Burton, J.W., Stein, M., Jensen, T.B.: A systematic review of algorithm aversion in augmented decision making. Behav. Decis. Mak. **33**, 220–239 (2020)
46. Grisold, T., Schneider, J.: Dynamics of Human-AI Delegation in Organizational Routines. In: ICIS 2023 Proceedings (2023)
47. Van Rijmenam, M., Logue, D.: Revising the 'science of the organisation': theorising AI agency and actorhood. Innov. **23**, 127–144 (2021)
48. Sundar, S.S.: Rise of machine agency: a framework for studying the psychology of human–AI interaction. J. Comput.-Mediat. Commun. **25**, 74–88 (2020)
49. Brandtzaeg, P.B., You, Y., Wang, X., Lao, Y.: "Good" and "Bad" machine agency in the context of human-AI communication: The Case of ChatGPT. In: Degen, H., Ntoa, S., and Moallem, A. (eds.) HCI International 2023. pp. 3–23. Springer Nature, Cham (2023)
50. Stahl, B.C., Rodrigues, R., Santiago, N., Macnish, K.: A European agency for artificial intelligence: protecting fundamental rights and ethical values. Comput. Law Secur. Rev. **45**, 105661 (2022)
51. European Commission: AI Act. https://digital-strategy.ec.europa.eu/en/policies/regulatory-framework-ai. Accessed 12 Dec 2024
52. Monterossi, M.W.: Liability for the fact of autonomous artificial intelligence agents. Things, agencies and legal actors. Glob. Jurist. **20**, 20190054 (2020)

Opportunities and Challenges of Artificial Intelligence in Hiring from a Recruiter's Perspective

Maja Dieckmann[1], Ralf-Christian Härting[1](✉), Joerg Bueechl[2], and Rosalie Schwörer[1]

[1] Aalen University of Applied Science, Beethovenstraße 1, 73430 Aalen, Germany
maja@dieckmann-aalen.de, {Ralf.Haerting, rosalie.schwoerer}@hs-aalen.de

[2] ESB Business School, Reutlingen University, Alteburgstraße 150, 72762 Reutlingen, Germany
joerg.bueechl@reutlingen-university.de

Abstract. Artificial Intelligence (AI) provides unprecedented possibilities to accelerate and improve the hiring process. However, the willingness of recruiters to use–and in turn–profit from AI is crucial. Perceptions of human resource (HR) experts on AI-based hiring has been explored further in an empirical study, which is partly shown in this paper. As the existing body of research is still small, an inductive approach was chosen. We conducted seven semi-structured expert interviews with recruiters from the German manufacturing industry. Based on the findings, we developed a conceptual model with testable hypotheses and practical implications. Noticeably, the recruiters both praised and criticised AI for providing a fact-based assessment of candidates. AI's increased efficiency was recognised as a major opportunity or even necessity. The most promising approach seems to be a tandem of human and algorithmic decision-making, each offsetting the shortcomings of the other.

Keywords: Artificial Intelligence · Digitization · Hiring · Opportunities · Challenges · Qualitative Research · Recruiters

1 Introduction

The world is changing at a rapid speed, as Artificial Intelligence (AI) is constantly shifting the conception of what is possible. Organisations recognise this enormous potential in all business areas [1]. Since AI has become mainstream, it has gone from "nice to have" to being a necessity. This is also the case in hiring. Here, AI may be a crucial asset in bringing people and work together.

One factor often left out of consideration is the view of those people who will be using the system: The recruiters. This is critical, as the success of a technology is connected to its users' satisfaction [2]. Several researchers address this topic and call for more

K. Węcel (Ed.): BIS 2025, LNBIP 554, pp. 18–29, 2025.
https://doi.org/10.1007/978-3-031-94193-1_2

research on employees' attitudes towards AI [3–5]. Park et al. (Datum) point especially to the lacking empirical studies in this area [5]. Lee seeks for more studies in real-life settings to investigate people's perceptions of human and algorithmic decision making [3]. Ore and Sposato suggest to replicate their qualitative study on opportunities and risks of AI adoption in hiring in different countries [6]. Thus, a need for further work in this regard arises. The goal of this paper is to address this gap by answering the following research question: *How do recruiters perceive the use of AI in hiring?*

A range of expert interviews with recruiters was conducted. This approach allowed for an in-depth investigation of individual views. The subjects of the interviews were seven recruiters from the German manufacturing industry. Using the Gioia Methodology [7], the findings were coded, structured and then moulded into a conceptual model. This work aims to extend the current research body, enriching the discussion on opportunities and challenges of AI usage from several angles. First and foremost, it shows a unique view of recruiters on work-place AI implementation. Influential topics, such as the debate between automation and augmentation [8, 9], "algorithm aversion/appreciation" [10, 11], and biases in hiring [12] are touched on. Furthermore, it paves the way for future scientific work by providing testable hypotheses. Lastly, propositions on how to implement AI into hiring practice were formulated.

In the hiring process, AI is used in three sets of activities: Outreach, Screening, and Assessment [13]. Outreach activities aim to reach and attract promising candidates. Here, screening efforts down the line can be minimised by only contacting candidates with a high person-job fit. Several works propose Machine Learning (ML) models that are able to identify these candidates by matching job requirements with candidate profiles [14, 15]. Another use case of AI in this first recruitment stage is to personalise job advertisements, as research suggests that individualised job ads stand out more [16].

The next step is to screen and filter the incoming applications, so only suitable candidates move on to the assessment stage. Even with a personalised and targeted attraction of candidates, there is still an enormous volume of application data to screen. Here, AI represents a solution to speed up the screening, while still evaluating each application with the same level of attention. Using text mining, AI can derive meaning from unstructured texts and thus analyse and score application documents like CVs, cover letters and accomplishment records [17]. Recent natural language processing models are also able to avoid the "information retrieval problem", the long-lasting inability of AI to detect synonyms [18].

Lastly, the prefiltered candidates are assessed further to find the most suitable one, typically by doing job interviews. To make the assessment phase more flexible and accessible, asynchronous interviews (AVIs) can be used. The applicant records themselves answering interview questions in audio and video and the recruiter can access the video whenever and wherever they want. To increase efficiency even further, AVIs can be analysed by ML models, extracting prosodic features (e.g. speaking style, rhythm, intonation), lexical features (e.g. use of unique words, filler words) and facial features (especially smiling) from AVIs and connecting them to personality traits important to the hiring decision [19].

After reviewing existing research, the study closest to our approach was conducted by Ore and Sposato [6]. They interviewed a total of ten active recruiters at a large US enterprise to investigate opportunities and risks of AI in hiring and analysed the responses thematically. Malik et al. [4] set out to illustrate the influence of AI-applications on employees' experiences of HRM. They conducted ten interviews with employees in different positions and hierarchy levels at an enterprise, developing AI applications. This inherent connection of the company with AI might have influenced the overall positive responses of the interviewees. Deducted from previous studies, Park et al. [5] developed a comprehensive scale to test people's attitude towards AI-application at work. Bankins et al. [22] and Lee [3] studied how people perceive the fairness of AI in human-based HRM decisions with experimental survey studies. Both used a combination of quantitative and qualitative analysis. However, their qualitative part lacks in depth, as they merely included free text survey responses instead of holding interviews.

Even though research on the perception of AI in HRM decisions exists, the field is still small, with few studies focusing on recruiters. Our study contributes to this area.

2 Research Design

The Gioia Methodology builds upon Grounded Theory [20] and is characterised by a clearly constructed process. One defining aspect is the "tandem reporting of both voices – informant and researcher" [7]. The so-called 1st order concepts contain the topics highlighted by the informants. In this study, they were extracted by open coding of the expert interviews. Then, the 2nd order themes form the theoretical structure, developed by the researcher. This step includes extant theoretical knowledge, which was acquired by a short but comprehensive literature research. At last, the 2nd order themes are abstracted further, forming aggregate dimensions. From these dimensions the data structure, as Gioia termed the conceptual model, is formulated [7]. This approach provides a clear separation of informants and researcher–purely inductive and theory-supported research. It was therefore selected to faithfully conduct this research.

The data was collected in the form of seven semi-structured expert interviews. The target group comprised experts with working experience in recruitment and selection. Decidedly, this group was not limited to recruiters who have already used AI in their job environment. This ensured a more representative picture of the current situation in German companies. The interviews were scheduled and held between 12th of April and 3rd of May 2024 and lasted 30 min on average. For reasons of practicability, they were conducted remotely via the video conferencing tool Zoom. Since it was the interviewees' native tongue, the interviews were all given in German language. Transcription and translation of the interviews were carried out using the AI-based online service TurboScribe.

Table 1 shows the seven participants and the characteristics of their respective companies. Notably, the interviewees are from manufacturing companies, or recruitment agencies and are specialised in the domain recruiting for these companies. The selection of the participants mirrors the fact, that the manufacturing industry in South-Germany plays a major role in Germany and is a driver for the European economy. Therefore, the emphasis of this study lies on the German manufacturing industry. In terms of age and gender the sampling was considerably homogeneous. The participants were bet ween 27 and 34 years old (avg. = 31 years; SD = 2.17) and 71.4% were female. Their working

Table 1. Overview of the Interviewees.

No.	Industry	Relevant positions	Working experience	Age	Gender
1	Optoelectronic industry	Recruiter	3 years	27	Male
2	Optoelectronic industry	Recruiter	4.5 years	34	Female
3	Recruitment agency	Recruiter	6 years	30	Female
4	Optoelectronic industry	Teamlead Recruiting & Talent Sourcing	7 years	32	Female
5	Technology consultancy	Deputy Head of Recruiting	7.5 years	33	Female
6	Recruitment agency	Teamlead Recruiting	7 years	32	Female
7	Automotive supplier	Personnel Officer	7 years	30	Male

experience in hiring ranged from 3 to 7.5 years (avg. = 5.71 years; SD = 1.51). This sampling was the result of personal contacts based on a comprehensive network in the manufacturing industry.

The coding process with MAXQDA was carried out sequentially alongside the interview process. Although not necessarily recommended by Gioia [7], testable hypotheses were formulated. They were chosen as an addition to this work, providing a starting point for future, quantitative research.

3 Results and Interpretation

The goal of the study is to show the perception of recruiters concerning the use of AI in recruitment and selection. As they will be on the frontline of AI-supported hiring, it is crucial to regard their experiences. Figure 1 shows the data structure and therefore the backbone of the emergent theory.

3.1 Opportunities of AI Use in Hiring

Improvement of Processes. One major benefit of AI, emerged in all interviewees, was the increased efficiency and speed of processes. "*Efficiency gain, I think, is the big advantage. If you use it correctly, AI can lead to more efficient processes for all participants.*" (R4).

This aspect was especially relevant to highly manual and repetitive tasks, e.g. "*maintaining Excel lists, reporting, appointment arrangements, screening, where it's really just about clicking a lot*" (R1). Furthermore, by automating not only single steps but also the

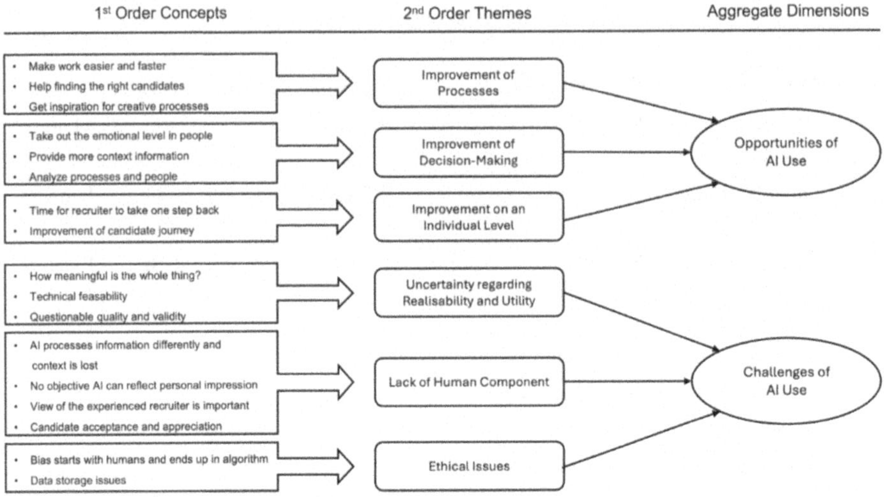

Fig. 1. Data structure.

transfer between them, the hiring process as a whole could be accelerated. As time pressure and a high administrative share were singled out as major challenges in recruiting, AI-powered efficiency increase may provide a solution.

In regard to efficiency increase, the term automation was frequently used. However, several recruiters were aware that automation is not necessarily connected to AI use. It is decidedly hard to differentiate these two in practice, without inside knowledge of the system. Therefore, the statements of the informants were generally treated as relevant to this study, even though some might not involve AI, strictly speaking. They nonetheless convey the sentiments towards characteristics and practical use cases connected to AI.

The *"classic lack of skilled workers"* (R1) was mentioned in numerous interviews as a major challenge. Therefore, active sourcing and employer branding have become extremely important. The informants named a range of use cases of AI that they already had in practice, at least partially. One AI tool frequently mentioned here improves job descriptions. But AI could also help recruiters to actively source candidates by recommending the right search terms for social media platforms, creating candidate pools and mining huge databases, e.g. job portals. Also, the first direct contact with candidates could be taken over by AI.

Generative AI, namely ChatGPT, was used by almost all recruiters for creative processes. Most of them pointed out the ability of generative AI to provide a first rough structure for various texts, e.g. phone calls, e-mails, public speeches. Nonetheless, they also noted the need to rework and adapt the output. *"Somehow, in more creative processes it can maybe offer a first starting point. To have a first prototype for a text, for an idea, for a representation or something like that."* (R5).

Also, the creation of images and videos for announcements or workshop material was mentioned. Being able to create exactly the image you need, would allow a whole new level of creative freedom. Additionally, they had lower concerns about copyright issues.

H1: Recruiters see a positive impact of AI use on processes in hiring.

Improvement of Decision-Making. Another improvement becoming apparent in the interviews was in terms of decision-making.

The recruiters acknowledged that as humans *"we are definitely biased"* (R1) and that personal preferences influenced hiring decisions. This might take place subconsciously and not in ill will. The belief that algorithms are generally the more objective decision makers has been reported in prior works [21, 22]. It has been termed as "machine heuristic" [23] or "AI appreciation" [11]. Noticeably, the recruiters in this study did not support AI making decisions completely on its own. They rather saw AI as a tool to uncover these implicit biases by providing a second, neutral opinion. *"And that's why an AI might be an addition to that, might not be that bad. Which can treat it more neutrally, to take out this emotional level in people."* (R3).

Another point that was brought up frequently, was the large data and knowledge base at the AI's disposal. As human capacities to store and remember facts are limited, AI could *"cover a lot more than what a person can at that point"* (R2). This would help in making fact-based decisions in several situations. The informants mentioned the use cases of ranking university degrees, providing background information for interviews and additional knowledge for company-specific characteristics.

While data analytics is widely seen as influential in all areas, notably only one recruiter directly addressed this topic. This study nonetheless reveals applications, already used in practice. *"In the personnel consulting agency we actually had a tool that could analyse: Where in the process do the candidates drop out? Many candidates drop out at the beginning, if it's too complicated. Where do they jump off?"*(R2).

This shows the possibilities for companies to question and improve their processes. In analysing applicant reactions, they could identify critical steps that might have led to a lower completion rate of the application process. Recruiter 2 also suggested using analytics to recognise behavioural patterns in applicants e.g. *"How flexible is she? How consistent?"* This could be achieved by looking at past behaviour and drawing conclusions for the future.

H2: Recruiters see a positive impact of AI use on decision making in hiring.

Improvement on an Individual Level. The last improvement becomes apparent in the individual perception of hiring. For recruiters this theme shows itself in a better working experience, for applicants in their candidate journey.

From a recruiter's view, the improvement is connected to the time saving attributed to automation. Several recruiters stated that they *"would like to invest more time in everything that happens between people"* (R1), if they had more time. Another big factor would be to invest in conceptual work and to improve existing processes. Lastly, one recruiter expressed the wish to use more time for personal education. Less time pressure through AI-infused automation could therefore lead to an improved working experience on various levels.

H3: Recruiters see a positive impact of AI use on individual recruiter experiences.

Candidates could also profit from this. On one hand, recruiters might be able to handle them with more individual care. As mentioned above, this was also desired from the recruiters' side. But they might also experience a better candidate journey because of AI's direct contact. Ad-hoc questions and problems could be addressed by chatbots, taking that kind of work off the shoulders of recruiters. This aspect was also addressed by Härting et al. in a qualitative study on emotionally sensitive AI applications [24]. They found their interviewees to highlight the possibilities of improved individual customer service, given the prerequisite that the applications deliver high-quality results.

H4: Recruiters see a positive impact of AI use on individual candidate experiences.

4 Challenges of AI Use in Hiring

Uncertainty Regarding Realizability and Utility. AI-implementation was a highly debated topic in many of the companies, the informants worked for. Therefore, they had first-hand experience with its difficulties. Several recruiters went as far as to question the utility of these applications altogether. This was in part connected to the need to rework or double-check evaluations. Then the efficiency gains from automating e.g. CV screening would begone. However, all recruiters showed some kind of reluctance to give AI complete decision power without human oversight. They feared that they would choose the wrong candidate or reject a more fitting one. As a consequence of recruiters not relinquishing important and complex decisions, the scope of action for AI would remain limited.

One recruiter questioned the practicality of data analytics in HR–People Analytics. This topic was also touched on by Angrave et al., who argue that even with insightful data at their disposal, HR practitioners lack the skill and knowledge to "ask the right questions" [25]. Without proper education in this regard, the usefulness of People Analytics would therefore be uncertain. *"And then there is always the question, does it make sense to save all this data at all? Or is it a flood of information with which nothing can be done in the end? (…) Then the question is, how practical is the whole thing?"* (R3).

While there are AI models successfully screening CVs in scientific [15, 26] as in practical environments, several recruiters doubted their functioning. According to them, an AI would need the documents to be structured in exactly the same way to be able to analyse them. Also, the notion prevailed that AI has not overcome the "information retrieval problem" [18]. This sheds a light on the "theory of machine" of the informants. Logg et al. coined this term, describing "people's lay perceptions of how algorithmic and human judgement differ in their input, process, and output" [11]. The informants' theory of machine is characterised by a rigid and strict input and processing. In processing information, many believed AI to follow an extremely binary approach. Either the key word is found in the CV, or it is not. They showed little trust that AI could grasp more complicated contexts. Conversely, the informants also stated AI to be capable of detecting patterns in data.

Overall, the recruiters rated the quality and validity of AI's output as *"questionable"* (R1). In their experience with AI applications, they often saw the need to check and adjust the results. This view is in line with the findings of several scientific works, studying people's perception of AI. The general undertone seems to be that algorithms can only provide the bare minimum of quality [26]. A study by Dietvorst et al. indicated that people are less likely to forgive errors made by algorithms compared to those made by humans [10]. Thus, mistakes detected by the recruiters might generally lower the opinion they have of AI's decisions.

H5: Recruiters doubt the realizability and utility of AI use in hiring.

Lack of Human Component. A high level of importance was attributed to the human capacities in hiring. All informants agreed in that sense. AI was generally believed to process information differently and with less leeway than humans. This is in line with the interviewees' "theory of machine" described above [11]. According to the informants, AI could mainly check off predefined hard skills. However, this approach would not fit the current notion in recruiting to prioritise personal competence and the candidate's ability to grow and develop. "*An AI is super fact-based, rational, all well and good, but recruiting is not fact-based and rational. So, not to that extent. It is too, of course, but not to that extent.*" (R4).

Regarding soft skills, team fit and other social aspects, all recruiters emphasised the importance of human assessment. An AI would not be able to read facial expressions, gestures and personality the way, a person could do. Recruiter 1 also used the term *"gut feeling"* to describe the basis on which he got his personal impression of a candidate. *"No objective AI or AI that imposes objective criteria can reflect my impression. If you think about team fit, social competence and so on."* (R1).

This criticism is a known phenomenon connected to algorithmic decision making. It is found under the terms of "lack of human touch" [27], "lack of emotional intelligence" [22] or "lack of personal contact/humanity" [28] in several qualitative studies. An important aspect to note is that the informants in this study support automation in administrative but not in human-centric tasks. Thus, this work underscores Lee's findings, which showed that the perception of AI is connected to the kind of task involved [3]. In mechanical tasks, human and algorithmic decisions were deemed as equally fair and trustworthy, but in human tasks the participants preferred human decision makers.

Aside from the inherently human ability to deal with social situations, the importance of the recruiters' competence was underlined. They had the education and experience necessary to thoroughly evaluate the candidate. This could be especially valuable with candidates, that do not have a classical career path or fit all the requirements. An AI that makes its assessment according to pre-set criteria might lack the flexibility to still take everything into account. *"I think the view of the experienced recruiter is important to say: 'ah, there is this one bit in the CV. Even if the CV may not be quite straightforward, this is exactly the interesting bit."* (R2).

Another point brought up, dealt with the candidates' view on AI tools. As recruiter 2 expressed, using a heavily AI-based process would rob the candidates of the possibility to get to know the culture of the company. Furthermore, little appreciation for the candidates would be signalled. Indeed, a quantitative study by Koch-Bayram and Kaibel concluded that candidates perceived algorithmic screening as a cost-reduction strategy, not as an improvement of screening quality. This significantly lowered the organisation's attractiveness [26].

H6: Recruiters see a negative impact in the lack of a "personal component" in AI use in hiring.

Ethical Issues The theme regarding ethics in AI use includes "algorithmic bias", discriminating phenomena in algorithms [12, 29], and data storage issues. The topic of algorithmic bias was highly discussed, and the interviewees showed a substantial level of knowledge and interest in this topic. Notably, several recruiters drew the connection between prevalent human biases and their occurrence in algorithms. "*To be honest, I don't think you can separate it at all. Because the algorithmic bias is created by the human bias.*" (R4).

A study by Cowgill et al. suggested that using training data sets which are unbalanced in some aspect, e.g. including significantly more male than female individuals, has the greatest impact on bias in AI decisions [12]. Several informants showed their understanding of this concept and mentioned that learning from past hiring data could lead to the repetition of patterns in the future. They mainly addressed this with the example of gender differences in specific job fields. *"And if it is, for example, a job that is very male dominated, where maybe the prejudice prevails, that rather only men can do it. (...) And you teach something like that, whether consciously or subconsciously, in some points to the AI."* (R3).

The findings suggest a paradoxical view of the informants–acknowledging aspired objectivity in algorithms as well as potential algorithmic bias. This phenomenon is also found in Bankins et al.'s study. Similarly, the informants displayed "AI appreciation" [11] as well as "AI aversion" [10], which is the preference of human decision makers over algorithmic ones.

H7: Recruiters see algorithmic bias in AI as a risk in hiring.

Last, several informants addressed the issue of data storage and protection. This is especially critical, as applicant data includes extremely personal information. When this information is stored in e.g. a company-wide Applicant Tracking System (ATS), all recruiters have access to it. But the problem would go even further. As the ATS is provided by a third party, the data is stored on their servers. Here e.g. the European General Data Protection Regulation (GDPR) must be complied with, which demands rigorous accounting of data flow and storage. Data security seemed to be the reason, why many AI implementation projects fail, as two recruiters reported.

H8: Recruiters see data storage and security as a risk of AI use in hiring.

5 Implications

From the data material, clear strategies for the practical use of AI in hiring were derived. The overarching theme is a *"symbiosis between AI and people"* (R4). The recruiters especially valued the qualitative part of their work, interviewing and dealing with people. Administrative tasks were seen as a necessary evil.

The first proposition for practitioners is therefore to consider automation for administrative tasks with little human contact. Examples are scheduling of interviews, sending out emails, or creating recurring documents. The use of AI in conducting interviews and direct contact with applicants was seen rather critically. Nonetheless, there is certainly a case to be made for its opportunities, in terms of efficiency as well as objectivity. Here, communication and collaborative evaluation with recruiters are crucial. Also, the support AI assessment has among applicants should be questioned and taken into account.

Secondly, there must be a balance between emotionality and neutrality. These two aspects symbolise the strengths as well as the weaknesses of AI and humans. While AI-based decision-making aspires to be completely objective, it lacks the emotionality also needed. Additionally, there is the risk of algorithmic bias, diminishing the advantage of neutrality. Human recruiters are strong in providing a human component and emotional intelligence, but they are subject to human cognitive bias and therefore lack in neutrality. The conclusion follows the work of Bankins et al. (2022), suggesting to constantly weigh human and algorithmic decision-making [22]. In important situations, they should work as a tandem, each offsetting the weaknesses of the other. This implies that AI assists human recruiters but does not function as the sole determinant in decision-making.

The second proposition for practice is therefore to first and foremost consider augmentation tools for qualitative tasks. In contrast to automation, which signifies the complete replacement of humans by machines, augmentation is defined as the collaboration of algorithmic and human decision makers [9]. Providing a second, objective view could mitigate human biases. These tools could also help to improve decisions in general and make work easier for recruiters. In the interviews it became apparent that additional support from AI tools would be welcomed by many.

The third proposition for organisations is to involve recruiters in the implementation process. The findings show that recruiters have a decided opinion on AI and its influence on hiring. It is therefore of utter importance to give employees the possibility to voice their concerns and in turn clear out misconceptions. User involvement has also been shown to lead to higher acceptance and satisfaction among end-users [2].

6 Conclusion

This study addresses the research gap regarding empirical works on the perception of AI [3–6] by giving an overview of recruiters' evaluation of AI use in hiring. The findings were encapsulated in a conceptual framework, displaying perceived opportunities and challenges. The recruiters highlighted possible improvements of processes, decision-making and individual experiences. AI could provide a solution to current challenges in hiring, namely the lack of skilled workers and high administrative share. On the other hand, the informants pointed out numerous challenges in AI implementation. Interestingly, they called the general usefulness and practicality into question. This might be influenced by their rigid "theory of machine" [11]. Another factor to be considered is the importance of the human component in hiring. Since both the informants and several similar studies [3, 22, 27, 28] concur on this point, it warrants serious consideration. Augmentation tools might be a suitable way to still ensure an objective and well-informed decision-making. In turn, with humans still reviewing AI, algorithmic biases might be prevented from causing harm. Another issue noted as highly influential in practice was data storage and security. Compliance with e.g. GDPR and the European AI Act should therefore be ensured.

The present study has some limitations. First, the number of informants interviewed is limited, so this work could be complemented with additional interviews. This study has already laid the groundwork, as several hypotheses were created based on the findings, therefore an additional quantitative study could enrich our dataset. Another limitation is the homogeneous structure of the sample. It mainly reflects a specific part of the German manufacturing industry. All but one are large enterprises; all situated in the industrial south of Germany. The sample of this study was also very homogeneous in terms of gender and age. Thus, another avenue of future research could be to compare the attitudes of different age cohorts and genders. All interviewed recruiters reported to have used AI –namely ChatGPT– in professional or personal use cases. However, few had working experience with tools specifically related to hiring e.g. CV screening or AVIs. As these applications were perceived as controversial in this study, an in-depth investigation might bring insightful results. This study could therefore be replicated with a focus on these tools, providing experience-based assessments. This approach might prove especially useful to practitioners seeking to implement AI-powered screening tools.

References

1. Kamen, M., Mahoutchian, T., Paytner, N.: Powering human impact with technology: technology contributes to making work better for humans and humans better at work™ (2023). https://www.deloitte.com/global/en/our-thinking/insights/topics/talent/human-capital-trends/2023/human-capital-and-productivity.html
2. Bano, M., Zowghi, D.: A systematic review on the relationship between user involvement and system success. Inf. Softw. Technol. **58**, 148–169 (2015). https://doi.org/10.1016/j.infsof.2014.06.011
3. Lee, M.K.: Understanding perception of algorithmic decisions: Fairness, trust, and emotion in response to algorithmic management. Big Data Soc. **5**(1), 1–16 (2018). https://doi.org/10.1177/2053951718756684
4. Malik, A., Budhwar, P., Patel, C., Srikanth, N.R.: May the bots be with you! Delivering HR cost-effectiveness and individualised employee experiences in an MNE. Int. J. Hum. Resour. Manag. **33**(6), 1148–1178 (2022). https://doi.org/10.1080/09585192.2020.1859582
5. Park, J., Woo, S.E., Kim, J.: Attitudes towards artificial intelligence at work: scale development and validation. J. Occup. Organ. Psychol. **97**(1), 1–32 (2024). https://doi.org/10.1111/joop.12502
6. Ore, O., Sposato, M.: Opportunities and risks of artificial intelligence in recruitment and selection. Int. J. Organ. Anal. **30**(6), 1772–1782 (2021). https://doi.org/10.1108/IJOA-07-2020-2291
7. Gioia, D.A., Corley, K.G., Hamilton, A.L.: Seeking qualitative rigor in inductive research. Organ. Res. Methods **16**(1), 15–31 (2013). https://doi.org/10.1177/1094428112452151
8. Maedche, A., et al.: AI-based digital assistants. Bus. Inf. Syst. Eng. **61**(4), 535–544 (2019). https://doi.org/10.1007/s12599-019-00600-8
9. Raisch, S., Krakowski, S.: Artificial intelligence and management: the automation–augmentation paradox. Acad. Manag. Rev. **46**(1), 191–210 (2021). https://doi.org/10.5465/amr.2018.0072
10. Dietvorst, B.J., Simmons, J.P., Massey, C.: Algorithm aversion: people erroneously avoid algorithms after seeing them err. J. Exp. Psychol. Gen. **144**(1), 114–126 (2015). https://doi.org/10.1037/xge0000033
11. Logg, J.M., Minson, J.A., Moore, D.A.: Algorithm appreciation: People prefer algorithmic to human judgment. Organ. Behav. Hum. Decis. Process. **151**, 90–103 (2019). https://doi.org/10.1016/j.obhdp.2018.12.005
12. Cowgill, B., Dell'Acqua, F., Deng, S., Hsu, D., Verma, N., Chaintreau, A.: Biased programmers? Or biased data? A field experiment in operationalizing AI ethics. In: EC '20: Proceedings of the 21st ACM Conference on Economics and Computation, pp. 679–681. Association for Computing Machinery (2020). https://doi.org/10.1145/3391403.3399545

13. Black, J.S., van Esch, P.: AI-enabled recruiting: what is it and how should a manager use it? Bus. Horizons **63**(2), 215–226 (2020). https://doi.org/10.1016/j.bushor.2019.12.001
14. Martinez-Gil, J., Paoletti, A.L., Pichler, M.: A novel approach for learning how to automatically match job offers and candidate profiles. Inf. Syst. Front. **22**(6), 1265–1274 (2020). https://doi.org/10.1007/s10796-019-09929-7
15. Qin, C., et al.: An enhanced neural network approach to person-job fit in talent recruitment. ACM Trans. Inf. Syst. **38**(2), 1–33 (2020). https://doi.org/10.1145/3376927
16. Pfiffelmann, J., Dens, N., Soulez, S.: Personalized advertisements with integration of names and photographs: an eye-tracking experiment. J. Bus. Res. **111**, 196–207 (2020). https://doi.org/10.1016/j.jbusres.2019.08.017
17. Campion, M.C., Campion, M.A., Campion, E.D., Reider, M.H.: Initial investigation into computer scoring of candidate essays for personnel selection. J. Appl. Psychol. **101**(7), 958–975 (2016). https://doi.org/10.1037/apl0000108
18. Dumais, S. T.: Latent semantic analysis. Annu. Rev. Inf. Sci. Technol. **38**(1), 188–230 (2004). https://doi.org/10.1002/aris.1440380105
19. Naim, I., Tanveer, M.I., Gildea, D., Hoque, M.E.: Automated analysis and prediction of job interview performance. IEEE Trans. Affect. Comput. **9**(2), 191–204 (2018). https://doi.org/10.1109/TAFFC.2016.2614299
20. Strauss, A.L., Corbin, J.M.: Basics of Qualitative Research: Techniques and Procedures for Developing Grounded Theory (2. ed. [repr.]. Sage, Newcastle upon Tyne (2005)
21. Araujo, T., Helberger, N., Kruikemeier, S., Vreese, C.H.: In AI we trust? Perceptions about automated decision-making by artificial intelligence. AI Soc. **35**(3), 611–623 (2020). https://doi.org/10.1007/s00146-019-00931-w
22. Bankins, S., Formosa, P., Griep, Y., Richards, D.: AI decision making with dignity? Contrasting workers' justice perceptions of human and AI decision making in a human resource management context. Inf. Syst. Front. **24**(3), 857–875 (2022). https://doi.org/10.1007/s10796-021-10223-8
23. Sundar, S.S., Kim, J.: Machine heuristic. In: Brewster, S., Fitzpatrick, G., Cox, A., Kostakos, V. (eds.) Proceedings of the 2019 CHI Conference on Human Factors in Computing Systems, pp. 1–9. ACM (2019). https://doi.org/10.1145/3290605.3300768
24. Härting, R.-C., Schmidt, S., Krum, D.: Potentials of Emotionally Sensitive Applications Using Machine Learning. In A. Zimmermann, R. J. Howlett, & L. C. Jain (Eds.), Smart Innovation, Systems and Technologies. Human Centered Intelligent Systems (Vol. 189, pp. 207–219). Springer Singapore. https://doi.org/10.1007/978-981-15-5784-2_17 (2021)
25. Angrave, D., Charlwood, A., Kirkpatrick, I., Lawrence, M., Stuart, M.: HR and analytics: why HR is set to fail the big data challenge. Human Resource Management Journal, 26(1), 1–11. https://doi.org/10.1111/1748-8583.12090 (2016)
26. Koch-Bayram, I.F., Kaibel, C.: Algorithms in personnel selection, applicants' attributions about organizations' intents and organizational attractiveness: an experimental study. Hum. Resour. Manag. J. 1–20 (2023). https://doi.org/10.1111/1748-8583.12528
27. Binns, R., van Kleek, M., Veale, M., Lyngs, U., Zhao, J., Shadbolt, N.: 'It's reducing a human being to a percentage'. In: Mandryk, R. Hancock, M. Perry, M., Cox, A. (Eds.), Proceedings of the 2018 CHI Conference on Human Factors in Computing Systems, pp. 1–14. ACM (2018). https://doi.org/10.1145/3173574.3173951
28. Haefner, F., Härting, R.-C., Bueechl, J.: Potentials and challenges of emotionally sensitive applications in Apprenticeship. Procedia Comput. Sci. **192**, 2606–2615 (2021). https://doi.org/10.1016/j.procs.2021.09.030
29. Lambrecht, A., Tucker, C.: Algorithmic bias? An empirical study of apparent gender-based discrimination in the display of STEM career ads. Manag. Sci. **65**(7), 2966–2981 (2019). https://doi.org/10.1287/mnsc.2018.3093

Identification of Challenges in the Economic Evaluation of AI Projects

Aleksandra Kopyto(✉) and Bartosz Wachnik

Warsaw University of Technology, Warsaw, Poland
aleksandrakrupa811@gmail.com

Abstract. Artificial intelligence (AI) plays a growing role in business, driving innovation, efficiency, and optimization. However, evaluating the economic feasibility of AI projects presents unique challenges. This paper explores key difficulties in assessing AI investments by conducting a critical literature review and case study analysis. The study highlights financial, organizational, and social factors impacting AI project evaluation. The research, based on case studies across various industries, identifies major economic assessment challenges, including the inability to isolate AI's specific impact on business performance, high data quality requirements, and compliance costs. Findings suggest that organizations require structured evaluation frameworks to better estimate AI's return on investment (ROI) and long-term economic impact. The study concludes that AI projects demand more refined assessment approaches that go beyond traditional financial metrics, incorporating ESG (Environmental, Social, Governance) factors and broader business value considerations.

Keywords: artificial intelligence · economic evaluation · AI projects

1 Introduction

AI is widely considered the latest general-purpose technology after the major technological innovations in information technology (IT), computers, and electricity [1]. According to a report by the McKinsey Global Institute, artificial intelligence could contribute to increasing global production growth by an additional $13 trillion per year by 2030 [2]. Such forecasts indicate the enormous potential of AI (Artificial Intelligence) in transforming various sectors of the economy, including manufacturing, health, transport and finance [3]. Fortune estimates that 84% of enterprises believe that implementing AI will provide a competitive advantage [4]. AI has also revolutionized the way they do business [5]. A recent study by MIT Sloan Management Review found that over 80% of organizations perceive AI as a strategic opportunity [6]. An important issue that must be considered in the rational design of IT systems based on artificial intelligence is the economic justification of such an undertaking throughout its life cycle, which includes the following phases: preparation, implementation, operation.

The research objective is to identify significant difficulties in the implementation of the economic analysis of IT projects involving the implementation of artificial intelligence in business, ex-ante and ex-post. They are distinguished mainly by the specificity

K. Węcel (Ed.): BIS 2025, LNBIP 554, pp. 30–38, 2025.
https://doi.org/10.1007/978-3-031-94193-1_3

concerning the ability to pose a business problem, the methodology of implementing the AI system, preparing data for training and testing, specificity in testing the results, the process of teaching AI algorithms. Considering the early stage of maturity of implementing artificial intelligence solutions in business, it is desirable to analyze and accumulate experience related to the use of artificial intelligence, which allows to increase the economic efficiency of individual business processes as well as entire business models.

2 Literature Review

Enterprises implementing artificial intelligence solutions expect benefits from such an investment in the form of increased business value, revenue growth, cost reduction and improved business efficiency [7]. According to Lech [8], research on the procedures for the economic evaluation of IT projects by economic organizations indicates the lack of commonly used methods in this area.

2.1 Methods of Economic Analysis of AI Investment Projects

As part of the literature study, the authors analyzed 20 scientific and research works. In 90% of them, scientists and practitioners used the quantitative ROI (Return on Investment) method and qualitative methods to analyze the economics of AI projects ex-post. Domin et al. [9] conducted research on the returns on investment in ethics and AI governance. In case of AI projects, the return on investment can come from several different sources, these are time savings, increased productivity, cost savings or increased revenues. According to Vayyavur [10] companies across various industries are increasingly investing in AI initiatives, driven by the potential for significant growth and improved operational efficiency. However, the reality of AI implementation is often challenging, causing many projects to fall short of expectations. AI projects help companies gain competencies in data, automation and predictive analysis can bring ROI in the form of long-term increase in competitive capabilities. Emerj [11] proposes its method for assessing the return on investment in AI, dividing ROI into measurable, strategic and building capabilities. According to a report conducted by Deloitte [12], companies that have implemented AI see a positive return on these investments. This study highlights the level of intangible benefits from AI solutions being implemented. To estimate the ROI of AI investments, Pandey, Gupta and Chhajed [13] propose three approaches: the market approach, the income approach (future cash flows) and the asset approach. In turn, Bevilacqua et al. [14] present three approaches to the economic evaluation of IT projects: the traditional approach, the intangible approach and the opportunity approach. In the case of AI ethics investments, the authors developed three paths to assess the impact of the investment: direct - through economic return, indirect - through the development of competences and reputation. In 2024, McKinsey Global Institute [15] conducted a study according to which the economic evaluation of AI projects consists of an analysis business value, an assessment of benefits in the value chain and operational KPIs. One of the methodologies for assessing the ROI and effectiveness of AI projects proposed by the McKinsey Global Institute is Cost-Benefit Analysis (CBA). CBA compares the costs of implementing and maintaining AI technology with the savings resulting from its implementation. Katragadda [16] has developed a methodology for assessing the ROI of AI implementation in customer service using a data-driven approach. This approach provides a solid framework for quantifying the multi-faceted impact of AI on customer service.

2.2 Challenges in the Economic Evaluation of AI Projects

Researchers and practitioners emphasize that investments in IT projects related to artificial intelligence are distinguished by the exceptional difficulty of identifying and quantifying costs and benefits, with intangible and indirect factors often playing a particularly significant role [17]. Studies on the economic evaluation of IT ventures are typically complex, and evaluators encounter numerous challenges, including the quantitative assessment of benefits and their accurate identification [18].

A key factor differentiating AI-related projects from other IT projects is the specific AI workflow. This process includes collecting, manipulating, and transforming data, which then serves as input for training, evaluating, interpreting, refining, and deploying complex mathematical models [19]. Additionally, isolating the specific impact of AI on business outcomes is difficult due to the parallel implementation of various solutions aimed at improving business processes. Botchkarev and Andru [20] highlight in their research that, while Return on Investment (ROI) is a widely used metric, it is particularly susceptible to errors, especially regarding future benefits, which are difficult to precisely estimate in the context of IT systems, and even more so for AI solutions. The difficulties in obtaining a high ROI from AI projects often stem from: lack of data, lack of appropriate skills, complex infrastructures [21].

3 Research Methodology

The authors employed a case study method, analyzing four AI-enabled IT system implementation projects. The objective of the case study is to expand the body of knowledge on the economic analysis of AI-based IT projects, particularly in understanding issues related to the precise estimation of costs and economic benefits. The authors aim to consider not only economic benefits but also non-economic aspects within the E - environmental, S – social responsibility, and G - corporate governance (ESG) framework. The ESG concept posits that enterprises should not focus solely on their financial interests but also ensure that their projects generate broader benefits for all stakeholders, local communities, and the environment.

The case study method was chosen as it enables the development of existing theories and provides explanations for underexplored methodologies in the economic evaluation of AI-based IT projects. This article focuses on the customer perspective throughout the entire lifecycle of an IT project, encompassing three phases: preparation, implementation, and operation. The selection of the case study approach was driven by two primary considerations [22]:

1. The early stage of theoretical knowledge development in the economic analysis of investment projects within a specific subset of IT projects—those involving AI system implementation.

2. The lack of established practical methods for conducting economic analysis in this domain.

The study is guided by the following research question: "*What key challenges have been identified in the ex-post economic analysis of AI-based IT projects in business?*".

The selection of studied cases was purposeful. According to B. Flyvbjerg [23], five main criteria guide case selection. Table 1 presents these criteria along with their application to this research.

Table 1. Criteria along with their application to research.

	Description in the Context of This Study
Data Availability	Ensured
Case Representativeness	Successfully completed projects utilizing various AI models
Case Diversity	The diversity of the cases analyzed is expressed in the selection of: • Different AI models • Client profile • Project implementation results
Criticality of the Phenomenon	The economic analysis of AI projects must address challenges in identifying costs and benefits from both economic and ESG perspectives
Metaphor for Understanding the Phenomenon	Implementing AI is like installing a high efficiency engine in an existing vehicle it improves processes and efficiency but presents challenges in compatibility, costs, and system adaptation

4 Research Findings and Interpretation

Table 2 presents the profile of companies X, Y, Z, and A, and AI project descriptions.

Table 3 presents detailing significant challenges identified in the ex-post economic analysis of AI-based IT projects. The difficulties were observed across the entire lifecycle of IT projects: preparation, implementation, and operation phases.

The research indicated that the economic analysis performed ex-ante and ex-post, although performed using the same methods, indicated that the ex-ante economic analyses indicated that the project profitability was lower from 75% to 170% than the economic analyses performed ex-post. Based on the findings in Table 3, the authors identified three primary factors contributing to challenges in the economic analysis of IT projects involving artificial intelligence implementation.

Table 2. Companies profile and characteristic of AI projects. Source: Own study

Companies	X	Y	Z	A
Client profile	Manufacturing firm	IT company	Service and trade	Law firm
AI project description	Detecting anomalies in X-rays to ensure aluminum casting quality	Risk identification in projects	Sales forecasting and customer segmentation for service personalization	Development of legal assistant for contract drafting and legal opinions
AI model	Neural network-based image recognition (Azure)	BERT/GPT for text, LSTM for sequences, Random Forest for structured data, K-means/DBSCAN for project clustering and pattern discovery	Time series forecasting and pattern classification using neural networks	LLMs

- Group 1. Very high uncertainty of identifying costs and revenues related to the implementation and operation of AI systems in the enterprise.

The ex-ante estimated costs differed significantly from the actual expenses incurred during project execution. The implementation process exhibited characteristics of research and development activities within the enterprise, implying that, at the project's inception, hypotheses were formulated regarding the economic benefits and costs associated with the anticipated outcomes.

- Group 2. Lack of a standardized methodology for the implementation of IT projects involving the implementation of artificial intelligence.

In all analyzed projects, the supplier did not provide a structured implementation methodology that would allow for the clear delineation of tasks, expected outcomes, and workload distribution. The absence of a standardized approach hinders the systematic execution of projects in accordance with best practices, significantly complicating the assessment of the reliability of the supplier's proposed budget.

- Group 3. Lack of knowledge among users of IT systems regarding the implementation and operation of artificial intelligence systems.

Users had great difficulty defining the problem being solved already in the preparatory phase, which resulted in limited possibilities for verification of the offer and later active participation in the project.

The analyzed projects utilized commonly available methods that are typically applied in IT project analysis [24]. The research revealed that companies did not conduct an economic analysis of artificial intelligence projects with a deep understanding of the

Table 3. Results of case study research. Source: Own study

	X	Y	Z	A
Preparation phase				
Duration of the preparatory phase	3 months	12 months	6 months	6 months
Identified difficulties in ex-ante approach	Challenges in project scoping, supplier selection, and contract negotiation	Challenges in project scoping, IP transfer, supplier selection, contract negotiation, and offer validation	Issues with project scoping, supplier selection, and contract agreement	Difficulties with project setup, supplier choice, and contract finalization
Implementation phase				
Duration	3 months	12 months	6 months	6 months
Identified difficulties in ex-ante approach	Difficulties in estimating data preparation, model/user training, and regulatory compliance costs (e.g. X-ray data, EU standards)	Difficulties with data timeliness, model selection and risk, AI impact isolation, and compliance cost estimation	Difficulties in adapting AI to trends, data scarcity, forecast uncertainty, and training/compliance cost estimation	Difficulties with LLM selection, document standardization, training costs, and EU compliance analysis
Operational phase				
Duration of the operational phase	5 years	1 year	1 year	1 year
Economic valuation method used	Payback Period – PB	Accounting Rate of Return – ARR	Net Present Value – NPV	Payback Period - PB

(continued)

Table 3. (*continued*)

	X	Y	Z	A
Comparison of economic analysis results ex ante and ex post	The ex-ante analysis showed that the project profitability was over 75% higher than the ex-post analysis	The ex-ante analysis showed that the project's profitability was over 60% higher than the ex-post analysis	The ex-ante analysis showed that the project's profitability was over 120% higher than the ex-post analysis	The ex-ante analysis showed that the project's profitability was over 170% higher than the ex-post analysis
Identified difficulties in ex-ante approach	Difficulties with SLA alignment, AI update costs, and process integration	Difficulties assessing unseen risks and high AI maintenance costs	Difficulties with SLA agreement and costly AI updates in dynamic environments	Difficulties validating rare events and high AI maintenance costs

specific nature of such initiatives. The findings indicate that AI implementation projects are characterized by high risk and uncertainty. In the analyzed cases, companies did not employ proven business procedures [25] for economic analysis of IT projects, as described in the literature, which could have been adapted for the evaluation of AI implementation projects.

5 Conclusion

Studies indicate a significant knowledge gap in the economic analysis of IT projects involving artificial intelligence implementation. As a result, economic analyses of such projects are often flawed, which may undermine the credibility of their execution. This diagnosis primarily stems from two key factors:

1. The high risk and uncertainty associated with implementing AI-based IT projects.
2. The lack of available knowledge and established business practices, both in terms of executing such projects and conducting their economic evaluation.

The authors suggest that future research on developing methodologies for the economic analysis of AI-driven IT projects should also incorporate guidelines from the ESG framework, which influence the sustainable value of projects, organizations, and investor assessments. Research and development efforts in this field should consider insights from analogous studies on IT project evaluations, such as the implementation of ERP, CRM, BI, and DMS systems. Criticism of currently available methodologies, techniques, and procedures for assessing the economic feasibility of IT projects should consider the principles of imperfect knowledge economics [26]. This approach emphasizes continuous, non-routine changes in the analyzed subject and acknowledges the inherent limitations in knowledge held by both theorists and practitioners, particularly in assessing project effectiveness.

Disclosure of Interests. The authors have no competing interests to declare that are relevant to the content of this article.

References

1. Lee, Y.S., Kim, T., Choi, S., Kim, W.: When does AI pay off? AI-adoption intensity, complementary investments, and R&D strategy. Technovation **118**, (2022)
2. Notes from the AI frontier: Modeling the impact of AI on the world economy. https://www.mckinsey.com/featured-insights/artificial-intelligence/notes-from-the-ai-frontier-modeling-the-impact-of-ai-on-the-world-economy/. Accessed 15 Oct 2024
3. Bughin, J., van Zeebroeck, N.: The promise and pitfalls of AI, McKinsey Global Institute (2018)
4. Murry, A.: Fortune 500 CEOs see A.I. As a big challenge (2017). https://fortune.com/2017/06/08/fortune-500-ceos-survey-ai/. Accessed 18 May 2020
5. Schneider, S., Leyer, M.: Me or information technology? Adoption of artificial intelligence in the delegation of personal strategic decisions. Manag. Decis. Econ. **40**(3), 223–231 (2019)
6. Ransbotham, S., Kiron, D., Gerbert, P., Reeves, M.: Reshaping business with artificial intelligence: closing the gap between ambition and action. MIT Sloan Manag. Rev. **59**(1) (2017)
7. Alsheibani, S., Cheung, Y., Messom, C., Alhosni, M., Winning AI strategy: six-steps to create value from artificial intelligence. In: Americas Conference on Information Systems (2020)
8. Lech, P.: Metodyka ekonomicznej oceny przedsięwzięć informatycznych wspomagających zarządzanie organizacją, Wydawnictwo Uniwersytetu Gdańskiego, p.192. Gdańsk (2007)
9. Domin, H., Rossi, F., Goehring, R., Ganapini, M., Berente, N., Bevilacqua, M.: On the ROI of AI Ethics and Governance Investments: From Loss Aversion to Value Generation (2024)
10. Vayyavur, R.: Why AI projects fail: the importance of strategic alignment and systematic prioritization. Int. J. Res. **11**, 386–391 (2024)
11. Applying Emerj's AI ROI Model to Assess the Upside of AI Projects. https://emerj.com/assess-the-roi-of-ai-projects/. Accessed 13 Oct 2024
12. ROI from AI: The importance of strong foundations. https://www2.deloitte.com/us/en/insights/industry/technology/artificial-intelligence-roi.html. Accessed 13 Oct 2024
13. Pandey, S., Gupta, S., Chhajed, S.: ROI of AI: effectiveness and measurement. Int. J. Eng. Res. Technol. (IJERT) **10**(5), 749–759 (2021)
14. Bevilacqua, M., Berente, N., Domin, H., Goehring, B., Rossi, F.: The return on investment in AI ethics: a holistic Framework. In: Hawaii International Conference on System Sciences (HICSS) (2023)
15. The state of AI in early 2024: Gen AI adoption spikes and starts to generate value. https://www.mckinsey.com/capabilities/quantumblack/our-insights/the-state-of-ai. Accessed 30 Nov 2024
16. Katragadda, V.: Measuring ROI of AI implementations in customer support: a data-driven approach. J. Artif. Intell. Gen. Sci. (JAIGS) (2024)
17. Willcocks, L., Lester, S. (eds.): Beyond the IT Productivity Paradox: Assessment Issues, pp.155. John Wiley&Sons, Chichester, Hoboken (1999)
18. Cypryjański, J.: Metodyczne podstawy ekonomicznej oceny inwestycji informatycznych przedsiębiorstw. Rozprawy i Studia/Uniwersytet Szczeciński **669**, 296 (2007)
19. Vial, G., et al.: Managing artificial intelligence projects: key insights from an AI consulting firm. Inf. Syst. J. **33**(3), 669–691 (2023)

20. Botchkarev, A.: Estimating the accuracy of the Return on Investment (ROI) performance evaluations (2014)
21. AI Return on Investment: How to Measure the Business Value of AI. https://aiexponent.com/measuring-the-artificial-intelligence-return-on-investment/. Accessed 13 Oct 2024
22. Yin, R.: Case Study Research: Design and Methods. Sage Publications, Thousand Oaks (1984)
23. Flyvbjerg, B., FiVe misunderstandings about case-study research, [in] qualitative research practice. In: Seale, C., Gobo, G., Gubrium, J.F., Silverman, D., pp. 420–434, Sage, London-Thousand Oaks (2004)
24. Wachnik, B.: Wdrażanie Systemów Informatycznych Wspomagających Zarządzanie, PWE (2016)
25. Ibidem
26. Aghion, P., et al. (eds.): Knowledge, Information, and Expectations in Modern Macroeconomics: In: Honor of Edmund S. Phelps. Princeton University Press, Princeton (2003)

Artificial Intelligence for Visualization

LumiViz: Automating Business Data Visualization with Generative AI

Szczepan Górtowski(✉) and Elżbieta Lewańska

The Poznań University of Economics and Business, 61-875 Poznań, al. Niepodległości 10, Poznań, Poland
{szczepan.gortowski,elzbieta.lewanska}@ue.poznan.pl

Abstract. The paper presents LumiViz, a system leveraging Generative AI to automate business data visualization processes. The tool processes user queries in natural language, retrieves relevant data, generates visualizations, and provides business-oriented result interpretations. The study introduces a layered system architecture comprising Security & Governance, AI, and Presentation layers, ensuring scalability and reliability. A Proof of Concept was developed and tested to assess the feasibility of GenAI-driven automation in visualization workflows. While AI successfully automates key stages of the process, user intervention remains essential for refining outputs. User evaluations demonstrate the system's usability, with an 88% satisfaction rate.

Keywords: Generative AI · Business Data Analysis · Large Language Models · System Architecture · Data visualization · NL2VIS

1 Introduction

Generative Artificial Intelligence (GenAI) is becoming increasingly relevant in the field of data analysis [20,22]. Its ability to generate text, images, and other forms of content makes it a valuable tool for automating data interpretation, visualization, and insights generation, improving efficiency in decision-making processes [4].

Analysis of business data is crucial for decision-making processes, especially in data-driven organizations. Effective data analysis enables businesses to identify trends, optimize operations, and develop strategic insights that drive competitive advantages. With the increasing complexity of datasets, businesses require advanced tools to process and interpret information efficiently, ensuring that decision-makers have access to accurate, timely, and actionable insights.

As the volume of business data continues to expand rapidly, organizations are turning to AI automation to streamline analysis processes, reduce reliance on large specialist teams, and accelerate internal reviews. Traditional Business Intelligence (BI) solutions can be time-consuming and often require users to have at least a basic knowledge of query languages (e.g., SQL) and analytical tools. The development of GenAI – particularly in the area of Large Language Models

K. Węcel (Ed.): BIS 2025, LNBIP 554, pp. 41–55, 2025.
https://doi.org/10.1007/978-3-031-94193-1_4

(LLMs) – opens new possibilities for automating the entire data analysis process. This process starts with formulating a question in natural language, retrieving relevant data, visualizing it, and ultimately interpreting the results in a way that is understandable to a broad audience.

The goal of this paper is to discuss the architectural elements necessary for designing and implementing a solution based on natural language queries, aimed at using GenAI to generate or improve analyses and visualizations. This tool should take into account various legal, organizational and technical challenges, such as issues arising from diverse business terminology, industry-specific vocabulary, and trust in AI algorithms. Additionally, this paper presents the challenges faced by organizations that implement such solutions.

Moreover, this study goes beyond the discussion of theoretical elements by proposing an example architecture of the natural language to visualization system (NL2VIS). A working Proof of Concept (PoC) has been developed to assess the practical effectiveness of AI-driven automation in business data analysis. The PoC undergoes practical evaluation through real-world testing, assessing its ability to automate business data analysis processes and improve efficiency.

This study poses the following research questions:

- Q1 What are the key technical components required for GenAI-based tools to automate business data visualization?
- Q2 Which stages of visualization process can be effectively automated using GenAI?
- Q3 What is the impact of using GenAI for automatic business data visualization on user experience and decision-making processes?

The structure of the paper is as follows. Section 2 presents related works in the area of automatic data analysis, as well as security and organizational aspects of integrating AI in company's infrastructure. Section 3 describes methodology followed in this research. Sections 4 and 5 introduce the LumiViz tool. In Sect. 6 system evaluation is presented. Section 7 concludes the paper and indicates directions of future research.

2 Related Works

Automatic Data Analysis

Analysis of business data is essential for decision-making processes in organizations and has been a central concern for researchers since the early development of management information systems. Three main phases of automation in business data analysis might be identify. The first phase was the advent of BI systems that require technical expertise and manual effort, limiting their accessibility to non-technical users. The second phase was the introduction of Machine Learning (ML) and Artificial Intelligence (AI) solutions, which improved efficiency by learning from historical data and generating insights dynamically, but still required specialized knowledge for effective implementation and interpretation.

With the emergence of GenAI in the past few years, the possibilities for streamlining business data analysis have expanded. The ability of GenAI to process vast amounts of unstructured and structured data, interpret context, and generate human-like responses opens new opportunities for automation and decision support in organizations.

The first area of GenAI applicability is creation of AI-driven data visualization. Multiple studies focus on this aspect. E.g., [17] developed a Chat2VIS system converts natural language prompts into Python code for data visualization. An example of domain-specific application is the FinFlier system [8], which enables the automatic creation of graphical overlays on charts using language models fine-tuned for the finance domain. FinFlier illustrates how leveraging domain knowledge can enhance the interpretability and trustworthiness of AI-generated results. The LADV system [16] uses deep learning to automate dashboard creation, allowing users to generate initial prototypes based on images and sketches. By leveraging models such as Faster R-CNN, this tool accurately identifies chart types and optimizes layouts based on visual design principles. Evaluation has shown that users were able to generate more design concepts in less time, increasing the efficiency of the creative process [16]. Despite its advantages, the LADV system has limitations, such as difficulties in recognizing non-standard charts and the lack of support for complex data relationships.

Secondly, GenAI is successfully applied to automate the whole data analysis process. Dhoni [6] highlights the significant value of AI tools in automating data workflows, where GenAI models can automatically generate queries and synthesize analysis results. Behesti et al. [1] propose ProcessGPT, an architecture for integrating data and knowledge graphs in GenAI-based systems. A key component of this system is the Process Knowledge Graph, i.e. a central repository of knowledge, enabling the mapping of technical data to business context. The study by Malacaria et al. [18] explores the adoption and maturity models of Generative AI within organizations, with a specific focus on governance frameworks. The research investigates how structured governance, particularly through the proposed Consensual Big Data Assessment System (CBDAS), can optimize AI adoption strategies.

Another research direction isuse AI to automatically translate natural language to SQL queries (NL2SQL). Research in this area has gained momentum due to advancements in neural networks and deep learning techniques [13,28], even before the GenAI era. Since 2017, numerous prototype systems based on transformers [25] have emerged, enabling the processing of user queries into complex SQL statements [29]. Yu [28] introduced TYPE-SQL, a model enhancing text-to-SQL conversion using type information. [29] contributed by developing the WikiSQL dataset to benchmark performance in cross-schema SQL query automation, improving accuracy in AI-driven database querying. However, generating SQL queries is only the first step; a crucial aspect is also the processing and visualization of data in a manner accessible to decision-makers [3].

Several papers provide broad reviews of GenAI's applications. [3] evaluates GPT-3's strengths, weaknesses, and ethical implications across various language tasks. [27] offers an overview of Generative AI's impact on Learning Analytics. [24] reviews the integration of AI and NLP in data visualization, while [23] surveys natural language interfaces (V-NLIs) for visualization generation.

LumiViz addresses a significant research gap by providing a comprehensive system that supports users throughout the entire visualization process. Unlike most AI-based visualization solutions, which typically focus on a single stage, LumiViz guides users from initial visualization design through data extraction to full visual representation with business insights. Additionally, the proposed system is tested in a real business environment and has been developed based on actual business needs, ensuring its practical applicability and effectiveness.

Security, Privacy and Ethics in AI Systems

Security-related aspects are crucial for every business application and might limit widespread adoption of such solutions [26]. For example, [26] discusses several issues regarding vulnerabilities in AI models that might occur at every stage of product life: corrupting training data, backdoor attacks, manipulations with model outcomes, etc. LLMs face jailbreak attacks, aimed at gaining access to sensitive data or bypassing security measures [14]. [7] lists such security problems as data leakage, model inversion, and membership inference attacks.

Privacy concerns include the process of collecting training data (e.g., if includes sensitive information that might lead to privacy breaches), lack of transparency (the "black-box" nature of some models makes it difficult to understand how they process information), and privacy leakage (models store conversation history that can be leaked) [1,26].

Ethical concerns also arise regarding the use of GenAI [10,11,26,27], the most important aspects include: the misuse of AI technology to distribute malicious software, adversarial attacks, the use of GenAI for plagiarism and cheating in education, generating harmful content, the presence of bias and discrimination in the data, the risks related to the fairness, transparency and accountability of the models, as well as the risk of amplifying the digital divide.

In the context of AI model interpretability, a critical issue is not only the lack of transparency in their functioning but also the biases in processed data. Data analysis requires consideration of both explicit schema (such as database structures or SQL queries) and tacit schema, which may reflect biases, cultural differences, or industry-specific contexts [2].

Although this research does not primarily focus on security, privacy, or ethical issues related to AI, these aspects were carefully considered during the LumiViz development. Since the prototype is deployed in a real-world business environment, it was designed with these concerns in mind to ensure compliance, reliability, and responsible AI usage.

Incorporation AI into Company's Infrastructure

GenAI offers transformational potential for businesses by increasing efficiency and fostering innovation [5,21]. However, its implementation comes with significant business challenges, including high costs, integration complexity, and ethical concerns [21]. Key technological obstacles include aligning AI initiatives with strategic goals and ensuring output quality [21]. Organizational challenges involve the need for skilled professionals, effective data management, and fostering a data-driven corporate culture [18]. To maximize benefits, businesses must navigate implementation complexity, prioritize ethical considerations, and adapt to workforce transformations through continuous skill development [21].

GenAI enhances personalization in data analysis, enabling precise recommendations based on user context. This improves decision-making and analytical efficiency in business as GenAI transforms raw text into structured data and enriches visualizations with explanatory analytics [27].

Advanced Questions&Answers (Q&A) systems like SmartQ can improve answer quality and response rates by expert question forwarding [15]. For a Q&A tool to function effectively, it must integrate with the existing IT infrastructure, including data warehouses, CRM and ERP systems. The diversity of data formats and sources poses significant challenges for consistent and rapid data processing in big data integration and analytics [12]. Additionally, adapting the system to industry-specific domains is often required [19].

For NL2VIS solutions such as LumiViz to function effectively, it is essential to provide the AI tool with a clear business context. Business analysts often rely on specialized terminology to define their analytical objectives, so incorporating these linguistic nuances ensures more accurate query interpretation. Additionally, separating AI modules from the data via data virtualization enhances security and streamlines system maintenance, allowing flexible integration and updates without exposing underlying data structures.

3 Research Methodology

To explore and organize the challenges of implementing GenAI in automated business data analysis, the Design Science Research (DSR) was chosen, as it is widely recognized and justified in the information systems literature [9].

Existing definitions and frameworks for GenAI-supported business data analysis serve as the Knowledge Base, forming the foundation of this research and methodology. On the other hand, business needs and trends define the Business Environment, ensuring alignment with real-world applications, particularly in integrating AI solutions into corporate infrastructure.

The study identifies key artifacts such as the LumiViz reference architecture, a model for data analysis supported by GenAI, and a prototype of LumiViz. These artifacts are iteratively developed, evaluated, and refined through the Rigor and Relevance Cycles. The Rigor Cycle ensures the refinement of artifacts based on insights from the Knowledge Base, particularly by leveraging existing GenAI applications, data visualization automation, and best practices. The Relevance Cycle ensures that the study addresses practical business needs through user evaluations and real-world testing, leading to iterative refinements.

The Design Cycle represents the core iterative process, bridging theoretical foundations (Rigor Cycle) with practical applications (Relevance Cycle). It involves the ongoing development, evaluation, and refinement of system modules and resources. Ultimately, the study contributes back to the Knowledge Base by defining requirements for automatic business data analysis and validating the evaluation results through empirical testing.

4 System Architecture

LumiViz is divided into three main modules (Fig. 1), supplemented by organizational and legal rules as well as best practices.

The Security & Governance module (S&G) ensures data security, performance monitoring, and solution management. The AI module, enriched with a database component, provides a fully functional query processing mechanism. Finally, the Presentation module serves business experts and analysts, enabling them to obtain business insights and assess the model through a feedback loop.

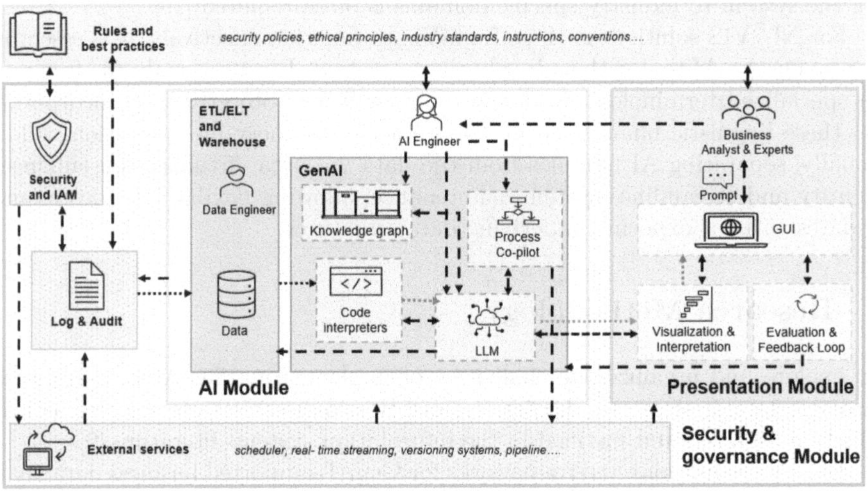

Fig. 1. LumiViz architecture

S&G is a fundamental component of the system, ensuring compliance with legal requirements, industry standards, and operational needs. It coordinates security policies, maintaining the platform's reliability and stability. Key guidelines are derived from *Rules and Best Practices*, encompassing security policies, ethical principles, and industry regulations.

To enhance data protection, Identity and Access Management (IAM) mechanisms regulate authentication and authorization. Commercial solutions enable centralized identity management and access verification, utilizing role-based access control or zero trust models based on organizational security needs.

This module also logs system activities for auditing purposes, allowing for security breach detection and performance monitoring of AI modules.

API integration ensures compatibility with enterprise tools and future scalability without overhauling the platform. The integration module supports data streaming for real-time transaction processing and schedulers for periodic tasks. Additionally, MLOps principles, including version control and automated testing , facilitate continuous model improvement.

ETL/ELT processes extract, cleanse, and standardize data from various sources before storing it in a data warehouse. Given its significance in the LumiViz system, the role of the Data Engineer has been highlighted. While not managing a traditional warehouse, the Data Engineer ensures data quality and updates, preparing it for AI applications by aggregating and enriching metadata.

At the core of LumiViz lies the GenAI module, a set of components designed to generate and interpret responses to natural language queries:

- The Knowledge Graph stores concepts, abbreviations, domain-specific terminology, and metadata, enhancing the system's ability to understand queries. It also includes predefined business metrics and corporate objectives. While partially integrated with the LLM, these elements remain distinct. Context provided by the Knowledge Graph helps to minimize GenAI hallucinations and to increase explainability.
- Code Interpreters (e.g., NL2SQL) translate user queries into database queries or analytical scripts.
- LLM, such as GPT, serves as the generative engine, producing structured responses and reports.
- Process Co-pilot supervises analytical workflows, acting as a set of agents evaluating queries and recommending next steps. It utilizes feedback loops and an internal scoring mechanism to refine LLM-generated insights.

The AI Engineer oversees model fine-tuning, explainability enhancements (XAI), and the Knowledge Graph updates. They determine retraining frequency, assess model performance, and collaborate with the Data Engineer to adapt the system's architecture to evolving data and business needs.

Users, such as Business Analysts or domain experts, can access a dedicated panel to submit prompts, configure analysis parameters, or review query history. The interface is available as a web or mobile application or can integrate with communication platforms.

Visualizing results through charts, dashboards, or tables is key to understanding complex data relationships. BI tools or programming libraries serve as ready-made components, allowing analysts to gain insights. If the XAI is active, users receive explanations on the model's decision-making process.

The Evaluation and Feedback Loop enables users to assess the relevance of insights through rankings or comments. This feedback helps recalibrate AI models, refine the Knowledge Graph, and update domain dictionaries, ensuring continuous learning and adaptation to organizational needs.

5 LumiViz Process Flow

Figure 2 presents the process flow of LumiViz system. For the clarity reasons only main components of the systems are presented (administrative modules and internal processing are skipped).

The main actors in the process are:

- End user – This is the main actor in the process. Usually, this person holds the position of business manager or data analyst. It is also possible that two different users cooperates in this process: the business expert defines information needs and visualization goals, while the technical-oriented data analyst will provide data requirements. The role of the end user is to provide requirements for the analysis, as well as validate the results at different stages. The end user is using GUI and communicates with the system with prompts written in natural language or images (e.g., hand drawn mock-ups).
- The visualization component (VC) is responsible for generating visualizations (single charts or whole dashboards), both based on the real data and mock-ups (which are just drafts built on synthetic data or without any data, just to show user-specific diagram types or layout of the dashboard).
- AI agent – a GenAI agent responsible for transforming user prompts into specific queries to retrieve data from the source or build a visualization. AI agent also clarifies user prompts by improving them with specific domain-related content, e.g. corporate jargon or the company's visual identification system, or ask the user for additional information. It also provides descriptions and insights written in natural language.
- ETL component (ETL) is to retrieve desired data set from available data sources and transfer it into the required structure.

System resources includes several different artifacts (not included on the Fig. 2 to increase readability). (1) The Knowledge Graph stores business domain context, as described in Sect. 4. (2) Data sources – As contemporary enterprises usually have a wide ecosystem of various IT tools, data sources are heterogeneous and decentralized. The information needs of business representatives can usually be met with data gathered in management information systems such as ERP, CRM, SCM, and others. Such systems register structured data about business transactions, as well as characteristics of different actors (e.g., business partners) and resources (e.g., materials, assets). However, external data sources might be required for certain data analysis, e.g., weather data, stock prices, currency exchange rates, events hosted in specific locations. (3) Direct inputs include all information provided by the end users in prompts. Those are usually one-time requirements for the analysis.

The process is divided into 4 phases.

Phase 1 – Mock-up design The goal of this phase is to set up initial requirements for the analysis: the type of visualization (e.g., type of the diagram, how it should be presented) and business requirements (e.g., what is the purpose of the visualization, who the audience is). The process is started by a user, who defines the problem. The input might be in the form of text (natural language

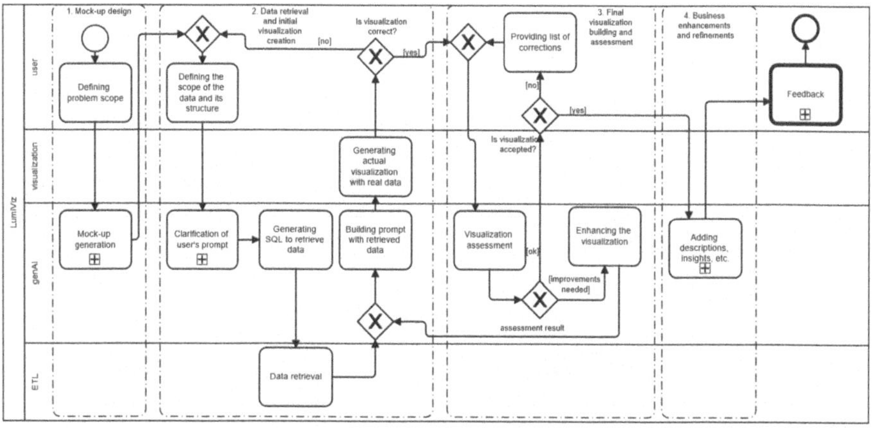

Fig. 2. LumiViz process flow in BPMN. Diagram created in https://demo.bpmn.io/

description) or image (sketch, image). The user prompts AI agent. In the "mock-up generation" subprocess, the AI agent adds business domain context from the Knowledge Graph, that includes: style guide, corporate visual identification system, list of acronyms and jargon used in the company, etc., and then prepares a prompt for the VC. At this point, only general mock-up of the final data visualization is prepared by VC (i.e., no actual data is included). This mock-up is presented to the user who might accept it (and proceed to phase 2) or create a list of refinements and prompt the AI agent to repeat this part of the process.

Phase 2 – Data retrieval and initial visualization creation The goal of this phase is to define and retrieve the data needed for the visualization: its scope, format, time range etc. This phase starts with the user who defines the scope and time restrictions for the analysis. As this prompt is built in natural language, users tend to use jargon, abbreviated terms, or inaccurate wording. Thus, the AI agent clarifies the prompt based on resources from the Knowledge Graph. In this subprocess, the AI agent might ask the user for additional clarification. Then, it determines the data sources and generates formal queries to retrieve data. Those queries are used by ETL to retrieve the final dataset, which is subsequently handed back to the AI agent, that prepares a prompt for the VC. The VC generates a visualization with real data. The user might restart this phase if the results are not correct.

Phase 3 – Final visualization building and assessment The goal of this phase is to create a final visualization. This phase starts with the visualization assessment performed by the AI agent against predefined visualization rules. If the assessment results are positive, then the visualization is presented to the end user for acceptance. If there are any errors or possible improvements, the AI agent prepares a list of corrections and rebuilds the prompt for the VC, a new version of the visualization is created, and the AI agent assesses it again.

Phase 4 – Business enhancements and refinements The goal of this phase is to enhance the analysis with business description and insights. After the visualization is accepted by the user, the AI agent adds natural language descriptions (e.g., labels, titles, full sentence description of the diagrams, natural language analysis of the insights, outliers, trends etc.). The length and style of this description are adapted to the analysis audience. The final results are again presented to the end user, who might prompt the system with the list of possible improvements or accept. The result of the example query is shown in Fig. 3.

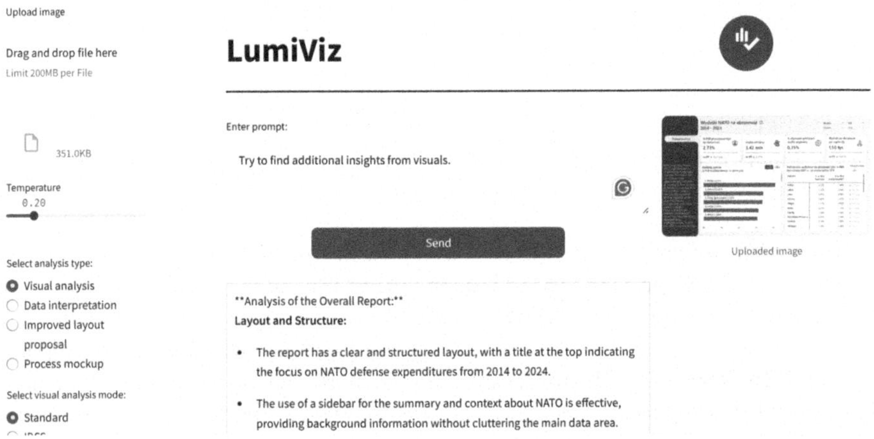

Fig. 3. Lumiviz GUI with example output

End of the process – Feedback There might be an additional feedback step at the end of the process. This is technically a separate business process, optionally initiated at this point. The Evaluation and Feedback Loop is described in Sect. 4.

6 Evaluation

The system functionalities are implemented as a Proof of Concept (PoC), meaning a fully integrated end-to-end solution is not yet available. Instead, individual modules, such as query interpretation and visualization analysis, were tested independently. Special attention was paid to the accuracy and relevance of the visualization, as it quickly provided measurable value and user feedback, forming a foundation for further system development.

The pilot implementation focused primarily on the visualization quality assessment module, which allows the evaluation of selected elements of the proposed architecture. A critical aspect was to ensure seamless communication between the initial chart generation processes and user activity tracking tools. This facilitated the practical validation of the integration of the AI module and the access control mechanisms within the S&G module. However, the full implementation of ETL/ELT orchestration and advanced MLOps practices, including model versioning and continuous maintenance in a production environment, remained beyond the scope of this paper and is designated for future research and development.

To evaluate the data visualization system, a survey was developed, with its core component comprising Likert scale questions, allowing respondents to express their level of satisfaction. The set of questions (1–7) addressed the intuitiveness of the interface, the readability of generated charts and dashboards, the extent to which the system met business needs, response time, flexibility in selecting and customizing visualizations, and the accuracy of AI-generated suggestions. Additionally, a "Comment" field was included, allowing respondents to provide more detailed feedback. A follow-up in-depth interview with each respondent was conducted. The questionnaire was distributed to 7 users of various seniority levels, with practical experience in creating visualizations. The number of respondents was limited due to the chosen method of evaluation. Their expertise in the data visualization allows to conduct in-depth interview and obtain valuable critical insights.

The responses indicate that most users positively rated the core features, particularly the ease of use and readability of charts. However, opinions varied regarding the speed of visualization generation as some users found the response time satisfactory, while others reported slower performance (which, as noted in in-depth interviews, might have been influenced by hardware constraints). Qualitative feedback also revealed that many respondents expect greater flexibility in styling and selecting report elements (e.g., custom colors, labels), but appreciated the ability to "evaluate" AI-generated visualizations before proceeding to the next phase. At the same time, it was noted that user preferences may differ, while the system enforces a high level of standardization. Several users, particularly those more experienced in creating charts, pointed out that while the system is a valuable aid for quick tasks and the initial phase of report creation, it requires advanced customization options for layout and styling at later phases to meet individual needs. The system was also appreciated for allowing less experienced users to receive rapid feedback on their visualizations.

A preliminary analysis of the results suggests that while the tool effectively performs key tasks in intuitive data visualization, there is room for further development in the areas of personalization, integration with corporate standards, and performance optimization.

	Unsatisfied	Neutral	Satisfied
Interface intuitiveness		0%	100%
Suitability to needs		18%	82%
Flexibility	9%	9%	82%
Clarity of results	9%	9%	82%
Generation speed	9%	9%	82%
Accuracy of suggestions		9%	91%
Overall satisfaction		0%	100%
Total	**4%**	**8%**	**88%**

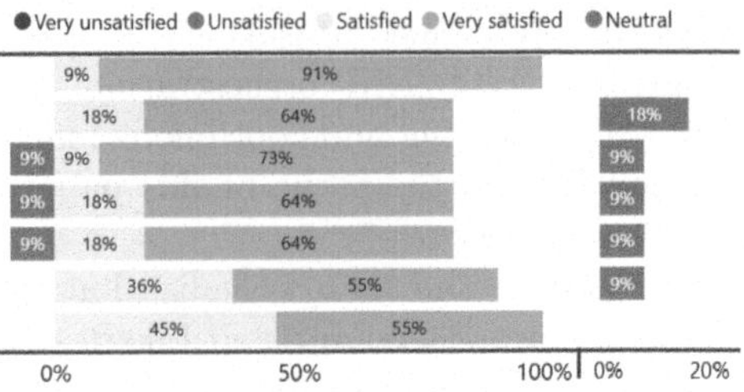

Fig. 4. Evaluation results

7 Conclusions

The paper presents a system that uses GenAI to automate business data analysis processes - from processing natural language queries, retrieving relevant data, and visualizing it, to provide business interpretations of the results. The study highlights key challenges related to both organizational issues (e.g., lack of trust in AI algorithms, diverse industry nomenclature, or competency gaps) and strictly technical aspects, such as integration with existing IT infrastructure.

In Sect. 4 elements of system architecture and organizational structure necessary for implementation were identified, which answers the first research question: *How effectively can GenAI-based tools automate business data visualization, and what are the key technical components required for their implementation?*. The proposed architecture consists of a Security & Governance module, an AI module, and a presentation module. Conducted research emphasizes that technical solutions, like automatic translation of natural language into SQL, must align with organizational arrangements, including clear role distribution (management, department managers, analysts, data engineers) and fostering an innovation culture with appropriate competency support.

In order to answer the second research question, *Which stages of visualization process can be effectively automated using GenAI?*, a process flow of LumiViz was modeled and analyzed (Sect. 5). The results show that automatization with GenAI can be successfully implemented at each stage of preparing visualization of business data, however, user's feedback and assessment are needed at the end of each phase of this process. This step-by-step approach gives the user the possibility to catch errors earlier and adapt the prompts accordingly.

The last research question relates to the user experience and the overall assessment of the LumiViz applicability in a real-world business scenario. The tool was tested by users on different seniority levels. The results of the evaluation are discussed in Sect. 6 but the overall users' satisfaction achieved 88%.

A preliminary analysis of the results suggests that while the tool effectively performs key tasks in intuitive data visualization, further improvements can be made in personalization, integration with corporate standards, and performance optimization in specific environments. It also allows less experienced users to receive initial feedback on their visualizations.

Future research should prioritize empirical testing of generative AI tools across various business sectors (e.g., e-commerce, manufacturing, logistics) to assess their effectiveness and develop best practices for integrating LLMs into existing data ecosystems. Additionally, advancements in automatic response verification tools are expected to improve the reliability of AI-generated insights by minimizing errors and mitigating the risk of model hallucinations.

Another critical area for exploration is the incorporation of GenAI into business process management systems. Specifically, further research should examine how integrating AI with knowledge graphs and real-time recommendation systems can improve decision-making quality and operational efficiency in dynamic business environments.

References

1. Beheshti, A., Yang, J., et al.: ProcessGPT: transforming business process management with generative artificial intelligence. In: Proceedings - 2023 IEEE International Conference on Web Services, ICWS 2023, pp. 731–739 (May 2023). https://doi.org/10.1109/ICWS60048.2023.00099
2. Berret, C., Munzner, T.: Iceberg Sensemaking: a process model for critical data analysis and visualization. IEEE Trans. Vis. Comput. Graph. **PP** (2022). https://api.semanticscholar.org/CorpusID:248084915
3. Brown, T.B., Mann, B., et al.: Language models are few-shot learners. Adv. Neural Inf. Process. Syst. **33**, 1877–1901 (2020). https://commoncrawl.org/the-data/
4. Bruckner, S., Turkay, C., Vrotsou, K.: Natural language generation for visualizations: state of the art, challenges and future directions **41** (2024). https://arxiv.org/abs/2409.19747v1
5. Chen, B., Wu, Z., Zhao, R.: From fiction to fact: the growing role of generative AI in business and finance. SSRN Electron. J. (2023). https://doi.org/10.2139/SSRN.4528225
6. Dhoni, P., Dhoni, P.S.: Exploring the synergy between generative AI, data and analytics in the modern age. Authorea Preprints (October 2023). https://doi.org/10.36227/TECHRXIV.24045792.V1
7. Feretzakis, G., Papaspyridis, K., et al.: Privacy-preserving techniques in generative AI and large language models: a narrative review. Information **15**, 697 (2024). https://doi.org/10.3390/INFO15110697
8. Hao, J., Yang, M., et al.: Finflier: automating graphical overlays for financial visualizations with knowledge-grounding large language model (December 2024). http://arxiv.org/abs/2412.06821
9. Hevner, A., Salvatore, M., et al.: Design science in information systems research. MIS Q. (2004). https://doi.org/10.5555/2017212.2017217
10. Hoffman, D.A., Arbel, Y.A.: Generative interpretation. SSRN Electron. J. (2023). https://doi.org/10.2139/SSRN.4526219
11. Inala, J.P., Wang, C., et al.: Data analysis in the era of generative AI (2024). https://arxiv.org/abs/2409.18475
12. Jovanovic, P., Nadal, S., Romero, O., Abelló, A., Bilalli, B.: Quarry: a user-centered big data integration platform. Inf. Syst. Front. **23**, 9–33 (2020). https://api.semanticscholar.org/CorpusID:218781877

13. Li, F., Jagadish, H.V.: Constructing an interactive natural language interface for relational databases. Proc. VLDB Endow. **8**, 73–84 (2014). https://doi.org/10.14778/2735461.2735468
14. Li, H., Guo, D., et al.: Multi-step jailbreaking privacy attacks on chatgpt. In: Findings of the Association for Computational Linguistics: EMNLP 2023, pp. 4138–4153. Association for Computational Linguistics (2023). https://doi.org/10.18653/v1/2023.findings-emnlp.272
15. Lin, Y., Shen, H.: Smartq: a question and answer system for supplying high-quality and trustworthy answers. IEEE Trans. Big Data **4**, 600–613 (2018). https://api.semanticscholar.org/CorpusID:54436126
16. Ma, R., Mei, H., et al.: Ladv: deep learning assisted authoring of dashboard visualizations from images and sketches. IEEE Trans. Vis. Comput. Graph. **27**, 3717–3732 (2021). https://doi.org/10.1109/TVCG.2020.2980227
17. Maddigan, P., Susnjak, T.: Chat2vis: generating data visualizations via natural language using chatgpt, codex and gpt-3 large language models. IEEE Access **11**, 45181–45193 (2023). https://doi.org/10.1109/ACCESS.2023.3274199
18. Malacaria, S., Grimaldi, M., et.al.: Business talk: harnessing generative ai with data analytics maturity. Int. J. Cybern. Inform. **12**, 01–10 (2023). https://doi.org/10.5121/IJCI.2023.120701
19. Molnar: Interpretable machine learning (2019). https://christophm.github.io/interpretable-ml-book/
20. Penko, I., Elitsa, P.: Enhancing business analytics through generative AI: integrating large language models with proprietary knowledge graphs for advanced data querying and visualization. In: Rajagopal, Goncalves, M., Zlatev, V. (eds.) Entrepreneurship, Innovation, and Technology. Palgrave Studies of Entrepreneurship and Social Challenges in Developing Economies, LNCS, pp. 239–256. Springer, Cham (2024). https://doi.org/10.1007/978-3-031-65314-8_10
21. Reznikov, R.: Leveraging generative AI: Strategic adoption patterns for enterprises. Modeling the Development of the Economic Systems, pp. 201–207 (March 2024). https://doi.org/10.31891/MDES/2024-11-29
22. Schwarz, J.: The use of generative AI in statistical data analysis and its impact on teaching statistics at universities of applied sciences. Teach. Stat. (2025). https://doi.org/10.1111/test.12398
23. Shen, L., Shen, E., et al.: Towards natural language interfaces for data visualization: a survey. IEEE Trans. Vis. Comput. Graph. **29**, 3121–3144 (2021). https://doi.org/10.1109/TVCG.2022.3148007
24. Uddin, M.K.S.: A review of utilizing natural language processing and AI for advanced data visualization in real-time analytics. Glob. Mainstream J. **1**, 34–49 (2024). https://doi.org/10.62304/IJMISDS.V1I04.185
25. Vaswani, A., Shazeer, N., et al.: Attention is all you need. Adv. Neural Inf. Process. Syst. **2017-December**, 5999–6009 (2017). https://arxiv.org/abs/1706.03762v7
26. Wang, Y., Pan, Y., et al.: A survey on chatgpt: ai-generated contents, challenges, and solutions. IEEE Open J. Comput. Soc. **4**, 280–302 (2023). https://doi.org/10.1109/OJCS.2023.3300321

27. Yan, L., Martinez-Maldonado, A.R., et al.: Generative artificial intelligence in learning analytics: contextualising opportunities and challenges through the learning analytics cycle. In: ACM International Conference Proceeding Series, pp. 101–111 (March 2024). https://doi.org/10.1145/3636555.3636856
28. Yu, T., Li, Z., et al.: Typesql: knowledge-based type-aware neural text-to-sql generation. In: NAACL HLT - 2018 Conference of the North American Chapter of the Association for Computational Linguistics: Human Language Technologies, vol. 2, pp. 588–594 (2018). https://doi.org/10.18653/V1/N18-2093
29. Zhong, V., Xiong, C., Socher, R.: Seq2sql: generating structured queries from natural language using reinforcement learning. arXiv preprint arXiv:1709.00103 (August 2017)

AI-Driven Video Avatar for Academic Support

Stanisław Kumor(✉), Mikołaj Maik, and Krzysztof Walczak

Poznań University of Economics and Business, Niepodległości 10, Poznań, Poland
{stanislaw.kumor,mikolaj.maik,krzysztof.walczak}@ue.poznan.pl

Abstract. This paper presents the implementation and initial evaluation of an AI-powered interactive video avatar designed to enhance student consultations in a university course. As educational institutions explore new ways to provide accessible and flexible academic support, we investigate the potential of AI avatars in consultation scenarios. The implementation combines video synthesis with natural language processing to create an interactive professor avatar trained on content from actual lectures. We evaluated the system with students through structured tasks focused on technical knowledge delivery, explanation capabilities, interaction and context management, and handling ambiguous queries. The assessment covered both educational effectiveness in sustaining meaningful dialogue and technical performance - including audiovisual synchronization and animation quality. Results demonstrate the avatar's effectiveness in conveying technical content and maintaining educational context during consultations. Technical performance was particularly strong in speech quality and audiovisual synchronization, demonstrating robustness even when addressing technical queries on multimedia topics. These findings highlight the potential of AI avatars to enhance student consultations in technical academic subjects.

Keywords: Educational technology · Student consultations · Interactive avatar · Conversational AI · HCI · NLP

1 Introduction

Artificial intelligence is reshaping educational practices across academic institutions. The integration of AI tools in education has progressed from basic instructional automation to intelligent tutoring systems that offer personalized learning experiences [23]. Institutions are particularly interested in AI solutions that extend teaching beyond traditional classroom hours, addressing the growing demand for flexible and accessible academic support-especially in online environments [13,22].

The application of AI in higher education has shown promising results across various domains [27]. However, a persistent challenge remains: enabling natural, engaging interactions that effectively replicate human-like teaching experiences

K. Węcel (Ed.): BIS 2025, LNBIP 554, pp. 56–69, 2025.
https://doi.org/10.1007/978-3-031-94193-1_5

[5]. Integrating AI into education also demands careful consideration of ethical implications and the development of inclusive frameworks to maximize benefits and minimize risks [14]. While text-based AI assistants offer valuable academic support, they often lack the personal connection and non-verbal cues that are essential to effective educational interactions [2].

Recent advances in video synthesis and natural language processing have opened new possibilities for creating more realistic and engaging AI-powered educational experiences. These technologies enable the development of virtual teaching assistants that combine visual presence with intelligent dialogue, offering potentially more natural and effective student interactions than traditional text-based chatbots [11]. Voice-based intelligent virtual agents-and systems such as IBM Watson-have also demonstrated significant potential in enhancing interactivity and providing personalized academic support [3,15].

This study evaluates the design and effectiveness of an AI-powered interactive professor avatar developed to assist with student consultations. The solution combines audio and video synthesis technology from Heygen with OpenAI's language processing capabilities to create an interactive and realistic virtual representation of the actual instructor, trained on content derived from 15 lectures delivered in a university class on Multimedia in Business. The research investigates whether AI-driven avatars can effectively supplement academic support by enhancing the flexibility and accessibility of student consultations.

Students enrolled in the course participated in the evaluation of the avatar through a structured set of tasks, including technical knowledge assessment, explanation capabilities, interaction and context management, and the handling of ambiguous queries. The evaluation addressed both the pedagogical effectiveness in explaining course concepts and sustaining meaningful dialogue with students, and technical aspects of the system-such as audio-visual synchronization and animation quality. This research explores several key questions: Can an AI avatar effectively assist professors in student consultations? How do students perceive and interact with a video-based AI lecturer? What are the current technical and practical limitations of AI avatar technology in educational settings?

The remainder of this paper is structured as follows. Section 2 reviews related work on the use of artificial intelligence in education, including adaptive learning systems, educational chatbots, and avatar-based technologies. Section 3 presents the implementation of the interactive avatar, describing the system architecture and its integration of video synthesis and language processing components. Section 4 outlines the evaluation methodology, including task design, participant details, and data collection procedures. Section 5 presents the results of the study, including technical performance, user interaction quality, and task completion times. Section 6 discusses the key findings and their implications. Finally, Sect. 7 concludes the paper and outlines directions for future research.

2 Related Works

The adoption of artificial intelligence in higher education is rapidly increasing, particularly in areas such as course automation, personalized learning, and text-based support systems. While educators initially played a limited role in AI research, their involvement has grown significantly in recent years [6,8,27]. Common applications include adaptive content sequencing, tracking of learning progress, performance prediction, and early identification of students who may require additional attention [8]. Research has been especially active in adaptive and personalized learning systems (54.5%), profiling and prediction (48.5%), and intelligent tutoring systems (21.2%) that dynamically adjust instructional content to individual learners [4]. Natural language processing and deep learning models-such as LSTM, BERT, and Bi-LSTM-have also been used in educational contexts, including automated essay evaluation systems (AES), though challenges remain in content interpretation and manipulation detection [20]. It is worth noting that, under the EU AI Act, automated systems for student assessment are classified as high-risk applications, requiring strict compliance with transparency, fairness, and oversight requirements [9].

Educational chatbots support learning through interactive quizzes and supplementary materials, offering immediate feedback and adapting the pace and scope of content to individual student needs [7]. AI has been shown to significantly improve mathematics performance (with a 20–30% increase in effectiveness) and to enhance student engagement in physics-although this does not always translate into better learning outcomes [16,21]. In academic writing, AI tools can assist with text organization, argumentation, and grammar; however, they still require human oversight, as they may introduce errors or generate false citations [10,24]. Integrating AI into education requires changes in teaching and assessment methods, with a growing emphasis on critical thinking and the evaluation of AI-generated content. Students recognize the opportunities offered by these technologies, but they face challenges in assessing reliability and express concerns about algorithmic grading [25]. Numerous risks have been identified in the use of generative AI in education, including inaccuracies, bias, disinformation, and privacy violations [26].

Avatars in education can be defined as digital representations of users in virtual environments, designed to support communication, collaboration, and interaction across a variety of educational contexts, including learning management systems (LMS), virtual classrooms, and simulation-based environments [18]. Research has demonstrated that the use of avatars can enhance student motivation and engagement by enabling more realistic interaction in digital settings. Students have been shown to assimilate knowledge more effectively and apply it in new contexts [12]. However, controlling avatars and navigating virtual spaces can be challenging and often requires additional technical support [12].

Integrating AI into education has been shown to enhance interactivity, although the ability to adapt to individual learning preferences remains limited [19]. Students have expressed a preference for hybrid models that combine AI tools with traditional teaching methods, citing accessibility and flexibility as key advantages [19]. Challenges include technical limitations, ethical concerns, and the need for seamless integration with existing educational practices [1,17]. The development of realistic voices and animations is widely recognized as a crucial requirement for enhancing the effectiveness of avatar-based educational systems.

To address the growing demand for flexible academic support and explore the potential of AI-driven interaction in educational contexts, we developed an interactive professor avatar designed to assist students outside regular office hours. The system integrates advanced video and audio synthesis with a large language model to enable natural, responsive, and realistic interactions. Unlike traditional chatbots or static learning platforms, our avatar simulates a lifelike representation of the actual course instructor, both visually and conversationally. The design focuses on providing personalized explanations, clarifying complex concepts, and supporting student engagement in a scalable and accessible manner. In the following section, we describe the architecture of the system, including its main components and the workflow enabling real-time interaction.

3 Implementation

The architecture of the interactive avatar is organized into two layers: the *front end* and the *back end*. The front end provides the user interface, enabling students to interact with the system. The back end handles communication with external services-namely OpenAI and HeyGen-and is responsible for interpreting user input, generating appropriate responses, and synthesizing audio and video output (Fig. 1). The system consists of four components:

- **The front-end component** is implemented using HTML, CSS, and JavaScript. Its primary function is to capture the audio signal from the user's microphone, transmit the audio file to the back-end layer, and receive the generated video stream, which is then presented to the user.
- **The back-end component** is implemented in Python using the Flask framework. It manages incoming requests, coordinates communication with external APIs, stores logs and session states, and appends system-level instructions (prompts) to user messages.
- **The OpenAI component** handles speech processing and language generation. The Whisper-1 model transcribes user speech into text, which is then refined for grammatical correctness using GPT-4o. A separate GPT-4o-mini instance is used to generate responses to the user's input.
- **The HeyGen component** manages interaction with the HeyGen platform API. It receives the generated response text and returns a video of the avatar delivering the response, synchronized with a synthetic voice.

The model's knowledge base has been populated with lecture materials stored in PDF format, comprising 780 slides and totaling approximately 51 MB. The model's access is restricted exclusively to this content, ensuring that all generated responses are grounded solely in the provided lecture material. This approach guarantees alignment with the original course content and eliminates the influence of external sources on the model's answers.

Each conversation is recorded in a JSON log, with each entry containing the conversation identifier, the user's transcribed message, and its grammatically corrected version (produced via an additional OpenAI API call). The log also stores any system instructions appended to the prompt, along with the final version of the message submitted to the model. If the file analysis procedure is invoked during the session, this is also documented. In addition, the model's response is saved together with detailed timestamps, including the query's initiation and completion times, as well as the total processing duration.

4 Evaluation

The evaluation of the AI-powered professor avatar focused on assessing its effectiveness in student consultations across multiple dimensions. This section presents the evaluation methodology, including the study design, task structure, data collection procedures, and performance metrics. The assessment framework was designed to examine both educational value and technical performance through structured one-on-one interactions with students enrolled in the Multimedia in Business course. The evaluation investigated the avatar's ability to convey technical content, explain complex concepts, maintain conversational coherence, and respond appropriately in varied consultation scenarios.

4.1 Study Design

The study participants were students enrolled in the Multimedia in Business course. Each participant engaged in a one-on-one consultation session with the AI avatar in a controlled environment (Fig. 2). The avatar was intentionally designed to represent the actual course instructor, providing continuity and familiarity in the learning experience. This design choice ensured educational consistency and allowed students to interact with a virtual version of a familiar educator. The individual sessions provided a standardized setting for evaluating the avatar's capabilities across all participants.

4.2 Tasks and Procedure

The evaluation process was structured into four distinct tasks, each designed to assess a specific aspect of the avatar's functionality. The first task evaluated the avatar's subject matter expertise in technical multimedia concepts. Students engaged in discussions involving complex calculations and theoretical understanding, including bitrate computations for HD video with H.264 compression, comparative analysis of MPEG-2 and MPEG-4 codecs, and the selection of appropriate color spaces for online video streaming applications.

The second task evaluated the avatar's ability to explain complex concepts effectively. Students were asked to select challenging topics from the course materials and engage in a structured dialogue with the avatar. This dialogue followed a careful progression from initial concept explanation to practical applications.

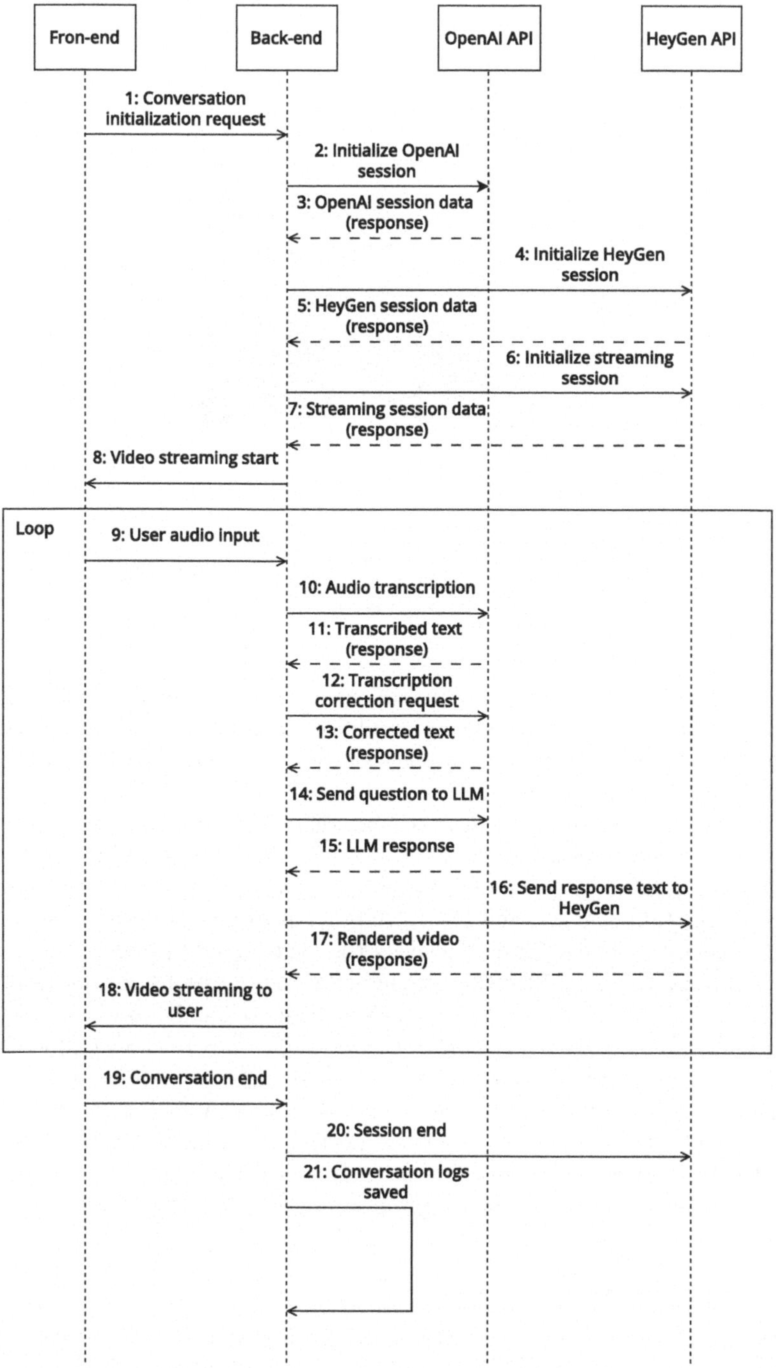

Fig. 1. System architecture

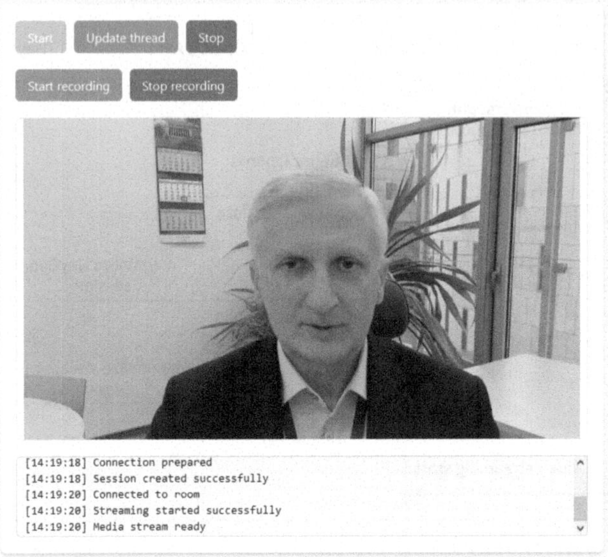

Fig. 2. Interactive avatar

The questioning sequence began with students requesting clarification on specific slide content, followed by requests for simplified explanations. The sequence then progressed to exploring practical applications and concluded with discussions of real-world implementation examples.

The third task assessed the avatar's interaction capabilities and contextual awareness. Students initiated discussions on lecture topics, then intentionally introduced unrelated questions to test the avatar's ability to manage conversation flow, maintain context, and return to previous topics after digressions. This task specifically evaluated how effectively the avatar could preserve the original discussion thread while addressing the interrupting questions appropriately.

The fourth task assessed the avatar's ability to handle problematic communication scenarios, with a specific focus on discussions related to office hours. Students presented the avatar with increasingly complex situations-starting with ambiguous queries about consultation times, followed by incorrect assumptions about scheduling, and concluding with logically contradictory statements. This task evaluated the avatar's capacity to maintain clear communication while managing uncertainty and responding appropriately to conflicting information.

4.3 Data Collection

The research methodology involved collecting data across multiple parameters during each interaction to evaluate the avatar's performance. These included quantitative measures such as task completion times and Likert-scale ratings of response quality and contextual coherence. The evaluation also tracked the avatar's ability to maintain conversation flow and appropriately handle diverse query types.

Student feedback was gathered after each task using questionnaires with 5-point Likert scales to evaluate specific aspects of the avatar's performance. The quality and effectiveness of the interaction were assessed immediately after each task. All timing data were captured using digital timers, with precise timestamps recorded at the start and end of each task.

4.4 Evaluation Metrics

The assessment focused on a quantitative evaluation of the avatar's educational performance. Response accuracy was measured by examining the avatar's ability to address technical subject matter, while clarity of explanations was rated by students based on their level of understanding. The ability to maintain conversational context was evaluated through the avatar's success in recovering original topics after interruptions. Effectiveness in managing contradictory information was assessed based on its responses to conflicting queries. All numerical results presented in this study represent arithmetic means calculated across all participants.

5 Results

The study analyzed data from 22 participants who engaged with the AI avatar through a series of structured tasks. The participant group had a consistent age distribution (mean: 23.2 years, standard deviation: 1.13) and reported a high level of prior experience with AI technologies (average: 8.0 out of 10). This demographic profile suggests that the evaluation was conducted with users who were well-acquainted with AI-based tools, providing informed and relevant feedback.

5.1 Task-Based Evaluation of Avatar Performance

Each task performed by participants targeted specific aspects of system performance: technical knowledge, explanation capabilities, interaction and context management, and handling of problematic scenarios. The results are graphically presented in Fig. 3. The following list summarizes the purpose, duration, and evaluation results for each task.

- **Task 1: Technical Knowledge Assessment**: This task had an average completion time of 295.5 s. Response coherence scored 4.18 out of 5, indicating strong performance in handling technical queries related to multimedia concepts.
- **Task 2: Explanation Capabilities Assessment**: This task required 350.9 s on average. The quality of explanations averaged 3.59 out of 5. The duration reflects the time needed for detailed concept breakdowns, while the score suggests that explanation clarity could be enhanced.

- **Task 3: Interaction and Context Management**: Context-switching exercises averaged 355.8 s, with high scores in transition fluidity (4.36/5) and context maintenance (3.95/5). These results demonstrate the avatar's ability to handle interruptions while preserving the conversation thread.
- **Task 4: Handling of Ambiguous Scenarios**: Ambiguity-handling tasks were completed in 192.6 s. The avatar's response to unclear or contradictory queries scored 3.90 out of 5, indicating effective management of challenging communication scenarios.

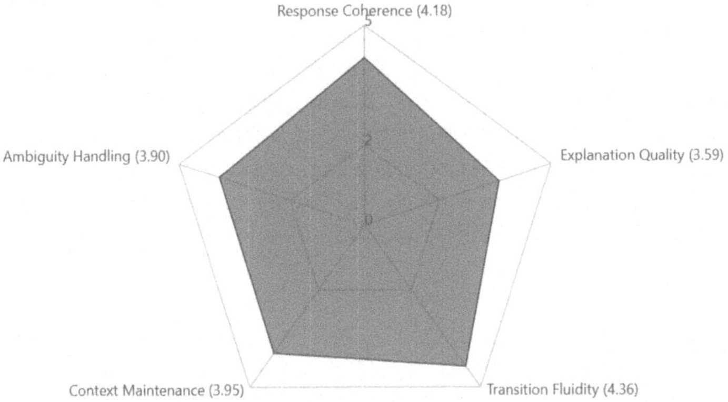

Fig. 3. Task Performance Metrics Chart

5.2 Technical Performance Assessment

The technical evaluation revealed distinct performance characteristics across multiple system components (Fig. 4). The speech synthesis module received a high rating of 4.50 out of 5, reflecting the overall quality of voice output. Avatar animation fluidity scored 3.68 out of 5, suggesting an acceptable level of motion realism, though with room for improvement. The highest technical score was achieved in audio-visual synchronization, which received 4.59 out of 5. This result confirms the system's effectiveness in maintaining precise alignment between spoken output and corresponding visual elements.

5.3 User Interaction Quality

Analysis of user interaction metrics revealed varying levels of performance across different dimensions (Fig. 5). The natural conversation flow received a score of 3.23 out of 5, indicating that while basic conversational functions operated effectively, interactions occasionally felt mechanical. Response adequacy was rated at 3.95, suggesting that the avatar generally provided relevant and contextually

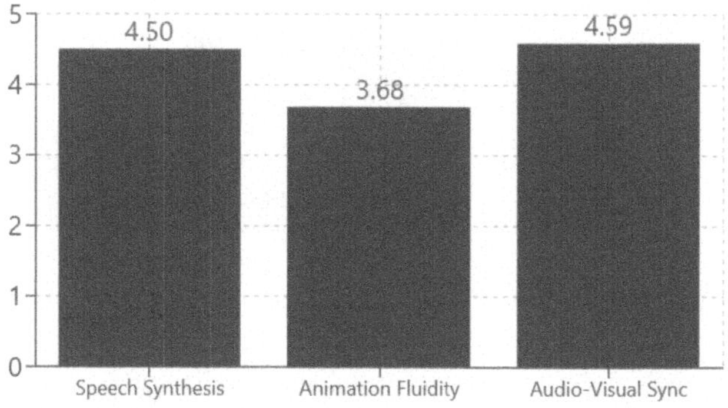

Fig. 4. Technical Performance Metrics

appropriate answers. Response completeness scored higher at 4.45, reflecting the avatar's ability to deliver thorough and comprehensive replies. The perceived embodiment of the professor's personality received a score of 3.59, indicating recognizable characteristics of educational authority. Overall consultation attractiveness was rated at 3.45, pointing to moderate user satisfaction and highlighting areas for further improvement.

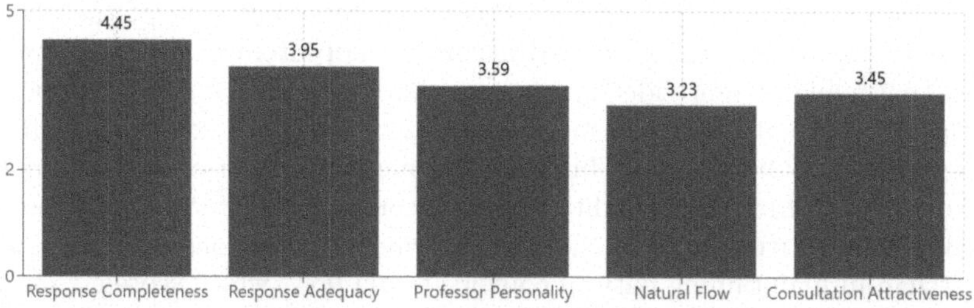

Fig. 5. User Interaction Quality Metrics

5.4 Task Completion Times

The temporal analysis revealed distinct patterns across different interaction types (Fig. 6). The technical assessment task had an average completion time of 295.5 s, indicating efficient handling of subject-specific queries. Explanation tasks required more time, averaging 350.9 s, which reflects the added complexity of breaking down and clarifying advanced concepts. Context management tasks showed similar temporal demands, with an average duration of 355.8 s. Ambiguity-handling tasks were completed most quickly, with an average time of 192.6 s. This shorter duration did not compromise performance, as reflected in the strong evaluation scores for this task type.

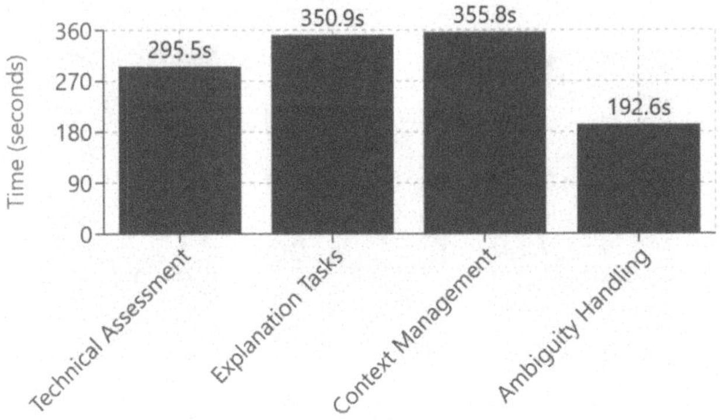

Fig. 6. Task Completion Times

6 Discussion

The evaluation results highlight several key aspects of the AI avatar's performance in the context of academic consultations. Among the technical metrics, speech synthesis (4.50/5) and audio-visual synchronization (4.59/5) demonstrated strong performance, indicating that the system delivers a high degree of realism in both voice and timing. In contrast, animation fluidity received a lower score (3.68/5), suggesting that the visual representation of the avatar's movements may reduce the perceived naturalness of interaction. These findings suggest that while current video synthesis technologies are capable of supporting essential educational dialogues, more complex or dynamic exchanges remain a challenge and may benefit from further refinement in avatar animation fidelity.

The analysis of response quality reveals a consistent pattern: the avatar performed more effectively on technical queries-achieving a response coherence score of 4.18-than in maintaining natural conversational flow, which scored 3.23. This disparity highlights the ongoing challenge of balancing accurate knowledge delivery with fluid, human-like dialogue in AI-based teaching assistants. The notably higher score for response completeness (4.45) compared to conversational flow further suggests that the system prioritizes informational accuracy over natural interaction style. This performance profile may be partially attributed to the constrained knowledge base, which was limited exclusively to lecture materials.

Task completion times offer additional insight into the avatar's operational efficiency. The longer duration required for concept explanations (350.9 s), compared to ambiguity-handling tasks (192.6 s), aligns with typical teaching dynamics, where in-depth explanation of complex topics naturally takes more time than responding to administrative or procedural queries.

Prompt engineering played a key role in constraining the avatar's responses to multimedia-related topics and in shaping the expected structure and tone of the answers. The high score in response accuracy (4.50) suggests that the structured knowledge base and prompt design were effective in guiding the model's output.

Student engagement data suggest that while the avatar effectively conveyed educational authority-reflected in a score of 3.59 for professor personality embodiment-the overall attractiveness of the consultation experience was rated slightly lower at 3.45. This observation raises important questions about how to balance professional distance with the creation of an approachable learning environment.

The technical knowledge assessment results (4.18) show that the avatar can effectively handle specific multimedia calculations and concepts. However, variation in explanation quality scores (mean: 3.59, standard deviation: 1.07) suggests that some technical topics may require additional context or alternative explanatory approaches within the avatar's knowledge base.

These findings suggest that future development should focus on enhancing animation fluidity, improving conversational naturalness, and refining the balance between formal instruction and engaging interaction. The results also indicate that current AI avatar technology is more effective for structured technical discussions than for open-ended educational dialogues.

7 Conclusions and Future Works

The implemented AI-powered interactive professor avatar demonstrates promising potential for supporting student consultations in education. Evaluation results indicate strong performance in technical knowledge transfer, particularly in audiovisual synchronization and speech synthesis quality. The system effectively maintains educational context during interruptions and accurately handles technical queries.

However, the study also identified areas for improvement, particularly in conversational naturalness and animation fluidity. These limitations currently affect the overall consultation experience, suggesting that the technology, while functional, requires refinement to enhance pedagogical effectiveness. Future efforts should focus on improving the natural flow of conversations to make interactions feel less mechanical and more engaging, and on refining animation quality to support more lifelike and dynamic interactions.

Further research should extend testing to other academic subjects and educational settings to validate the generalizability of findings and ensure robust performance across contexts. Expanding multilingual capabilities could support international students by improving the technical accuracy of terminology in multiple languages and promoting more inclusive engagement.

Longitudinal studies could help track avatar performance over time and assess long-term educational impact. Additional priorities include load testing under multi-user conditions, comparative studies on avatar design and its influence on student engagement, and exploring hybrid approaches to balance hallucination prevention with broader knowledge access. Finally, examining the impact of knowledge base file formats and structures on avatar performance may help optimize content organization for interactive consultations, supporting the more effective deployment of AI avatars in education.

Disclosure of Interests. The authors declare no competing interests relevant to the content of this article.

References

1. Arkün-Kocadere, S., Çağlar-Özhan, Ş: Video lectures with ai-generated instructors: low video engagement, same performance as human instructors. Int. Rev. Res. Open Distrib. Learn. **25**(3), 350–369 (2024)
2. Arya, R., Verma, A.: Role of artificial intelligence in education. Int. J. Adv. Res. Sci. Commun. Technol. (2024). https://doi.org/10.48175/ijarsct-19461
3. Benedetto, L., Cremonesi, P., Parenti, M.: A virtual teaching assistant for personalized learning. arXiv preprint arXiv:1902.09289 (2019)
4. Bond, M., et al.: A meta systematic review of artificial intelligence in higher education: a call for increased ethics, collaboration, and rigour. Int. J. Educ. Technol. High. Educ. **21**(1), 4 (2024)
5. Chen, S., Xu, X., Zhang, H., Zhang, Y.: Roles of chatgpt in virtual teaching assistant and intelligent tutoring system: opportunities and challenges. In: Proceedings of the 2023 5th World Symposium on Software Engineering (2023). https://doi.org/10.1145/3631991.3632024
6. Chen, X., Xie, H., Zou, D., Hwang, G.J.: Application and theory gaps during the rise of artificial intelligence in education. Comput. Educ. Artif. Intell. **1**, 100002 (2020)
7. Chen, Y., Jensen, S., Albert, L.J., Gupta, S., Lee, T.: Artificial intelligence (AI) student assistants in the classroom: designing chatbots to support student success. Inf. Syst. Front. **25**(1), 161–182 (2023)
8. Crompton, H., Burke, D.: Artificial intelligence in higher education: the state of the field. Int. J. Educ. Technol. High. Educ. **20**(1), 22 (2023)
9. EU Regulation: Artificial intelligence act. The Official Journal (OJ) of the European Union (2024/1689). Accessed 01 Apr 2025
10. Fyfe, P.: How to cheat on your final paper: assigning ai for student writing. AI Soc. **38**(4), 1395–1405 (2023)
11. González, L.A., Neyem, A., Contreras-McKay, I., Molina, D.: Improving learning experiences in software engineering capstone courses using artificial intelligence virtual assistants. Comput. Appl. Eng. Educ. **30**, 1370–1389 (2022). https://doi.org/10.1002/cae.22526
12. Han, E., Bailenson, J.N.: Lessons for/in virtual classrooms: designing a model for classrooms inside virtual reality. Commun. Educ. **73**(2), 234–243 (2024)
13. Joshi, V., Tripathi, S.K.: Optimizing higher education using artificial intelligence. Int. J. Multidiscip. Res. (2024). https://doi.org/10.36948/ijfmr.2024.v06i03.19721
14. Kamalov, F., Calonge, D.S., Gurrib, I.: New era of artificial intelligence in education: towards a sustainable multifaceted revolution. Sustainability (2023). https://doi.org/10.3390/su151612451
15. Katsarou, E., Wild, F., Sougari, A.M., Chatzipanagiotou, P.: A systematic review of voice-based intelligent virtual agents in EFL education. Int. J. Emerg. Technol. Learn. **18**, 65–85 (2023). https://doi.org/10.3991/ijet.v18i10.37723
16. Kretzschmar, V., Seitz, J.: Ai-based learning assistants: enhancing math learning for migrant students in German schools. In: 2024 IEEE Conference on Artificial Intelligence (CAI), pp. 144–149. IEEE (2024)

17. Moolman, J.H., Boyle, F., Walsh, J.: Revolutionising engineering education: creating photorealistic virtual human lecturers using artificial intelligence and computer generated images (2023)
18. Nasir, S.M., Hashim, M.E.A.H., Nee, C.K., Kamaruddin, N.H., Wibawanto, W., Hanapiah, M.L.H.M.: Enhancing student engagement in learning management systems through exploration of avatars in virtual classrooms: a systematic (2024)
19. Pang, C.C., Zhao, Y., Yin, Z., Sun, J., Mogavi, R.H., Hui, P.: Artificial human lecturers: initial findings from Asia's first AI lecturers in class to promote innovation in education. arXiv preprint arXiv:2410.03525 (2024)
20. Ramesh, D., Sanampudi, S.K.: An automated essay scoring systems: a systematic literature review. Artif. Intell. Rev. **55**(3), 2495–2527 (2022)
21. Robledo-Rella, V., Toh, B.Y.: Artificial intelligence in physics courses to support active learning. In: Proceedings of the 2024 10th International Conference on e-Society, e-Learning and e-Technologies (ICSLT), pp. 68–75 (2024)
22. Sajja, R., Sermet, Y., Cwiertny, D.M., Demir, I.: Platform-independent and curriculum-oriented intelligent assistant for higher education. Int. J. Educ. Technol. High. Educ. **20** (2023). https://doi.org/10.1186/s41239-023-00412-7
23. Shekhar, G., D'souza, R., Fernandes, K.: AI-driven contextual virtual teaching assistant using rasa. In: Proceedings of the 21st Annual Conference on Information Technology Education (2020). https://doi.org/10.1145/3368308.3415442
24. Utami, S., Karnedi, K.: Enhancing students' writing paragraphs through canva magic ai. Leksika: Jurnal Bahasa, Sastra dan Pengajarannya **18**(2), 105–116 (2024)
25. Walczak, K., Cellary, W.: Challenges for higher education in the era of widespread access to generative AI. Econ. Bus. Rev. **9**(2), 71–100 (2023)
26. Walczak, K., Cellary, W.: Navigating risks: Inaccuracies, bias, disinformation, and privacy in educational AI. In: Corbeil, J.R., Corbeil, M.E. (eds.) Teaching and Learning in the Age of Generative AI Evidence-Based Approaches to Pedagogy, Ethics, and Beyond, pp. 163–194. Routledge (2025). https://doi.org/10.4324/9781032688602-11
27. Zawacki-Richter, O., Marín, V.I., Bond, M., Gouverneur, F.: Systematic review of research on artificial intelligence applications in higher education-where are the educators? Int. J. Educ. Technol. High. Educ. **16**(1), 1–27 (2019). https://doi.org/10.1186/s41239-019-0171-0

Think Outside the Box – and Between Them: A Visual Guide to Smarter Technological Development

Dorothea Schneider(✉)

Chair of Machine Tools Development and Adaptive Controls, Institute of Mechatronic Engineering, TUD Dresden University of Technology, 01062 Dresden, Germany
dorothea.schneider@tu-dresden.de

Abstract. The rapid evolution of digital and AI-assisted technologies presents significant challenges for technological development, particularly in complex contexts such as cyber-physical systems (CPS). Addressing these challenges requires iterative approaches to navigate wicked managerial problems where analytical methods fall short. This paper introduces the Technology Inquiry Reference Model (TIRM), a lightweight framework that bridges problem and solution spaces in technological development. Supported by the Technology Inquiry Canvas (TIC) and the Technology Inquiry Process (TIP), the artifact integrates relevance and rigor to enable informed, collaborative development. The TIRM introduces a novel triadic approach that centers the developer as an empowered actor in complex design tasks. This contribution advances the field by operationalizing relevance, rigor, and human agency through integrated visual and process tools.

Keywords: Canvas · Human-centered Knowledge Management · Design Science

1 Chartering the Landscape: Challenges and Opportunities in Technological Development

The technological environment is evolving at an unprecedented pace, with advancements such as artificial intelligence (AI)-assisted systems offering substantial opportunities while introducing challenges such as reliance on robust data management, overwhelming complexity, and increasing cybersecurity risks [13]. This rapid evolution demands a rethinking of technological development tasks, as the convergence of hardware, software, and AI blurs the boundaries between information technology and operational technology. Systems are increasingly evolving into cyber-physical systems (CPS), requiring developers to navigate a wide spectrum of socio-technical considerations [9]. Beyond addressing systemic challenges, the ensemble artifact presented in this work emphasizes the central role of the human developer. In contexts characterized by complexity and uncertainty, the ability of developers to act with confidence, purpose, and creativity is crucial. The proposed approach aims to provide more than structured guidance:

K. Węcel (Ed.): BIS 2025, LNBIP 554, pp. 70–83, 2025.
https://doi.org/10.1007/978-3-031-94193-1_6

it seeks to enable users through accessible knowledge, empower them with practical tools, and engage them by fostering motivation, agency, and a sense of contribution. This perspective positions the artifact as a human-centered support system for informed and sustainable technological development. This kind of development support fosters innovativity as companies are called upon to aim for ambidexterity, to find the balance between exploring and exploiting at the same time [12].

While a range of established frameworks exist to support technology development, notably DSR, HCD, and co-design models, these approaches often emphasize either scientific rigor or user-centeredness without directly addressing the support needs of developers themselves. The human actors tasked with operationalizing complex socio-technical innovation often lack structured tools that balance theoretical grounding with practical adaptability. Recent literature (e. g. [2, 14, 18]) increasingly calls for more interactive, lightweight, and context-sensitive tools that empower development teams. The TIRM responds to this gap by integrating knowledge-making, usability principles, and inquiry-based co-creation into a developer-centered ensemble artifact.

The challenges of technological development in this context often qualify as wicked managerial problems: ill-defined, unique, complex, and intangible. These problems cannot be addressed solely through analytical methods due to infinite solution possibilities, complex interdependencies, and boundary conditions. Instead, heuristic and iterative approaches are required to explore and refine viable solutions [18]. Such challenges are particularly pronounced in innovation-driven, knowledge-intensive, and project-based development contexts, as seen in manufacturing, customer-specific machinery, plant engineering, and the design of cyber-physical production systems (CPPS). These challenges are further exacerbated in resource-constrained settings, such as start-ups or small and medium-sized enterprises (SMEs), which face limited personnel, expertise, and financial resources [6].

This paper addresses the interdisciplinary and multi-criteria challenges of the "*development and integration of complex cyber-physical technologies in the industrial domain*" (**cTI-Dev**) by presenting a structured approach to support technological development. The proposed approach is built upon three core elements: a lightweight conceptual model, a set of design principles, and a visual inquiry tool. These components work together to ensure both relevance (alignment with application contexts) and rigor (adherence to knowledge foundations), allowing for developers to address complex development tasks more effectively and collaboratively. By bridging the problem space (e.g., project requirements) and solution space (e. g., technological options), this approach provides a flexible framework for tackling development problems.

The research is guided by the following overarching research question:

RQ: What constitutes a developer-centered support system for the **cTI-Dev**?

This question is further divided into the following sub-questions:

*RQ1: Which conceptual model describes a support system for the **cTI-Dev**?*

*RQ2: How can a visual inquiry tool be designed to support the **cTI-Dev**?*

*RQ3: Which guidelines should accompany the visual inquiry tool for **cTI-Dev**?*

A central aim of this research is to not only improve the structure of the solution process for technological development tasks but also to enhance the role and experience of the human developer. Existing frameworks such as DSR focus on generating scientific design knowledge, while Human-Centered Design (HCD) prioritizes user needs but offers limited actionable guidance for developers themselves. Usability criteria and utility theories provide valuable outcome-oriented benchmarks but fall short of guiding the development process in real-world complexity. The ensemble artifact introduced here seeks to bridge this gap by enabling, empowering, and engaging the developer in their role. Enablement is achieved through contextualized knowledge management that supports understanding of tasks; empowerment stems from methodical, tool-supported agency; and engagement emerges when developers feel capable, supported, and motivated to actively drive the development process and contribute back to the knowledge system. This aspiration is aligned with the vision of Industry 5.0, which emphasizes sustainable, resilient, and human-centric innovation in complex industrial systems [15].

The remainder of this paper is structured as follows: Sect. 2 outlines the methodology, Sect. 3 reviews foundational approaches, Sect. 4 presents the ensemble artifact and Sect. 5 describes the artifact's instantiation, implementation, and testing through intervention cycles. Finally, Sect. 6 provides an outlook on future research and potential enhancements.

2 Navigating Complexity: A Design Science Research Approach

This work aims to design an ensemble artifact for supporting technological development, following the Design Science Research (DSR) methodology proposed by Peffers et al. [17]. The research process began with problem identification and motivation, drawing insights from literature, norms, and standards in engineering, as well as prior research and interactions with companies. Based on this foundation, we defined theory-driven research goals and developed an initial draft of the artifact.

To refine the artifact, we adopted the iterative Action Design Research (ADR) approach articulated by Sein et al. [23] and expanded by Mullarkey & Hevner [16]. ADR comprises intervention cycles each involving "Problem formulation", "Creation of an artifact", "Evaluation", "Reflection", and "Learning". In each cycle, a preliminary version of the artifact is instantiated, implemented, and tested in a realistic setting. Feedback from these evaluations informs revisions, initiating subsequent cycles. Mullarkey & Hevner [156] recommend completing at least four of these cycles, focusing on problem diagnosis, solution design, implementation, and artifact evolution. Currently, the described research project has completed the diagnosis and design cycles and is progressing through the implementation and evolution phases. In the bigger picture, this iterative testing in example scenarios further refines the DSR artifact, ensuring effectiveness and relevance for eventual adoption by the broader community, as envisioned by Peffers et al. [17].

Venable [25] highlights the role of design theories, particularly **utility theories**, in connecting the problem space, encompassing issues and relationships such as aggregation, generalization, and causality, with the solution space, which defines the characteristics and technologies of proposed solutions. Utility theories ensure alignment between these spaces by linking technological concepts with the problems they address, demanding effectiveness, efficacy, and efficiency.

This alignment can be operationalized by applying **usability criteria** defined by Wiemer [27], such as understandability, explainability, transferability, and interoperability. These criteria, considered in both the problem and solution spaces, enable a targeted approach to achieving a good fit, fostering technologies tailored to their intended applications. Evaluating this fit involves assessing both the importance and maturity levels of the respective criteria in the problem space (application needs and constraints) and the solution space (solution approaches).

The lightweight conceptual model developed in this work integrates these perspectives by conceptualizing the solution space as technological options and the problem space as use cases, problems, and challenges. Tools and heuristics provided by the ensemble artifact facilitate the alignment of these spaces, promoting the development of context-specific, user-centered technologies. This approach particularly draws on Roschnik [18] contributions, which emphasize the value of **lightweight conceptual models and visual inquiry tools** in addressing wicked problems.

3 Foundational Perspectives: Insights from Existing Models and Frameworks

Numerous models and frameworks aim to support technological development by addressing specific phases or contexts [2, 4]. For example, domain-specific approaches like REMODEL [10] excel in addressing targeted phases of development processes, complementing the proposed ensemble artifact. The Technology Inquiry Reference Model (TIRM) is grounded in **Hevner's Three Cycle View of Design Science Research** [7], which emphasizes the interplay between the Relevance Cycle, Rigor Cycle, and Design Cycle. The Relevance Cycle addresses real-world application contexts by identifying problems, defining requirements, and evaluating whether artifacts improve their environment. The Rigor Cycle ensures scientific validity by integrating knowledge from existing theories, methods, and artifacts, thereby connecting research with established expertise. The Design Cycle focuses on iterative artifact creation and evaluation, synthesizing inputs from the other cycles. Together, these cycles form a dynamic process that aligns practical relevance with scientific rigor, as originally inspired by Simon's The Sciences of the Artificial for a broad range of design artifacts [24]. While the Design Science Research (DSR) methodology provides a robust structure for constructing and evaluating artifacts, it traditionally focuses on scientific excellence rather than directly supporting the human developer in their practical task execution. The ensemble artifact presented in this work expands this scope by aiming to enable (through structured knowledge), empower (through practical competence), and engage (through motivational agency) the human actor involved in development tasks. This lens reinforces the human-centered potential of DSR and ADR by embedding support not just for artifact development but also for those who perform the development.

Human-Centered Design (HCD) further strengthens the artifact's foundations. As defined by ISO DIS 9241-210:2019 [11], HCD emphasizes usability, accessibility, and satisfaction by aligning system design with user needs and iterative evaluation. Models such as the Double Diamond Model and the Understand-Create-Evaluate Cycle support iterative user-centered processes for understanding, creating, and refining solutions [14].

Lastly, Wiemer's [27] criteria for broader usability, like usefulness, suitability, integrability, and interoperability, extend the usability claim to general technological development. These criteria emphasize clarity, domain relevance, compatibility with existing systems, and smooth interaction with external stakeholders. Using these criteria ensures a strong fit between development tasks and the solutions they aim to provide, not only in AI-related development, but in general technological development as well. Therefore, the concept and the application of the defined criteria should be broadened to a definition of **Usable Technological Design (UTD)**. Abiding by this concept, developers can create technological systems that are not only functional but also user-friendly, relevant, and effectively integrated into real-world application settings.

While existing approaches such as DSR, HCD, and UAI focus on process structure, usability, and system performance, they often fall short of fully supporting human actors as creative, decision-making agents within technological development. The **Enable–Empower–Engage perspective** introduced in this paper aligns closely with the vision of Industry 5.0 [15], which reframes industrial innovation around human agency, resilience, and sustainability. This conceptual framing is in particular inspired by patient empowerment in digital health, where tools and platforms enable patients to understand their condition (enablement), make informed choices (empowerment), and become active participants in their care (engagement) [8, 22]. This triadic approach reframes the developer not merely as a facilitator of user-centered design but as a central, supported actor whose creative capacity and contextual knowledge are essential to success. Thus, the artifact translates abstract goals into concrete developer support, bridging a critical missing link in existing models.

4 The Ensemble Artifact: Bridging Relevance and Rigor in Technological Development

As Simon [24] describes Design Science Research (DSR) as a suitable approach for developing technological artifacts and as these technological artifacts progress intensely towards being information systems (IS) as much as physical artifacts, it seems an intuitive consideration to draw from the extensive body of knowledge on DSR concerned with the development of IS artifacts to benefit the development of technological artifacts. Namely, Hevner's [8] approach to ensure relevance and rigor in scientific IS artifact development seems generally applicable to any kind of artifact development, scientific or entrepreneurial, as *rigor* in the sense of state-of the-art insight on all *relevant* factual and methodological aspects of the artifact is an essential prerequisite for any advanced development. What is more, the interplay of the two aspects offers a whole new approach of converging complex technological solutions. To allow for this, the power of concepts like the effectuation approach [20], interactive design formats [1], interactive external knowledge representation as described by Yasuoka et al. [28] and inquiry-based knowledge retrieval are crucial elements. They allow holistic knowledge development and by this lastly inquiry-based systems development [26] and by this open up the opportunity to cease knowledge input and user input in combination. Accordingly, these explicit and implicit knowledge components then can add up to more than just the sum of the parts, but instead draw from formerly hidden, much broader resources and create entirely new insight by externalizing contextual, implicit and experience-based knowledge directly drawn from the interaction with the user. This is especially important as conventional information sources usually do not provide the complete domain knowledge as needed in real-life development contexts, as it is often completely specific to one particular subdomain, company, project or even use-case or data set context. This disparity can be overcome only by providing knowledge interactively and by directly involving the users and developers of complex cyber-physical systems.

4.1 Design Principles for a System Supporting Inquiry-Driven Technology Development

In Design Science Research (DSR), design principles (DPs) serve as a means to articulate prescriptive design knowledge in a structured and precise manner, detailing how the mechanisms of a technology or approach contribute to achieving specific objectives [5]. As outlined by Chandra et al. [3], design principles should define the actions enabled by the use of an artifact, describe the material properties that facilitate those actions, and specify the boundary conditions under which the principle applies by formulating a DP in the following format: "Provide the system with [material property–in terms of form and function] in order for users to [activity of user/group of users–in terms of action], given that [boundary conditions–user group's characteristics or implementation settings]."

The design principles of the Technology Inquiry Reference Model (TIRM) are systematically derived from the Knowledge Making principles K1 to K6 as can be found in [21]. These principles emphasize structuring, labelling, and utilizing knowledge in development contexts and serve as a foundational framework for addressing interdisciplinary challenges and ensuring accessible knowledge transfer. Each TIRM design principle builds upon its corresponding Knowledge Making principle ensuring a seamless integration of these concepts into the artifact. To provide one example for these six transformations: the Knowledge Making principle K1 (Open up new possibilities) inspires T1 (Relevance Pull), which provides mechanisms for interactive input from the problem space, including collecting, clustering, and prioritizing potential use cases and problems. This alignment ensures that the TIRM leverages the creative, structured, and iterative strengths of the Knowledge Making framework to address the demands of technological development. These design principles thus serve a dual purpose: structuring system development and scaffolding human capability.

By enabling developers through access to structured, contextual knowledge, empowering them with actionable methods and heuristics, and engaging them through co-creative, interactive processes. Each TIRM design principle, which are described in Table 1, builds upon its corresponding Knowledge Making principle ensuring a seamless integration of these concepts into the artifact.

4.2 A Lightweight Conceptual Model for Integrating Knowledge and Application

The Technology Inquiry Reference Model (TIRM) as shown in Fig. 1 visually represents the interplay between application relevance, system development, and knowledge rigor. The central **System Development** section represents the creation of a technological

Table 1. Design Principles Technology Inquiry.

Headline	Provide system with [property]	In order for users to [activity]	Given that [boundary condition]
T1: Relevance Pull	Features to enable interactive input from the problem space	Gather relevant development needs and potentials	Diversity of user roles and their varying levels of familiarity with the solution options
T1.1: Inspire	*Features to enable the interactive description of the problem space and ist problems and potentials*	*Think creatively about potential development opportunities*	*Users may lack initial clarity or motivation in poorly defined problem spaces or complex solution spaces*
T1.2: Collect	*Means to systematically collect relevant data and user inputs*	*Capture essential problem-space information*	*Potentially biased information because of resource constraints, fragmented data availability and specifics of involved users*
T1.3: Cluster	*Means for clustering and organizing collected data*	*Identify patterns and themes*	*Variability in data and information quality, accessibility and variability of formats*
T2: Rigour Accumulation	Workflows to collect and structure pieces of information	Create a comprehensive repository of relevant knowledge	Limited time and cognitive capacity of users to process large amounts of unstructured data
T3: Rigour Consolidation	Features to structure, connect, and abstract pieces of information	Build coherent and actionable insights	Given the user's preconditions to obtain a balance of potent solutions and manageable effort
T4: Rigour-Relevance-Match	Means for the contextualised assessment of potential solutions	Align insights with the specific use case context	Potentially incomplete information and need to make assumptions
T4.1: Prioritize	*Features to conduct a priorisation procedure*	*Focus on the approaches with the highest potential for success*	*Biased view of user roles and the need to assume some stakeholder's expectations and system states*
T4.2: Specify Requirements	*Means to articulate and refine specific requirements, prerequesites and preferences*	*Narrow down development goals*	*Biased view of user roles and the need to assume some stakeholder's expectations and system states*
T5: Rigour Push	Interactive mechanisms to connect relevant information between problem and solution spaces	Iteratively refine both problem understanding and proposed solutions	Highly specific information needs and fragmented communication and knowledge gaps across interdisciplinary teams
T5.1: Inform	*Features to inform users of relevant data and insights, especially methodological knowledge*	*Enhance developed solutions*	*Heterogeneity of prior knowledge of the users*
T5.2: Substantiate	*Features to enrich developed solutions further by using additional information*	*Generate and operationalise contextual knowledge*	*Need and difficulty of tackling various knowledge types like e.g. tacit knowledge or procedural knowledge*
T6: Rigour Feed	Workflows to re-feed the knowledge base with information from interactive interchange	Ensure continuous learning and improvement	Time and resource constraints
T6.1: Contextual Knowledge Combination	*Process to include produced solutions and insights into knowledge base*	*Generate new insights and approaches*	*Time and resource constraints in terms of a scarcity of skilled personnel to maintain the knowledge base*
T6.2: Heuristic Solution Seed	*Process to re-iterate solution finding and refining process for drafted solution ideas*	*Motivate iterative development process*	*Time and resource constraints in terms of a scarcity of skilled personnel to reevaluate the solution ideas*

system and is divided into "Develop and Build" and "Test and Deploy" phases, linked by iterative feedback loops. This process is supported by the **Field of Application** on the left, ensuring relevance by identifying and addressing user needs, and the **Knowledge Base** on the right, providing rigor through the integration of scientific expertise and technological resources.

The **Field of Application** is inspired by Hevner's [8] Relevance Cycle, focusing on identifying problems, defining requirements, and evaluating whether artifacts improve their contextual environment. The **Knowledge Base** relates to Hevner's [8] Rigor Cycle, emphasizing the role of domain expertise, existing theories, and methods in ensuring innovative and scientifically grounded solutions. The **System Development** section reflects Hevner's [8] Design Cycle, which synthesizes inputs from the other two cycles into iterative artifact creation and evaluation. Connections between these sections are represented by labeled arrows that depict the iterative and interdependent nature of the process. Requirements flow from the Field of Application to the System Development section, while knowledge provision flows from the Knowledge Base. Feedback loops that ensure usability insights and contextual knowledge are reintegrated into the Field of Application and Knowledge Base, respectively. The TIRM encapsulates these iterative relationships through its core principles: The interaction between the"System Development" section and the "Knowledge Base" is grounded in the approach of knowledge making as described in Sect. 4.1 in its adaption to technological systems development in general and can therefore be described as the principle of "**Systems Knowledge Making (SyKnoMa)**".

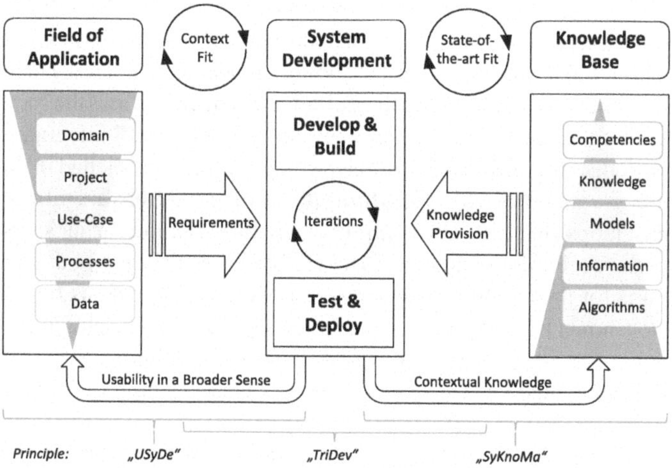

Fig. 1. Technology Inquiry Reference Model (TIRM).

Similarly, the interaction between the "System Development" section and the "Field of Application" can be described as the principle of "**Usable Systems Development (USyDe)**", as it is based on the concept of Usable Technology Design (UTD) as described in Sect. 3. The iterative design and building approach in the "System Development" section can be described as the principle of "**Triple-i-Development (TriDev)**", as it is constituted as being interactive, inquiry-based and informed. This lightweight conceptual model has been assessed guided by Roschnik [18] central work providing a framework and criteria for the evaluation of conceptual models. Accordingly, it demonstrates coherence, parsimony, and relevance. Necessary future validation through domain expert feedback will refine the model further, ensuring its adaptability to diverse technological contexts.

4.3 Operationalizing Technology Inquiry: A Visual Inquiry Tool and Guidelines for Use

The Technology Inquiry Canvas (TIC), as can be seen in Fig. 2, operationalizes the TIRM as a visual inquiry tool, guiding users through creative and informed technology development. Structured as a matrix, the TIC divides the process into four quadrants, as illustrated in Fig. 2: **Collect Use Cases & Problems** (Field of Application), **Collect Options & Solutions** (Knowledge Base), **Connect Knowledge to Options & Solutions** (Test and Deploy), and **Connect Solutions to Use Cases & Problems** (Develop and Build). Each quadrant corresponds to a section of the TIRM, ensuring alignment with its conceptual structure. To support "Collecting" and "Connecting" activities, the TIC includes two triangular frames: the **Cluster** frame in the upper half, which organizes collected ideas and information, and the **Prioritize** frame in the lower half, which refines selected elements for further exploration.

The Technology Inquiry Process (TIP) flowchart, as depicted in Fig. 3 as an overlay above the canvas, operationalizes the Technology Inquiry Canvas (TIC), illustrating the sequence and relationships of activities involved in the process. The TIP starts with the activity "**Inspire**", serving as the entry point to the process, followed by "**Collect**", where data and insights related to the problem and solution spaces are gathered. Next, "**Cluster**" organizes these inputs into structured groups, and "**Prioritize**" refines the focus by selecting the most critical elements for further exploration. "**Connect**" integrates prioritized ideas and knowledge into actionable insights, which are further enriched in the "**Inform**" stage. The process concludes with "**Substantiate**", where solutions are refined and elaborated. Feedback loops ensure iterative improvement by revisiting earlier steps as needed. To implement the Technology Inquiry Process (TIP) a variety of methods can be chosen. The **suggestions** for the quadrants are depicted in Fig. 3. Suggestions for the "Cluster" element are not shown in the figure. They include suggestions like classification by topics, departments, user roles, or timelines or Affinity Diagramming. Suggestions for the "Prioritize" frame involve various techniques such as Dot Voting or Weighted Scoring Models.

All suggestions are intended to provide inspiration and flexibility; in practice, only one or a combination of methods may be chosen based on the specific needs of the project. Moreover, any other suitable methods that align with the process and context can be applied, ensuring the TIP remains adaptable to diverse real-world settings. By integrating the TIRM, TIC, and TIP, the ensemble artifact serves as both a practical tool and a conceptual framework, empowering interdisciplinary teams to navigate complex development tasks effectively.

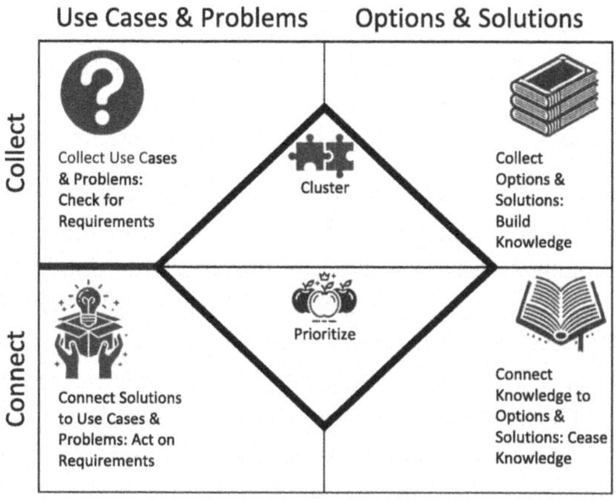

Fig. 2. Technology Inquiry Canvas (TIC).

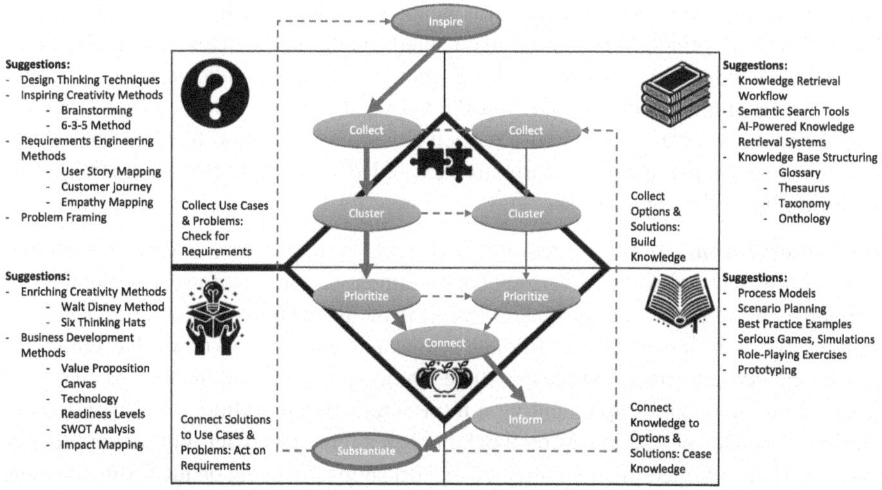

Fig. 3. Technology Inquiry Process supported by the Technology Inquiry Canvas and Suggestions for each quadrant.

5 From Concept to Reality: Iterative Testing and Refinement of the Artifact

The iterative evolution of the artifact follows the Action Design Research (ADR) methodology, as outlined in Sect. 2, through active testing with users during interventions. Three interventions have been completed, with the first two conducted in experimental settings and the third in a real-world scenario with an SME. Each intervention identified flaws in the artifact, addressed in subsequent development cycles. Currently, two additional interventions with SMEs are underway, alongside a new test series in a different application domain, addressing open-source hardware designs in engineering. For the interventions, the Technology Inquiry Reference Model (TIRM) was instantiated as a graphical representation depicting the company's perspective, the knowledge retrieval process, and their interconnection. All interventions focused on the thematic field of developing and evolving digital assistance systems with cybersecurity considerations.

First Intervention: In a simulated role-play setting carried out with two researchers enacting practitioners with a background in technology development, the "SyKnoMa" principle was implemented as a note-and-link-based textual database containing solution space details aimed to provide context-specific insights. The "USyDe" principle was implemented as semi-structured interviews to gather problem space data (e.g., company, user, and solution details). The "TriDev" principle was implemented by matching interview findings with database content. Internal (workshop leaders and knowledge maintainers) and external (participant) feedback indicated that the interview guidelines were too rigid, and the database could not adequately support development due to unforeseen questions, highlighting the need for greater flexibility and contextual specificity.

Second Intervention: Conducted with engineers in a real-world technological research environment, the "USyDe" principle was adapted to an interactive workshop format. The "SyKnoMa" principle remained unchanged, while the "TriDev" principle linked database knowledge to workshop results. Participant feedback indicated that the workshop was inspiring and informative, but the knowledge base content was at times too specific and not entirely aligned with participants' needs. Internal feedback noted that the demand for highly specific information exceeded what could be anticipated during the knowledge collection process.

Third Intervention: Implemented in a real-world setting with six workshop participants, who were employees of various backgrounds within a medium-sized enterprise developing automation systems, this intervention included significant updates. The "SyKnoMa" principle introduced additional synthesis steps in the knowledge collection process, including creating synthesis articles, redesigning the linking process using an expanded ontology, and implementing a more interactive workflow for database curation. The "USyDe" principle was realized through a four-stage user interaction process, including a kick-off meeting, two interactive workshops (focusing on idea collection and use-case development), and a results discussion. External knowledge provision occurred through formats such as impulse lectures, topical canvases, serious play, and interactive documentation. The "TriDev" principle supported development by co-creating innovative, inquiry-based solution ideas enriched with facts and methodologies from the knowledge base, which were then integrated back into the knowledge base as enriched use cases and contextual information. Documentation was provided in a digital wiki format, including workshop results, details of the enriched solution, a subset of relevant knowledge base content, a SWOT analysis, and suggested next steps.

Despite rating the experience and outcomes as extremely helpful, participants noted a lack of clarity about the overall process and the purpose of specific activities, as well as uncertainty about available options for organizing the process. Based on these insights, the TIC canvas and TIP flowchart were added to the ensemble artifact to enhance transparency and process comprehension and to further stimulate creativity.

Future interventions will incorporate these refinements. The ongoing SME interventions aim to validate the updated artifact, while the new test series in open-source hardware design will explore its adaptability to different application domains.

6 Outlook

This article presents the Technology Inquiry Reference Model (TIRM), a lightweight conceptual framework for supporting informed, interdisciplinary technological development. Accompanied by the Technology Inquiry Canvas (TIC) and the Technology Inquiry Process (TIP) it integrates principles of usability, knowledge making, and inquiry-based development. The approach fosters relevance, rigor, and collaboration in addressing complex development tasks. The main contribution lies in providing a flexible, modular framework that can guide practitioners and researchers alike in tackling technological innovation challenges across diverse contexts. This orientation aligns with the principles of Industry 5.0, reinforcing the shift toward human-centered, sustainable, and resilient development frameworks that respect and leverage human capabilities in symbiosis with digital technologies.

Feedback from intervention cycles highlights both opportunities and challenges. Participant bias emerged as a significant factor, emphasizing the need for balanced stakeholder representation. Carefully designed interaction formats are key to addressing these issues. Additional critical challenges are the influence of possible researcher biases, and the paradox of evaluating the concept itself instead of evaluating the mere implementation. Another critical factor is the ongoing maintenance of the knowledge base, requiring participants to engage in regular interaction phases and iterative updates to sustain the process over time.

Ultimately, the ensemble artifact not only fosters methodological alignment and rigorous design but also promotes human development contexts. This orientation resonates with emerging hybrid approaches like the Double Diamond Model [14] and co-creation canvases [4], which emphasize iterative exploration. However, the TIRM advances these models by tightly integrating structured knowledge management and developer empowerment into one cohesive framework. Its triadic Enable–Empower–Engage structure thus provides not only process scaffolding, but also personal agency and cognitive support for development practitioners navigating wicked problems. It positions the developer and crucial stakeholders as central actors navigating complex development tasks, paving the way for more sustainable, satisfying, and scalable technological innovation.

Future work includes completing the final intervention cycles of the current Action Design Research (ADR) project and extending testing into new domains, such as open-source hardware development, hopefully contributing to resolving potential existing biases. The TIRM will undergo further evaluation by domain experts according to Roschnik's [18] framework to refine its components and applicability. Major envisioned enhancements of the ensemble artifact involve digitizing interaction formats for remote execution and to follow the suggestion of Roschnik [19] of exploring full digitalization of the process to leverage advanced functionalities, such as automated knowledge management and interactive virtual tools. These enhancements aim to broaden the framework's scalability and accessibility.

Lastly, testing the approach in varied organizational contexts, including larger companies and entirely different domains, will provide valuable insights into its adaptability and generalizability. The suggested ensemble artifact and its elements are not supposed to be a rigid construct, but to serve as an inspiration to draw from and to be further evolved and refined by the scientific and practitioner community.

Disclosure of Interests. The authors have no competing interests to declare that are relevant to the content of this article.

Data Availability. The Technology Inquiry Reference Model, the Technology Inquiry Canvas and the Technology Inquiry Process with the respective design principles presented in this study are open access, DOI: https://doi.org/10.5281/zenodo.14760079.

References

1. Becker, F., Siemon, D., Robra-Bissantz, S.: Smart participation design: prescriptive knowledge for bottom-up participation. Commun. Assoc. Inf. Syst. **51**(1), 484–508 (2022). https://doi.org/10.17705/1CAIS.05121
2. Cerqueira, C., Sequeira, M., Varajão, J., Sousa Pinto, C., Veloso, M.: Canvas for IT/IS—a literature review and a framework. ICEB 2024 Proceedings (Zhuhai, China), p. 61 (2024). https://aisel.aisnet.org/iceb2024/61
3. Chandra, L., Seidel, S., Gregor, S.: Prescriptive knowledge in is research: conceptualizing design principles. In: Proceedings of the 8th Hawaii International Conference on System Sciences, pp. 4039–4048 (2015). https://doi.org/10.1109/HICSS.2015.485
4. Garmann-Johnsen, N.F., Olsen, D.H., Eikebrokk, T.R.: The co-creation canvas. Proc. Comp. Sci. **181**, 189–197 (2021). https://doi.org/10.1016/j.procs.2021.01.120
5. Gregor, S., Kruse, L.C., Seidel, S.: The anatomy of a design principle. J. Assoc. Inf. Syst. (2020)
6. Grooss, O.F., Presser, M., Tambo, T.: Balancing digital maturity and operational performance. Proc. Comp. Sci. **200**, 495–504 (2022)
7. Hevner, A.R.: A three cycle view of design science research. Scand. J. Inf. Syst. **19**(2), 87–92 (2007)
8. Hickmann, E., Richter, P., Schlieter, H.: All together now – patient engagement, patient empowerment, and associated terms in personal healthcare. BMC Health Serv. Res. **22**(1), 1116 (2022). https://doi.org/10.1186/s12913-022-08501-5
9. Hoffmann, M., et al.: Developing industrial CPS: a multi-disciplinary challenge. Sensors **21**(6) (2021). https://doi.org/10.3390/s21061991
10. Ferreira, A., Hunt, S., Claypool, S., Yuan, W.: Promoting Open Source Models in the Danish Manufacturing Industry. Worcester Polytechnic Institute (2017)
11. ISO DIS 9241-210. (2019). Ergonomics of human system interaction–Part 210: Human-Centred Design. International Organization for Standardization
12. Kassotaki, O.: Review of Organizational Ambidexterity Research. Sage Open **12**(1) (2022). https://doi.org/10.1177/21582440221082127
13. Kayan, H., et al.: Cybersecurity of industrial cyber-physical systems: a review. ACM Comput. Surv. **54**(11s), 1–35 (2022). https://doi.org/10.1145/3510410
14. Kochanowska, M., Gagliardi, W.R.: The Double Diamond Model. In: Perspectives on Design II. Springer International Publishing (2022)
15. Leng, J., et al.: Industry 5.0: prospect and retrospect. J. Manuf. Syst. **65**, 279–295 (2022). https://doi.org/10.1016/j.jmsy.2022.09.017

16. Mullarkey, M.T., Hevner, A.R.: An elaborated action design research process model. Eur. J. Inf. Syst. **28**(1), 6–20 (2019)
17. Peffers, K., Tuunanen, T., Rothenberger, M.A., Chatterjee, S.: A design science research methodology for information systems research. J. Manag. Inf. Syst. **24**(3), 45–77 (2007). https://doi.org/10.2753/MIS0742-1222240302
18. Roschnik, A.:. A Framework Proposal to Evaluate Conceptual Models Framing Wicked Managerial Concepts. ECIS 2022 Research Papers, p. 67 (2022)
19. Roschnik, A., Jaccard, D., Monnier, S., Missonier, S.: Approaching Visual Inquiry Tools through the Lens of Systems Thinking. ICIS 2024 Proceedings, p. 7 (2024)
20. Sarasvathy, S.D.: Effectuation: Elements of Entrepreneurial Expertise. Edward Elgar Publishing, Cheltenham (2008). https://doi.org/10.4337/9781848440197
21. Schneider, D., Kusturica, W.: Towards a guideline affording overarching knowledge building in data analysis projects. Bus. Inf. Syst. 49–59 (2021)
22. Schlieter, H., et al.: Designing virtual coaching solutions. Bus. Inf. Syst. Eng. **66**(3), 377–400 (2024). https://doi.org/10.1007/s12599-024-00871-w
23. Sein, M.K., Henfridsson, O., Purao, S., Rossi, M., Lindgren, R.: Action design research. MIS Q. 37–56 (2011)
24. Simon, H.A.: The Sciences of the Artificial (3. ed.). MIT Press, Cambridge (2008)
25. Venable, J.R.: The role of theory and theorising in design science research. In: Proceedings of the 1st International Conference on Design Science in Information Systems and Technology, pp. 1–18 (2006)
26. Wen, Y., et al.: Integrating augmented reality into inquiry-based learning approach. Ed. Tech Res. Dev **71**, 1631–1651 (2023). https://doi.org/10.1007/s11423-023-10235-y
27. Wiemer, H., et al.: Need for UAI. Multimodal Technol. Interact. **7**(3), 27 (2023)
28. Yasuoka, M., Nakatani, M., Taoka, Y., Hamaguchi, N.: Interactivity matters. In: Proceedings of the 57th Hawaii International Conference on System Sciences (2024)

Artificial Intelligence for Language Processing

Interview Study to Determine the Potential of Large Language Model Usage in IT-Consultancy

Moritz Jehnert[1], Christoph Meyer[2], and Kurt Sandkuhl[1,3](✉)

[1] University of Rostock, 18059 Rostock, Germany
moritz.jehnert@outlook.de, kurt.sandkuhl@uni-rostock.de
[2] MSG for Automotive, Braunschweig, Germany
Christoph.Meyer@msg.group
[3] Jönköping University, 55111 Jönkoping, Sweden

Abstract. Generative AI and large language models (LLM) are considered in many industry and service sectors as promising technology for automating routine tasks. The potential of LLM has also been noticed in the area of consulting services, but there is not much work in the scientific body of knowledge. With a focus on IT consultancy (ITC), the paper aims to contribute to a better understanding of how LLM can be applied from the perspective of the consultants. An interview study with two interlinked series of interviews, in total 20 experts in IT consultancy and AI experts, is conducted. The main contribution of our work is (1) a systematic literature analysis of LLM use in IT consultancy, (2) results of two series of expert interviews, and (3) a feasibility study of LLM use in an ITC activity considered by the interviewees as a time-consuming and promising area for LLM usage. The findings of the study indicate the substantial potential of LLM in ITC for automating or increasing the efficiency of a number of tasks, but, so far, no substantial changes in ITC value creation processes.

Keywords: Generative AI · IT-Consultancy · Large Language Model · LLM · Potential

1 Introduction

Generative artificial intelligence (AI), language technologies, and large language models (LLM), such as OpenAI's ChatGPT, have motivated a lot of research into potential application areas, new business models, or entirely new products and services based on AI [12]. Generative AI and, in particular, LLM are considered in many industry and service sectors as promising technology for automating routine tasks, with the perspective of substantial improvement of productivity (see, e.g., [11] and [8]). First investigations and experience reports from applying LLM in engineering confirm this technology's potential. Examples are LLM use in construction industries [19], factory organization [20], and manufacturing [21].

K. Węcel (Ed.): BIS 2025, LNBIP 554, pp. 87–100, 2025.
https://doi.org/10.1007/978-3-031-94193-1_7

The potential of LLM has also been noticed in the area of advisory and consulting services, and triggered studies into the possibility of applying LLM in consultancy. Consultancy is defined as the process in which a party within a professional area assists the client to achieve a stated outcome typically manifested as information, recommendation, or an artifact [2]. Although some market studies have been performed (e.g., [5]), there is not much work on LLM potential in consultancy in the scientific body of knowledge (see Sect. 3.2). The focus of this paper is on IT consulting (ITC), which encompasses strategic consulting, implementation services, and operational services [14]. Digital transformation and the introduction of AI into organizations have resulted in a growing demand for ITC and the need for increasing the efficiency of ITC activities [13].

The aim of this paper is to contribute to a better understanding of how LLM is and can be applied in ITC from the perspective of the consultants. What ITC activities are suitable for LLM usage, what outcomes are expected by the consultants, and what do consultants see as the main purpose of LLM usage? Due to the lack of empirical material available in the scientific literature, we decided to conduct an interview study with two interlinked series of interviews as the primary source of empirical data. The main research question (RQ) in this context is: *In what areas of ITC do LLM have the potential to assist consultants, automate tasks, or improve the quality of ITC results?* The main contribution of our work is (1) a systematic literature analysis of LLM use in IT consultancy, (2) results of two series of expert interviews, and (3) a feasibility study of LLM use in an ITC activity considered by the interviewees as time-consuming and promising area for LLM usage.

The paper is structured as follows: Sect. 2 briefly presents the overall research approach. Section 3 provides the background and related work for our research, including a systematic literature analysis. Section 4 contains the interview study design and results. Section 5 has a focus on the feasibility study. Section 6 summarizes the work and discusses future research.

2 Research Approach

The research presented in this paper is part of a research program that aims at developing technical and methodical support for LLM use in information systems design (ISD) and engineering. While previous work focused on specific areas of ISD, like conceptual modelling [17] or quality control [16], this paper addresses the overall development process of ISD from the perspective of IT consulting and the associated value creation. The research question presented in the previous section is broken down into several sub-questions: What previous scientific work is visible in publications about using LLM in IT consultancy? For which activities did IT consultants observe or expect benefits from LLM usage? For what activities can AI experts confirm LLM potential?

Several research methods are applied to investigate these RQs in a research process depicted in Fig. 1. The first step is to investigate the state of research as published in scientific body of knowledge by performing a systematic literature review (SLR) according to [9], which is described in 3.2. As the SLR shows a lack of publications and empirical investigation, expert interviews are prepared by determining the different tasks included in IT consultancy. This argumentative-deductive part of our research is presented in Sect. 3.3. It follows an interview study among experts in IT consultancy to determine, for which tasks they observed or expect LLM potential. This results in short descriptions of consultancy tasks and their underlying processes with expected LLM potential for each process step. The second part of the interview study is performed among AI experts to validate if the expectations of the IT consultants can be confirmed. The methodology for the two expert studies is presented in detail in Sect. 4. The final step is a feasibility study for LLM usage in one of the tasks.

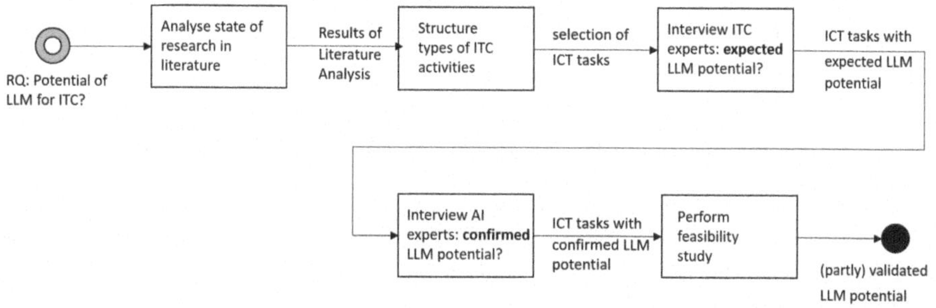

Fig. 1. Research process

3 Background and Related Work

3.1 Generative AI and Large Language Models

The launch of ChatGPT by OpenAI in November 2022 marked a pivotal milestone in the integration of artificial intelligence into everyday applications, significantly enhancing capabilities in diverse problem-solving scenarios. Built on the GPT-3 architecture, ChatGPT leverages a massive 175 billion parameters and is trained on an extensive dataset comprising nearly a trillion words. These datasets include sources such as Common Crawl, WebText2, internet-based books, and the English-language Wikipedia [3]. The model learns by predicting subsequent tokens based on contextual input, enabling it to produce coherent and contextually relevant text.

Large Language Models (LLMs) like GPT-3 are trained using a task-agnostic approach [3], allowing them to be easily adapted through in-context learning via natural language prompts at runtime. This flexibility promotes rapid experimentation and prototyping without the need for retraining the model. ChatGPT is capable of performing a variety of tasks, including summarization, translation, and grammar correction, making advanced AI tools widely accessible. This accessibility brings both opportunities and challenges, particularly in domains like education where concerns around academic integrity have emerged.

The growing field of prompt engineering has become essential in optimizing LLM performance, as discussed by [10]. It focuses on crafting prompts that effectively instruct the model to execute specific tasks. However, prompt design currently lacks standardized methodologies and is often guided by iterative trial and error [22]. This has led to the creation of domain-specific languages such as Impromptu, aimed at supporting structured and platform-independent prompt development.

Recent scholarly discourse highlights the importance of developing meta-models and software tools to enhance prompt engineering. These tools facilitate prompt reuse across different contexts-such as the prompt chainer [23]-and aid in managing prompt complexity in a way similar to traditional programming paradigms. Additionally, incorporating expert knowledge into prompt design processes is seen as vital for improving the validity and effectiveness of interactions with LLMs [6].

3.2 Systematic Literature Analysis on LLM in IT-Consultancy

The current state of research on the application of Large Language Models (LLMs) in IT consultancy was examined through a systematic literature review (SLR), following the six-step methodology outlined by Kitchenham [9]. The initial step involves formulating research questions (RQs) to guide the review. In this study, the primary RQ is: What prior academic work has been published on the use of LLMs in IT consultancy?

The second step, defining the search space, requires selecting relevant literature databases. For broad coverage across information systems and computer science, we chose Scopus, IEEE Xplore, and AISeL. Step three, the population phase, involves constructing the search string and conducting searches within the selected sources. Our search string-detailed in Table 1-includes the terms "LLM" and "IT consultancy," as well as their synonyms and related terms like "generative AI." The search was executed on January 28, 2025, and the resulting number of publications is also reported in the table.

In step four, paper selection is carried out based on predefined inclusion and exclusion criteria, followed by a manual screening process. We included only studies that specifically address the practical application of LLMs in IT consultancy. Papers that merely contained both keywords or discussed LLMs as a future prospect were excluded.

The fifth step, data collection, focuses on extracting information from the selected papers that directly relates to the research question. Finally, in step six, the extracted data is analyzed and interpreted to formulate a comprehensive response to the research question.

The search in Scopus returned 54 hits, but none of them was about the use of LLM or generative AI in IT consultancy. 21 of the 54 hits were excluded as they were from the medical field; another four from law consulting. Many of the remaining hits used the term consulting in their abstract with the meaning of "asking", "questioning", "visiting" or "exploring", but were neither relevant

Table 1. Results of the literature analysis

Query	Scopus	No. of AISeL	Hits Xplore	Relevant
(LLM OR "Large Language Model" OR "Generative AI") AND (consulting OR consultancy)	54	29	2	1
Total				1

from an LLM or consultancy perspective. Four papers addressed LLM use in IT-related topics but focused on cybersecurity, blockchains, tuning LLM models, or knowledge management. Thus, no relevant papers were identified.

Search in IEEE Xplore returned 29 results, 18 of them were also among the Scopus results. Among the remaining 11 results, four were on student training, five from the medical area and two about fine-tuning LLMs. Only one result about generative AI in software engineering [7] is relevant for our work as ITC often prepares software engineering projects and, thus, the areas have overlaps. This paper is based on experiences from a single case study and the expectations of the authors, two experienced software engineers. The paper addressed LLM usage in software development (e.g. for code generation) and for documentation tasks, but not for consultancy.

AISeL returned two results. One addresses LLM in teaching and the other AI transparency, i.e., both are not relevant for our research.

In addition to the above SLR, we also performed a search on Google Scholar that returned an investigation on LLM potential based on a case study [15]. The author concludes that business models of ITC might change due to LLMs affecting methods and client demands and LLMs will increase the productivity and versatility of individual consultants. Furthermore, there is research on specific tasks in the IT area that are also part of ITC, for example, the use of co-Pilot in software engineering [24] or of ChatGPT in information system design for UML modeling [4]. However, these results do not address the integrative perspective and essential value creation of consultancy work.

3.3 IT-Consultancy

Becker et al. [1] define consulting as a service to produce value by supporting companies in developing and improving their performance. IT consultancy is a consulting sector with a growing market size [13] that can be divided into segments [14], such as advisory services (strategic, governance, and architecture consulting), implementation services (system integration or software development), and operational services (maintain and manage production systems).

To identify ITC activities suitable for LLM usage, a breakdown structure of ITC activities is required that is more detailed than the above market sectors. The textbook 'Consulting: Knowledge for Strategy, Process and IT Consulting' [18] identifies essential consulting tasks that are also used in IT consulting and can be divided into general project management tasks and specific tasks for IT projects. Project management tasks reflect the different phases of projects to fulfill customer requirements efficiently and effectively:

- Project acquisition: Acquisition begins with the creation of a proposal that analyses the customer's needs and outlines a strategy for project implementation. This includes definition of objectives, the drafting of a project plan, and the assessment of risks and costs. Based on the proposal, a formal offer is prepared that fulfills legal requirements and includes cost calculations.
- Project planning: Once the contract has been awarded, the project plan is drawn up, including schedule, milestones, risk management, and resource allocation. Communication and quality control mechanisms are established.
- Project monitoring: During project implementation, progress is monitored on the basis of the project plan, risks, and problems are identified at an early stage, and the plan is adjusted if necessary.
- Project closure: Finally, project results are analyzed, documented, and handed over to the customer. An ex-post analysis is used for quality assurance and to derive improvements for future projects.

The specific tasks for IT projects focus on requirements management, release planning, and various documentation tasks.

- Requirements management: Customer requirements are collected, structured, and categorized in close coordination with stakeholders, taking risks and project goals into account. In agile development (e.g., the SCRUM process), user stories are used to document customer requirements in clear language, improve communication, and define acceptance criteria. They facilitate prioritization, iteration planning, and the integration of changes.
- Releases: Requirements are planned and implemented. In agile development methods, such as SCRUM, requirements are implemented in sprints and deployed at the end of a sprint. Release planning defines goals and responsibilities, prioritizes backlogs, and creates feasible plans.
- Release notes, which document the benefits of new features and improvements, are part of the delivery strategy and serve developers, stakeholders, and users. Releases and release notes ensure efficient, structured, and customer-orientated project implementation.
- IT system and security documentation: this documentation is used to manage information on system architecture, configuration, operation, and SLAs. It facilitates maintenance, compliance, and optimization and includes security guidelines, emergency plans and risk mitigation measures.
- User manual: the manual contains an inventory of the functionality of the developed systems and how to use them for different user groups. Typical usage scenarios and user interaction can also reflect test cases and form the basis for test data.
- Customer handbook: Documents service level agreements, supplier management, and cloud sourcing. The aim is to identify risks and organize collaboration efficiently. It also contains IT service management documentation, including guidelines, processes, and procedural instructions.

4 Expert Interviews

Two series of interviews were conducted on the basis of different guidelines. The guidelines were tested in pilot interviews before the start of the interview series. In the first series, IT consultants were asked about the extent to which generative AI is already used. The aim is to determine the experiences and expectations with a focus on making work easier. In addition, the expectations are to be determined in which areas the consultants expect AI to simplify their everyday work in the future. Taking into account the results of the interviews with the IT consultants, the second series is conducted with the AI experts. In these interviews, the focus is on the technical capabilities of LLM. The aim is to clarify how the technology is developing and how the AI experts assess the expectations of the IT consultants, i.e. which expectations are justified and which are unrealistic.

All expert interviews were conducted as semi-structured interviews via online video calls. The online format made it easier to find appointments with the interview partners and to record the conversations so that the interviews could be transcribed for evaluation and subjected to a qualitative content analysis.

Turboscribe.ai[1] was used for the transcription in the first step. As the results showed incorrectly transcribed words and errors in the presentation of the dialogue, the transcripts were revised by hand. Furthermore, ChatGPT was used to support the coding process. First, the transcribed interviews were summarised by ChatGPT to identify the main themes and key statements. Specific questions were then posed to the model to gain deeper insights into certain aspects of the interviews and extract relevant key points. The data generated in this way was checked manually in a further step by comparing the ChatGPT outputs with the original transcripts and checking for consistency and completeness.

The evaluation took into account how much experience the individual experts have in order to be able to weigh their statements accordingly.

4.1 Interviews with IT Consultants

In preparation for the interviews, it was necessary to acquire a sufficiently large number of experts who were willing to share their expertise. The focus was on people who work full-time as IT consultants. A total of 14 IT consultants working for five different IT companies were recruited. Three of the five companies belong to the same group of companies as the majority of the interviewees. The participants have different positions and corresponding levels of experience (between 2 and 25 years), such as Senior Business Consultant, IT Consultant, Senior Project Manager, Consultant, Lead Business Consultant or Junior Business Consultant. Figure 2 visualizes the experience profiles of the IT consultants.

In order to give the IT consultants the opportunity to prepare, an overview of the consulting tasks was sent out alongside the guidelines. In the interviews, the IT consultants were asked to focus primarily on their core competencies so that their assessment of whether LLMs can be used or not is based on sound knowledge and knowledge of the relevant tasks and subtasks.

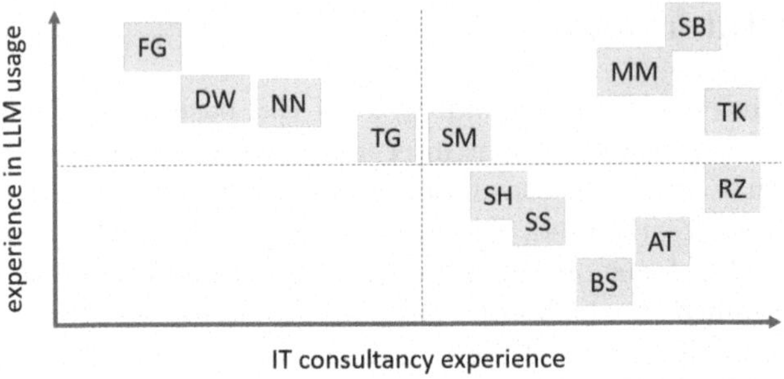

Fig. 2. Experience profile of the IT consultants in the first interview series

IT-Consultant Interview Results. As a result of the consultant interviews, basic process models were developed for the ITC activities described in Sect. 3.3 which illustrate the IT consulting tasks contained in the activities. These models were derived from the interviewees' descriptions of their activities and the expected or observed AI deployment options. Nine IT consultants made statements about all of the activities, five consultants focused on six activities only. For each activity, at least ten statements were available. The statements were summarised and no contradictory statements were discovered. The resulting models (Figs. 4 and 5 show examples) illustrate the observed or expected AI potential on the one hand and form the basis for the interviews with the AI experts on the other. In the models, the process steps have three different colours. For the process steps that are marked in green, at least eight of the consultants already use generative AI or they expect profitable support in the future. A yellow colour means that the technology is known, but the current implementation status of AI is not sufficient to support the tasks. If yellow is used, at least four experts expect for the future that generative AI can be used sensibly for yellow process steps. In addition, the colour yellow indicates that the respective process step has been addressed by only a few experts and therefore the general opinion of consultants is unclear. For all process steps marked in red, none of the consultants sees any possibility of supporting them with generative AI. Finally, the models include comments indicating if a process step involves time-consuming work. A total of ten process models were created; for brevity reasons, only two were selected and presented in the following: proposal preparation and creation of release plan. The activities included in the other process models and how interviewees see LLM suitability are summarised in Fig. 3.

Proposal Preparation. Proposal preparation includes the steps for preparing the proposal and the preparation of the formal offer (see Fig. 4). All interviewed consultants can imagine extensive LLM support in this area. For the first task, understanding the customer tender, the consultants are unanimous. Gathering information from documents is a time-consuming task. In addition, ten consultants were aware that the LLM are able to answer questions about documents

ITC activity	Suitability of tasks for LLM support		
	Observed	Expected	Not suitable
Project Acquisition	analyze customer requirements, prepare project structure, prepare written offer	Resource and competence planning, prepare project presentation	
Project Planning	generate milestone & deliverable plan	Develop documentation strategy, risk identification, strategy to minimize risks	Resource and technology planning
Project Monitoring	quality control, inform stakeholders, adjust project plan	Cost control, progress monitoring	
Project Closure	prepare final report	Provide project data	Delivery to customer, project retrospective
Requirements	prepare user stories	Understand customer demands, specify requirements	
Release Planning	assign user stories and increments to sprints, assign staff to sprints based on capacity	Analyse milestones plan, determine important increments	Prioritize backlogs
Release Notes	analyze content of increments, prepare release notes, publish release notes		
IT system documentation	enable retrieval of relevant content	Document security guidelines, write security handbook	Analyze norms, define security guidelines, define emergency measures
User manual	prepare test cases and test data, prepare user manual, enable retrieval of relevant content	Define test strategy	Document application configurations
Customer handbook	document customer relation, enable retrieval of relevant content	Analyse customer profile, define customer relation, define collaboration, agree on tasks	

Fig. 3. ICT activities and their suitability for LLM according to the observations and expectations of IT consultants

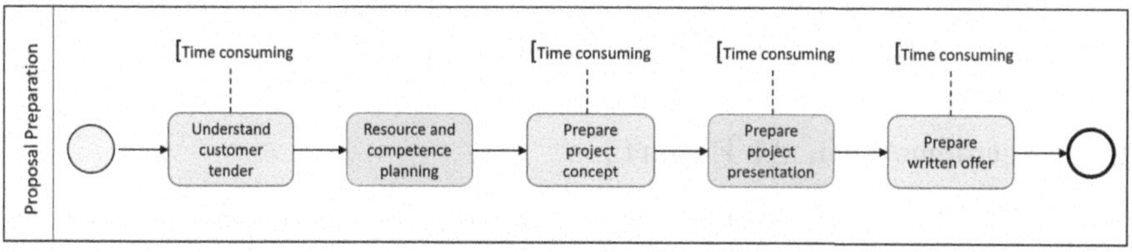

Fig. 4. Basic tasks of project planning as derived from IT consultant interviews

and summarise them. The second process step, internal resource and competence planning, was only mentioned by four consultants. Only these four consultants see support possibilities through LLM, which explains the yellow color. The creation of a written project concept is, again, a time-consuming task. Ten consultants see potential for the use of generative AI, particularly in the written formulation of the concept, which is why this is marked in green. The consultants do not agree on the general concept work, including modeling, and there is a great deal of skepticism regarding the use of LLM for this. All consultants agree that the automated creation of a project presentation from texts would bring added value because this task is time-consuming. However, nine consultants agree that it must be possible to define some formal requirements the LLM output has to fulfill, such as the predefined mandatory structure, layout and visualizations of time plans and economic aspects. LLM is currently already able to create presentations from texts. However, these presentations do not even begin to fulfil the requirements. The last process step considered is the creation of the written offer. The aim of this proposal is to win projects from customers, which is why it must be of a very high quality. This is why ten consultants see this task as a 'time waster' due to the many rounds of review. All consultants see a possible integration of LLM in this process step. This ranges from smaller support measures, such as spell-checking, to the complete generation of offer documents.

Creation of the Release Plan. As part of release planning, it is determined which increments must be implemented in which releases and at what times. To do this, the process steps analyse milestone plan and determine important increments must be carried out in order to determine the relevant increments from the backlog. Combining the process steps is a time-consuming task because it involves a lot of searching. The colour yellow comes from the fact that the six consultants assume that the AI would have to learn a project in order to assist with these tasks. Based on data from previous projects, eight consultants hope that AI will be able to carry out the process step of assigning increments/user stories to sprints. In addition, the same consultants hope that the AI will be able to learn how efficient a team is and be able to carry out the process step observe team efficiency. Using this information, five consultants are sKeptical that the generative AI will be able to prioritise the backlog process step. Nevertheless, the consultants point out that human control is essential to ensure that the prioritisation of the sprint backlog is carried out properly.

4.2 Interviews with AI Experts

A total of six AI experts working for four different IT companies were recruited for the second interview series. All experts have been using AI in their daily operations and also acted as consultants to enterprises exploring the possibility of AI and LLM use. The experts have on average more than 10 years of experience in the IT sector and occupy different positions, such as Lead IT Consultant, Senior Business Consultant or Head of IT innovation management.

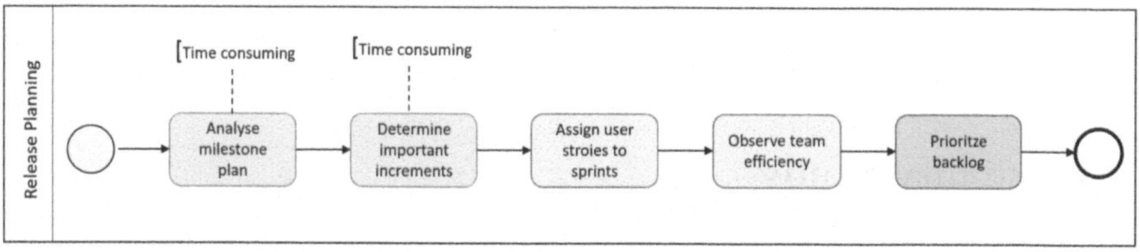

Fig. 5. Basic tasks of release planning as derived from IT consultant interviews

Results of the AI Expert Interviews. A key area for the use of generative AI is the structuring and summarisation of customer data. AI can process large amounts of information quickly and filter out relevant data, which is particularly helpful when creating formal offer documents and proposals. In addition, generative AI supports the generation of texts and content for these documents by creating templates that can be further refined and customised by the consultants. This reduces manual effort and speeds up the creation process considerably. Another area of application is support for project planning and monitoring. Generative AI can help with the creation of milestone plans and the monitoring of project progress. Based on legacy data, it develops initial drafts for the plans, which then need to be further refined by humans. The AI can also analyse data and recognise patterns, allowing potential risks and deviations to be identified at an early stage, enabling proactive project management.

Eliciting requirements and creating user stories are also time-consuming tasks where generative AI shows considerable potential. The AI can extract and structure requirements from various documents and identify gaps between the requirements. It also supports the creation of user stories from the requirements and the integration of these stories into management tools. This facilitates the creation of detailed and precise requirements that serve as a basis for development. However, it is emphasised that the results must be thoroughly checked and validated to ensure accuracy. In the distant future, it is possible that the story generation process will be fully automated. Support for text generation and wording improvement was also emphasised. This capability is particularly useful for creating release notes and writing documentation. Four of the experts from the field of artificial intelligence have expressed that complete automation of the generation of release notes based on generation of release notes based on user stories could be possible in the near future. Significant potential for the use of generative AI is also recognised for the creation of documentation. It can generate content that can subsequently be edited and facilitate the summarisation of documents, which is an advantage for any type of documentation.

To summarise, the analysis of the expert interviews shows that generative AI offers a wide range of potential applications in IT consulting. It can increase efficiency and reduce the workload by automating routine tasks and structuring complex information. However, human review and validation remains essential to ensure the quality and accuracy of results. Generative AI is seen as a valuable tool that supports, but does not completely replace, IT consultants. However, the use of generative AI can save a significant amount of working time.

5 Feasibility Study

The feasibility study, as the final step in our research process, had the purpose of investigating if it is feasible to fully automate time-consuming tasks in IT consultancy and, in this context, to gather information on how much effort could be saved by using LLM. From the activities categorized by consultants and AI experts as both suitable for LLM support and time-consuming, we selected the creation of user stories. User stories are used in requirements management to specify required functionality from a user perspective. AI experts expressed the expectation that LLM can be used to fully automate the generation of user stories if the specification of functional requirements to a software system has been completed. For the feasibility study, we used ChatGPT-4o and started with a document describing the functionality of an accounting system as input. When developing the prompt, we used several prompt patterns (e.g., persona and template pattern [22]) and stepwise extended the prompt to include the required structure of the user stories, the request to include acceptance criteria, and to define test cases. The final version of the prompt is shown in Fig. 6.

> You are an IT consultant in the area of requirements management. Your task is to subdivide the uploaded document into requirements and write user stories. Do not omit any requirements. Also use the information from the appendix included in the document. Write user stories for every requirement. The stories should consist of an introductory sentence, the acceptance criteria numbered in a list and a short description. Write down the introductory information as a sentence. Only use the content from the document and ist appendix. Do not add any content.
>
> A user story should have the following format:
> [Title of the user story] [Role of the user] [Description of the story]
> The following acceptance criteria must be met: [Numbered list of acceptance criteria]

Fig. 6. Prompt to generate user stories from description of functionality)

The generated user stories were checked by a subject matter expert and perceived as very useful, without hallucinations but with improvement potential. Examples of shortcomings were that not all parts of the input document were processed or that some of the acceptance criteria needed extensions. The expert estimated that between 80 and 90% of the efforts for creating user stories could be saved by LLM. The exercise was repeated with other functional specification documents and the final prompt, which confirmed this impression.

6 Conclusion, Limitations and Outlook

The aim of our study was to better understand the potential of LLM in IT consultancy. The first finding is that IT consultants use LLM for various tasks in their daily work and expect more LLM support in the future. Intuitively, this result was expected, but so far, there was no evidence in the scientific literature. The second finding is that the way LLMs currently are used does not indicate the redefinition of value-creation processes in ITC or innovations in the business model. LLM usage focuses on automating routine tasks, such as documentation, or increasing efficiency in time-consuming tasks, such as specifying user stories. None of the consultants reported on enterprise-level support from their employers to use LLM in a systematic way. There seems to be much room for better integration of LLM into the consultancy value chain.

The research design described in Sect. 1 has consequences with respect to limitations of our work: The expert interviews reflect observations, experiences and opinions of human experts, which are subjective in contrast to objective measurements. However, in the absence of instruments for exact measurements, this way of data capturing created a rich dataset that can be used to explore and understand the phenomenon better. Not all ITC activities and tasks were included in the investigation, but we had to focus on a subset suitable for an interview study. The tasks included in the investigation can be positioned in the area of implementation services and do not cover strategic advisory and operational services well. The process descriptions derived from the first interview series (IT consultants) simplify the tasks to be performed and are not necessarily complete, but are suitable to structure the investigation.

The above limitations resulting from the research design also indicate possibilities for future work. A larger number of interviewees, in combination with a quantitative investigation into the extent of LLM use and the expected effects on reducing efforts and increasing quality, would create a more holistic picture. Furthermore, the ITC areas not covered in this study (e.g., advisory and operational services) should be investigated. Although the feasibility showed promising results, a more detailed investigation into the potential and limitations of LLMs is needed. In specific client cases or application domains, LLMs might produce reports that are too generic or contain bias, which is not acceptable for ITC consultancy and has to be avoided, for example, by specifically designed prompts.

References

1. Becker, S.V., Aromaa, E., Eriksson, P.: Client-consultant interaction: the dynamics of and conflicts in value co-creation and co-destruction. Int. J. Serv. Technol. Manag. **21**(1–3), 40–54 (2015)
2. Biech, E.: The Business of Consulting: The Basics and Beyond. Wiley & Sons, Hoboken (2011)
3. Brown, T., et al.: Language models are few-shot learners. Adv. Neural Inf. Process. Syst. **33**, 1877–1901 (2020)
4. Cámara, J., Troya, J., Burgueño, L., Vallecillo, A.: On the assessment of generative AI in modeling tasks: an experience report with chatgpt and uml. Softw. Syst. Model. **22**(3), 781–793 (2023)
5. Chui, M., et al.: The Economic Potential of Generative AI. McKinsey and Co., New York (2023)

6. Du, R., et al.: Rapsai: accelerating machine learning prototyping of multimedia applications through visual programming. In: Proceedings of the 2023 CHI Conference on Human Factors in Computing Systems, pp. 1–23 (2023)
7. Ebert, C., Louridas, P.: Generative AI for software practitioners. IEEE Softw. **40**(4), 30–38 (2023). https://doi.org/10.1109/MS.2023.3265877
8. Eloundou, T., Manning, S., Mishkin, P., Rock, D.: GPTs are gpts: an early look at the labor market impact potential of large language models. arXiv preprint arXiv:2303.10130 (2023)
9. Kitchenham, B., Brereton, O.P., Budgen, D., Turner, M., Bailey, J., Linkman, S.: Systematic literature reviews in software engineering-a systematic literature review. Inf. Softw. Technol. **51**(1), 7–15 (2009)
10. Liu, P., Yuan, W., Fu, J., Jiang, Z., Hayashi, H., Neubig, G.: Pre-train, prompt, and predict: a systematic survey of prompting methods in natural language processing. ACM Comput. Surv. **55**(9), 1–35 (2023)
11. Makridakis, S.: The forthcoming artificial intelligence (AI) revolution: its impact on society and firms. Futures **90**, 46–60 (2017)
12. Mariani, M.M., Machado, I., Nambisan, S.: Types of innovation and artificial intelligence: a systematic quantitative literature review and research agenda. J. Bus. Res. **155**, 113364 (2023)
13. Oesterle, S., Buchwald, A., Urbach, N.: Investigating the co-creation of it consulting service value: empirical findings of a matched pair analysis. Electronic Markets 1–27 (2022)
14. O'Mahoney, J., Markham, C.: Management Consultancy. Oxford University Press, USA (2013)
15. Roima, A.: The disruption of the it consulting business by large language models: an analysis of implications. Master Thesis, Tampere University (2023)
16. Sandkuhl, K.: Llm-assistance for quality control of llm output. In: International Conference on Business Informatics Research, pp. 36–50. Springer (2024)
17. Sandkuhl, K., Barn, S.B., Barat, S.: Full paper: neural text generators in enterprise modeling: can chatgpt be used as proxy domain expert? In: Information Systems Development, Organizational Aspects and Societal Trends (ISD2023 Proceedings). Association for Information System (AIS) (2023)
18. Scheer, A.W., Köppen, A.: Consulting: Wissen für die Strategie-, Prozess-und IT-Beratung. Springer-Verlag, Cham (2013)
19. Taiwo, R., et al.: Generative artificial intelligence in construction: a delphi approach, framework, and case study. Alexandria Eng. J. **116**, 672–698 (2025). https://doi.org/10.1016/j.aej.2024.12.079
20. Tsushima, Y., Yamamoto, S., Ravankar, A.A., Luces, J., Hirata, Y.: Task planning for a factory robot using large language model. IEEE Robot. Autom. Lett. (2025). https://doi.org/10.1109/LRA.2025.3531153
21. Wang, X., Anwer, N., Dai, Y., Liu, A.: Chatgpt for design, manufacturing, and education. Procedia CIRP **119**, 7–14 (2023)
22. White, J., et al.: A prompt pattern catalog to enhance prompt engineering with chatgpt. arXiv preprint arXiv:2302.11382 (2023)
23. Wu, T., Jiang, E., Donsbach, A., Gray, J., Molina, A., Terry, M., Cai, C.J.: Promptchainer: chaining large language model prompts through visual programming. In: CHI Conference on Human Factors in Computing, pp. 1–10 (2022)
24. Ziegler, A., et al.: Measuring github copilot's impact on productivity. Commun. ACM **67**(3), 54–63 (2024)

Sentiment Analysis of Wikipedia Articles About Companies: A Comparison of Different Models

Milena Stróżyna, Włodzimierz Lewoniewski, Izabela Czumałowska(✉), and Aleksandra Wojewoda

Department of Information Systems, Poznań University of Economics and Business, Al. Niepodległości 10, 61-875 Poznań, Poland
izabela.czumalowska@ue.poznan.pl

Abstract. Wikipedia is a widely recognized, collaboratively edited encyclopedia and a key information source for journalists, investors, and consumers researching companies. A crucial quality criterion of Wikipedia articles is the Neutral Point of View (NPOV), but ensuring neutrality can be challenging due to editor biases. Therefore, it is essential to verify whether NPOV is met. One possible approach to achieve this is by assessing the sentiment of Wikipedia articles.

This paper presents a comparative study of different approaches for sentiment analysis of Wikipedia articles about companies. We used lexicon-based (TextBlob, VADER) and transformer-based models (RoBERTa-based, DistilBERT-based models and PySentimiento). Our main objectives was to determine the differences between sentiment in various Wikipedia articles and how publicly available models apply to such texts. First, we described the data collection process, including article selection and preprocessing steps to ensure content relevancy. We also proposed three methods to calculate an aggregation score for each article. Next, we presented the overall results across all selected articles, showcasing how each model performs in sentiment analysis. Additionally, we defined subsets of articles belonging to different quality classes and separate industries and compare the sentiment for these groups.

Results of this study shows that sentiment in Wikipedia articles varies depending on the industry and chosen model. Additionally it provides a practical framework for researchers and practitioners seeking to implement a sentiment analysis that may be applied in the case of longer texts. Our findings contribute to a better understanding of how each model evaluates sentiment in Wikipedia articles about companies.

Keywords: sentiment analysis · Wikipedia · companies · TextBlob · VADER · RoBERTa · DistilBERT · PySentimiento · Wikidata

1 Introduction

Throughout its extensive history as a collaborative online encyclopedia, Wikipedia has established itself as one of the most widely recognized sources

K. Węcel (Ed.): BIS 2025, LNBIP 554, pp. 101–115, 2025.
https://doi.org/10.1007/978-3-031-94193-1_8

of information online. Thanks to its easily accessible and open-editing model, it attracts millions of readers seeking quick, comprehensive overviews on a vast range of topics, including information about companies. In fact, Wikipedia articles can serve as a starting point for anyone looking for information about history, products, services, and notable achievements of an organization.

One of the important quality criteria for Wikipedia articles is Neutral Point of View (NPOV). This policy underpins the articles' credibility, guiding the collaborative effort to ensure accurate and balanced coverage of myriad topics. NPOV dictates that articles should be written without personal bias, presenting all significant viewpoints fairly and proportionately. However, in practice, ensuring neutrality can be challenging. Editors come from varied backgrounds and may bring their own biases to the editing process. Therefore, it is crucial to determine whether the NPOV standard is met. One way to do this is by analyzing the sentiment of Wikipedia articles.

Today the sentiment of publicly available information, such as Wikipedia articles, may have a significant impact on perceptions, decision-making, and societal behaviors. Sentiment analysis, which aims to categorize content as positive, neutral, or negative, might be used to automatically interpret vast amounts of digital content that is published online every day. The insights provided by such analysis are relevant to businesses, academia, and the general public alike, fostering a deeper understanding of public discourse and its implications. Sentiment analysis helps identify reputational risks, track trends in public opinion, and evaluate the credibility of information.

One of the common goals of sentiment analysis is to gain valuable insights into consumer opinions, public reactions to events, and market trends [1,7,14]. That is why it often focuses on short-form texts, such as those from social media platforms [2] as well as customer reviews [29], and product feedback [9]. These texts, while rich in real-time data, pose unique challenges due to their conciseness and informal language, highlighting the need for advancing methodologies to apply sentiment analysis effectively to other types of content, including longer and more structured texts like Wikipedia articles. This article aims to address this challenge by conducting sentiment analysis on a selected set of English Wikipedia articles that describe companies from various industries. The objective is to evaluate how commonly used, state-of-the-art methods for sentiment analysis can be applied in this context.

This study is guided by the following research questions:

- What prevailing sentiment patterns can be identified in Wikipedia articles about companies?
- Can or should commonly used sentiment analysis methods be applied to assess the sentiment of such articles?
- To what extent are sentiment evaluations consistent across different state-of-the-art methods when applied to structured and detailed content like Wikipedia articles?
- How does the quality of a Wikipedia article affect the sentiment analysis outcomes?

– How does the outcome of sentiment analysis vary depending on the industry sector of the described company?

The novelty of the paper stems from applying sentiment analysis to assess sentiment and thereby verify if NPOV is maintained in Wikipedia articles. The main contributions of this papers concerns the following points. First, it compares five state-of-the-art methods for sentiment analysis with regard to their applicability and performance in evaluating the neutrality in Wikipedia articles. Second, it proposes three methods to calculate an aggregated sentiment score for Wikipedia articles, offering a practical framework that might be used for sentiment assessment of longer texts. Finally, it analyzes how the sentiment of Wikipedia articles about companies varies depending on both the industry and the chosen model.

The remainder of this article is organized as follows: Sect. 2 reviews related work, Sect. 3 describes the methodology used, Sect. 4 presents and discusses the results, and Sect. 5 provides the conclusion and summary.

2 Related Work

2.1 Neutrality in Wikipedia

Several academic studies have explored the challenges and controversies surrounding the Neutral Point of View (NPOV) policy on Wikipedia. One of the works [24] found that ambiguity is inherent in Wikipedia's policymaking process. The NPOV principle is often interpreted flexibly by editors, which leads to conflicts in defining neutrality. Some studies have developed tools to detect bias in Wikipedia articles. For example, the study by Hube [13] focused on detecting biases in Wikipedia articles using different factors, including language style. Another work by [23] proposed machine learning models for detecting linguistic bias patterns. However, to the best of our knowledge, there was no research that would focus on assessment of Wikipedia articles' neutrality by applying sentiment analysis.

Companies may attempt to edit or influence their Wikipedia pages to present themselves in a more favorable light, sometimes leading to conflicts between corporate representatives and independent editors. Roessing & Einwiller [27] found that Wikipedia articles about companies differ significantly between English and German versions, with domestic versions often being more detailed and containing more criticism.

2.2 Sentiment Analysis

Sentiment analysis, a crucial aspect of natural language processing, aims to determine the emotional tone or attitude expressed in a piece of text. This is achieved by employing various methods, including rule-based or lexicon-based approaches, supervised and unsupervised machine learning algorithms, and deep learning techniques. These methods categorize the sentiment into different groups, most commonly as positive, negative, or neutral [31].

An analysis of the related studies on sentiment analysis shows that the most commonly used tools and methods include TextBlob, VADER, e.g., [5,7,9,14,15, 19]. Some studies also employed AFINN [2], as an auxiliary tool for data labeling. The second group of methods are machine learning-based methods such as SVM [26], CNN-LSTM [26], Naive Bayes [11], and KNN [11]. For thematic analysis and feature extraction, approaches like LDA (topic modeling) [1] and GloVe (word embeddings) [30] were utilized.

Recent development of transformer-based architectures and emergence of Large Language Models (LLMs) have significantly enhanced the capabilities of sentiment analysis. One example is transformer-based models, such as BERT (Bidirectional Encoder Representations from Transformers) and its variants, commonly referred to as BERT-like models. Unlike traditional methods, presented above, that rely on predefined lexicons or shallow linguistic features, LLMs excel at understanding complex language structures, such as idiomatic expressions and context-dependent meanings. This makes them particularly effective in applications like sentiment analysis [28]. As a result, certain BERT variants have been fine-tuned specifically for sentiment analysis task, e.g., DistilBERT, roBERTa, or PySentimiento (see Sect. 2.3). They were applied in studies like analysis of crisis-related texts [26], review analysis [18], opinions analysis [17], tweets analysis [16], or data labeling [3].

In the related studies, the most frequently processed datasets consist of textual content from social media, particularly Twitter (e.g., tweets about COVID-19, airlines, technological products, or telecommunications services), Reddit and user reviews. Typically, authors focused on analyzing opinions and sentiments associated with specific topics (e.g., the COVID-19 pandemic [1], the Russia-Ukraine war [26], or products and services of selected companies [14]).

Applications of sentiment analysis included: data labeling [2,3,19], public opinion analysis, e.g., in the context of events like the COVID-19 pandemic [1,19] or crisis [26], improving services (analyzing product and service reviews [7,15]), or comparison of different sentiment analysis methods [11,29].

In summary, existing studies in sentiment analysis have most often focused on content shared on social media, such as short texts like tweets, user reviews, and comments. These types of texts are typically concise and centered around specific topics. However, there is a notable lack of research analyzing longer texts, such as Wikipedia articles or news articles. Our study aims to address this research gap and expand sentiment analysis to more complex and longer textual datasets.

2.3 Models for Sentiment Analysis

In this section, we briefly described the sentiment analysis models used in our research. These models include two lexicon-based approaches – VADER and TextBlob – and three transformer-based models – RoBERTa, DistilBERT and PySentimiento.

TextBlob[1] is a Python library designed for working with human language. It builds upon the NLTK library, which is another tool for language processing [2]. When provided with a sentence, TextBlob analyzes it to determine two key aspects: polarity and subjectivity. Polarity scores range from -1, indicating extremely negative sentiment, to 1, indicating extremely positive sentiment. Subjectivity measures how subjective or opinionated the text is. In our research, we focus only on the polarity score.

VADER (Valence Aware Dictionary and sEntiment Reasoner)[2] is a sentiment analysis tool that utilizes a dictionary and a set of rules to understand the sentiment conveyed in social media text. It can accurately extract sentiments from text, as well as emotions represented by text, emojis, abbreviations, and specific terms [10].

The authors of VADER used a combination of qualitative and quantitative methods and also empirically validated sentiment lexicon for English as a *gold-standard* to train VADER. The model was specifically trained to be attuned to microblog-like contexts.

VADER provides two types metrics: *the positive/negative/neutral score*, which is, respectively, the proportion of positive, negative, and neutral sentiment in the text (the percentage of words in each category), and *the compound score*. The latter, being a normalized, aggregated sentiment, combines the positive, negative, and neutral scores into a single value ranging from −1 (most negative) to +1 (most positive). Unlike other sentiment scores, it weighs the intensity of words and phrases rather than just categorizing them, thus reflecting the overall sentiment intensity of the text. Its value can be used to classify text as positive (value $>= 0.05$), neutral (value between −0.05 and 0.05), or negative (value $<=$ −0.05). In our study, we used both types of VADER's metrics.

PySentimiento[3] is a Python toolkit developed for four types of NLP tasks: sentiment analysis, emotion detection, hate speech detection, and irony detection. It is based on several pre-trained transformer models (e.g., RoBERTa, BERT), fine-tuned for analyzing social media text. To achieve this, representative datasets in Spanish, English, Portuguese and Italian languages were used [25]. Unlike traditional lexicon-based approaches, it allows for a more context-aware classification of text.

In our study, we use only the sentiment analysis module for English, which provides categorical sentiment classification (positive, neutral, and negative) along with probability scores.

DistilBERT, more precisely distilbert-base-multilingual-cased-sentiments-student model[4] but referred to as DistilBERT later in the article, is a multilingual sentiment analysis model based on BERT. It was trained to classify text into positive, negative, and neutral sentiment while being computationally efficient

---

[1] https://textblob.readthedocs.io/en/dev/.
[2] https://vadersentiment.readthedocs.io.
[3] https://github.com/pysentimiento/pysentimiento.
[4] https://huggingface.co/lxyuan/distilbert-base-multilingual-cased-sentiments-student.

in terms of both processing power and memory usage. DistilBERT achieves this efficiency by reducing the model's size by 40% compared to the original BERT while still maintaining a remarkably high level of performance, retaining approximately 97% of BERT's original accuracy [6]. The model used in our research was trained on a dataset consisting of multilingual sentiment-labeled tweets. Similar to PySentimiento, it provides sentiment classification into positive, neutral, and negative along with probability scores.

RoBERTa, more precisely Twitter-roBERTa-base model[5] but referred to as RoBERTA later in the article, is a RoBERTa-based transformer model designed specifically for sentiment analysis on Twitter and social media text. It was trained on the TweetEval benchmark, which includes a large corpus of tweets (58 million) labeled for various NLP tasks, including sentiment analysis [4]. The reported classification accuracy measures an F1 score of 0.834, with precision of 0.830 and recall of 0.838 [20]. Although the model was trained for various NLP tasks, our research focuses on sentiment analysis. Similar to the two previous models, it provides sentiment classification into positive, neutral, and negative along with probability scores.

3 Methodology

According to the Design Science Research framework [12], our research adheres to structured guidelines, clearly addressing relevance, rigor, and design contributions. Our primary artifact is the proposed sentiment analysis framework, which includes selecting and evaluating sentiment analysis methods (lexicon-based: TextBlob, VADER; transformer-based: RoBERTa, DistilBERT, PySentimiento), as well as developing three aggregation methods for sentiment scoring. We conducted evaluation through comparative analysis using a comprehensive dataset of Wikipedia articles about companies. The dataset was categorized by article quality and industry. This allowed to highlight industry-specific sentiment trends and the correlation between article quality and sentiment outcomes.

3.1 Dataset Preparation

We began our analysis by creating a list of English Wikipedia articles related to companies. To ensure relevance, we aligned our selection with WikiProject Companies[6] - a dedicated initiative aimed at improving and highlighting articles on notable companies. This alignment acted as a quality filter, as the project focuses on articles that are likely to be informative and well-structured.

From this initial collection, we identified a total of 58,673 articles. To further refine the dataset, we applied specific filtering criteria. First, we retained only articles that had undergone content quality assessments. Next, we removed approximately 1,000 articles that were stand-alone lists.

[5] https://huggingface.co/cardiffnlp/twitter-roberta-base-sentiment.

[6] https://en.wikipedia.org/wiki/Wikipedia:WikiProject_Companies.

The sentiment analysis models selected for our research require text inputs to be of a given size. Therefore, we segmented the content of articles into smaller parts (referred to as segments later in the article) considering:

- Character limit: each segment was constrained to no more than 2000 characters (including spaces) to meet the input limits set by the selected models.
- Maintaining contextual integrity: the segmentation process was carefully designed to avoid arbitrarily breaking paragraphs or sentences. Paragraphs exceeding the 2000-character limit were split, ensuring that the semantic and syntactic context was largely preserved.

With a quality-assured set of articles, we proceeded to categorize the dataset into different groups based on their quality assessments. To this end, we adopted the quality groups defined by the Wikipedia quality grading scheme[7], which categorizes articles from highest to lowest quality: Featured Articles (FA-class), Good Articles (GA-class), B-class, C-class, Start, Stub. This classification also allowed us to control for quality-related factors in further analyses, particularly when comparing sentiment trends across different levels of content quality.

To add another layer of contextual relevance, we aligned each Wikipedia article with specific industries. This was achieved by mapping the articles to their corresponding Wikidata items and leveraging the semantic representations contained within them [21,22]. Due to paper length limitations, we selected industries associated with the largest number of companies described on English Wikipedia.

It should be noted that we used in this study lexicons as they were assumed and implemented in the six selected sentiment analysis methods. Our analysis did not involve an in-depth examination of these lexicons. Therefore, the findings should be interpreted with this limitation in mind.

3.2 Implementation

For implementation of TextBlob and VADER, we accordingly used TextBlob and vaderSentiment Python libraries. In the case of DistilBERT and RoBERTa we used Transformers Python library and their implementation available at HuggingFace. For PySentimiento its direct implementation available on HuggingFace was used.

Since TextBlob, unlike other models, provides a sentiment value ranging from –1 to +1, we adopted the assumption proposed by [15,29,30] to make its results comparable with those of other models: scores <–1;–0.05> are classified as negative sentiment, (–0.05;0.05) as neutral sentiment, and <0.05:1> as positive sentiment.

As mentioned in the previous section, due to input limitations, articles were divided into smaller segments before being processed by each model. Consequently, the outputted sentiment scores corresponded to individual text segment

[7] https://en.wikipedia.org/wiki/Wikipedia:Content_assessment.

rather than entire articles. Therefore, we proposed three methods for aggregating the sentiment of individual text segment derive the overall sentiment of an Wikipedia article:

1. **Method 1:** Maximum value – sentiment classifications of individual text segment are treated as votes; the overall sentiment of the article is determined by the class (positive, neutral, negative) that receives the majority of votes from all considered segments of the Wikipedia article;
2. **Method 2:** Weighted voting – the overall sentiment of the article is calculated similarly to the Method 1, but each segment's vote is additionally weighted based on its text length – the longer a segment is, the greater its sentiment influence on the overall score of the analyzed Wikipedia article;
3. **Method 3:** Probability multiplied by length – unlike the other two methods, all three probability scores (for each sentiment class) of each segment are taken into consideration; each probability is further weighted by the segment's length. As a result, the overall sentiment assessment consists of three values, one for each sentiment class. These values are then normalized to maintain a similar range as at the segment level, ranging from 0 to 1 on a continuous scale for each sentiment class.

The calculation of the overall sentiment value for a given Wikipedia article using Method 3 can be represented by the following equation:

$$\tilde{p}(j) = \frac{\sum_{i=1}^{N} \ell_i \, p_i(j)}{\sum_{i=1}^{N} \ell_i}, \quad \forall j \in \{positive, neutral, negative\}, \tag{1}$$

where:

- ℓ_i - the length of a text segment i,
- $p_i(j)$ - the probability score for one of the sentiment class j in a segment i,
- N - the total number of segments in a given Wikipedia article.

The above equation ensures that the overall score for each sentiment class is a weighted average of the segment-level probabilities, where the weight corresponds to the segment's length, and that the final values fall within the $[0, 1]$ interval.

In order to calculate the overall sentiment for each group of Wikipedia articles (selected based on the quality class or industry), additional level of aggregation was performed. This aggregation depends on the applied method:

1. For Method 1: as at the article level – votes were counted and the overall sentiment for the group was determined by the dominant sentiment class across all articles within the group,
2. For Method 2: the same approach as for Method 1,
3. For Method 3: having three separate values assigned to each sentiment class for each article, the average value was calculated within the group.

Despite proposing three different methods, during the research we discovered that Method 1 gives results very similar to Method 2. Therefore, Method 1 is ommitted from the results presented in the following section.

4 Results

4.1 Sentiment Patterns in Wikipedia Articles

The first analyzed aspect concerned the prevailing sentiment patterns in Wikipedia articles about companies. To this end, we compared the results provided by five selected models and using the three aggregation methods presented in the previous section. The results of sentiment analysis of all articles in the dataset are presented on Fig. 1. We distinguished between two metrics provided by VADER: the first was calculated based on positive, negative, and neutral scores (labeled on the figures as VADER), while the second based on the compound score (labeled as cVADER) – see Sect. 2.3.

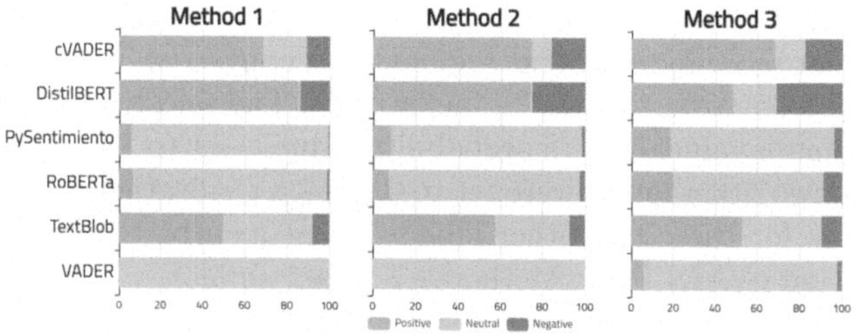

Fig. 1. The overall results of sentiment analysis for all considered Wikipedia articles using different models and methods. Extended version available at [8]

The results indicate that, in general, regardless of the method or model used, the majority of articles were classified as neutral or positive. However, some differences were observed across the different models. The highest agreement was found between transformer-based models (RoBERTa and PySentimiento). In both cases, the neutral class is dominant, with a small share of the positive class and a practically marginal negative class. Comparable results were obtained by VADER. In this case, the dominance of the neutral class is even greater, with Methods 1 and 2 yielding 100% neutrality, and Method 3 nearly 92%.

In the case of the results by Textblob and cVADER, the dominant class is positive, followed by neutral, with cVADER showing a lower proportion of neutral in favor of positive.

The sentiment provided by DistilBERT deviates significantly from the other models. Regardless of the method used, the dominant class is positive, with neutral having the smallest share. For Methods 1 and 2 almost no articles were classified as neutral.

When focusing on the consistency of sentiment evaluations across different models, we can conclude that such coherence can be observed in RoBERTa, PySentimiento, and VADER – all of which indicate that the sentiment of the analyzed articles is predominantly neutral. In contrast, models such as DistilBERT, TextBlob, and cVADER tend to lean more towards a positive sentiment. Given these differences in evaluations, it is difficult to definitively determine which method is best suited for sentiment analysis of longer texts, such as Wikipedia articles. However, considering Wikipedia's encyclopedic nature, which strives for objectivity and neutrality, it can be assumed that the results by RoBERTa, PySentimiento, and cVADER align most closely with these expectations.

Finally, we compared the three proposed aggregation methods. Methods 2 and 3, which take into account the length of the analyzed text, lead to slightly different results compared to Method 1. This is likely due to the weighing based on the text length, which influences the final article classification, especially in cases where longer sections have a distinct sentiment. In the case of Method 3, a decrease in the proportion of articles classified as neutral can be observed compared to Method 1. For most models, this decrease corresponds to an increase in the proportion of articles classified as positive.

It is important to note, that in the case of TextBlob and cVADER, the continous sentiment score for each text segment was converted into one of the sentiment classes ("positive", "neutral", "negative"). The selected class was asigned with the maximum value, i.e., its probability value is set to 1.0, while the probabilities for the two other labels were set to 0.0. As a result, there is no probability values available for these two other labels. Consequently, the results presented for TextBlob and cVADER for Method 2 & 3 are similar.

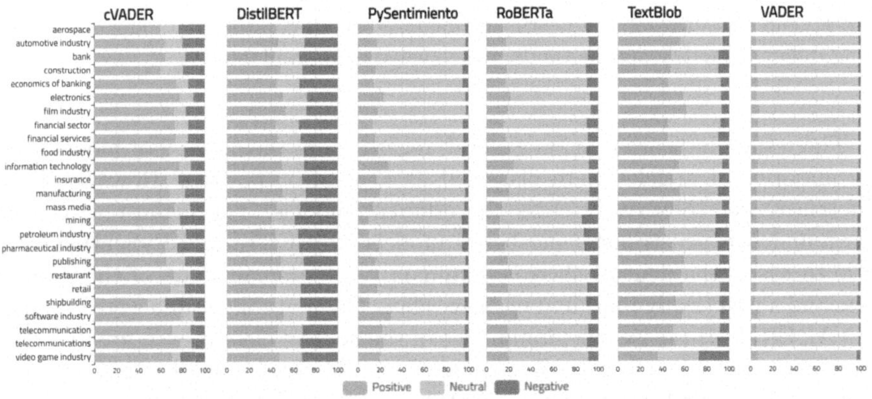

Fig. 2. Sentiment analysis results for Wikipedia articles about companies from different industries, using various models and Method 3. Extended version available at [8]

4.2 Sentiment Patterns Depending on the Company's Industry

The second aspect of our analysis focused on studying if sentiment varies depending on the industry sector of the company described in the Wikipedia article. The results are presented on Fig. 2. This analysis was conducted exclusively using the results form Method 3.

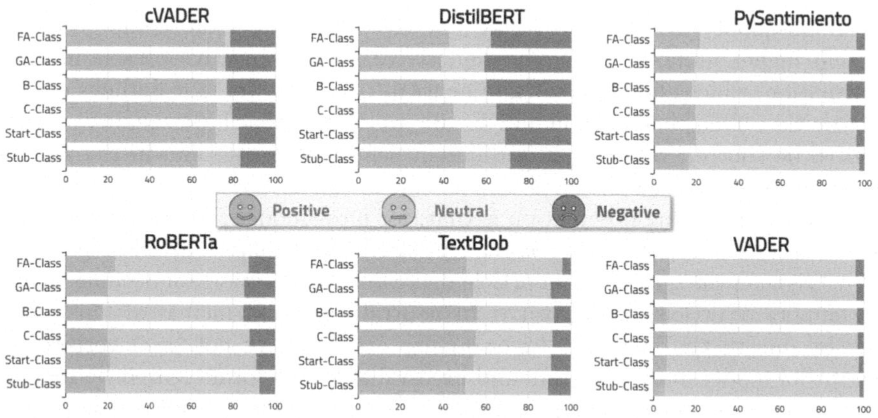

Fig. 3. Sentiment analysis results for Wikipedia articles about companies from different quality classes, using various models and Method 3. Extended version available at [8]

General observations showed significant variability between models. For instance, PySentimiento and VADER predominantly classify texts as neutral, while cVADER, DistilBERT, RoBERTa, and TextBlob provide more balanced distributions among positive, neutral, and negative sentiments.

The findings indicate also that there is some variation in sentiment depending on the industry sector. There are some industries that, regardless of the applied model, exhibit a higher percentage of positive sentiment, e.g., software, information technology, or film industry. This may be due to frequent reporting on innovations, financial successes, or advancements in these fields which can be reflected also in Wikipedia articles. On the other hand, there are industries that, regardless of the applied model, exhibit a higher percentage of negative sentiment, i.e., petroleum, mining, or pharmaceutical industry. This may be related to controversies and environmental issues associated with these sectors, and some information about them may be incorporated into articles.

Industries with a predominance of neutral sentiment include e.g., mass media, financial services, and automotive industry. This may suggest that information presented in articles about companies from these sectors is conveyed in an objective manner, with a neutral tone.

4.3 Sentiment Patterns Depending on the Article's Quality Class

In the final part of our analysis, we examined whether the quality class of an article (see Sect. 3) affects sentiment outcomes (see Fig. 3). This analysis was conducted exclusively using the results from Method 3.

The findings show that, for most of the applied models, the proportion of articles with a neutral tone increases as article quality decreases. This observation is counterintuitive and surprising, since one would generally expect that the higher quality of an article, the more neutral its tone. It is generally expected that high-quality articles (FA-Class and GA-Class) are more comprehensive, contain higher number of sources, and strive for an objective presentation of information, which should result in a neutral tone. However, our findings do not confirm this. This phenomenon may be due to the fact that lower-quality articles tend to be shorter, resulting in a different sentiment score than in the case of longer, high-quality articles.

Another surprising observation is that, for most models, the highest proportion of negative sentiment is recorded in the highest-quality articles (FA- and GA-Class). As the quality of an article decreases, this proportion declines, while the number of articles with positive or neutral sentiment increases. This trend may be due to the fact that lower-quality articles have not been verified by Wikipedia moderators or editors and are available in a form created by regular users. In the case of company descriptions, these users may include employees or fans of the company, who may be inclined to present the company in a positive or neutral manner while omitting negative aspects.

5 Conclusions and Future Works

This paper presented a comparative study of various approaches for sentiment analysis of Wikipedia articles about companies. To this end, five different models for sentiment analysis were applied, including lexicon-based (TextBlob, VADER) and transformer-based (DistilBERT, RoBERTa, PySentimiento). Moreover, we proposed three different methods for aggregating sentiment scores across an article's segments to assess its overall sentiment.

The results of our analysis show that the selected models for sentiment analysis vary in their sensitivity to different sentiment classes. While RoBERTa-based model, PySentimiento, and VADER indicate that most articles exhibit a neutral tone, which aligns with Wikipedia's mission of objectivity, DistilBERT-based model and cVADER tend to generate more positive sentiment.

Our insights show that when selecting a model for sentiment analysis of Wikipedia articles, one should consider that the majority of articles, especially high-quality ones, exhibit a neutral tone. Therefore, models such as RoBERTa, PySentimiento, or VADER, may be well-suited for this type of analysis. Additionally, when in comes to a method for aggregating sentiment scores for separate paragraphs of the article, methods that take text length into account (Methods 2 and 3 in the case of our analysis) may better reflect the actual tone, particularly in articles with many sections of varying sentiment.

An industry-based analysis of Wikipedia articles showed that certain sectors tend to have a more positive sentiment, while others more frequently exhibit a negative tone. The comparison of different quality classes, in turn, revealed two interesting findings. Firstly, a decreasing proportion of articles with a neutral tone as article quality increases. This contradicts Wikipedia's editorial principles, assuming that the highest-quality articles are more carefully edited and adhere more closely to Wikipedia's standards. Secondly, the highest proportion of negative sentiment was recorded for higher-quality articles.

Finally, when it comes to comparison of sentiment results provided by lexicon-based models and transformer-based model, the latter should theoretically perform better on more complex articles. In the case of our study, RoBERTa and PySentimiento demonstrated classification of the majority of articles as neutral, similarly to VADER. This shows that both types of models can be considered for analyzing sentiment of Wikipedia articles.

Our study shows that there is a need for better contextualization of models and their adaptation to the nature of the analyzed content, such as encyclopedic articles, which are characterized by greater neutrality. Standard models for sentiment analysis, such as those applied in the article, often overemphasize positive or negative tones, which can be misleading when analyzing Wikipedia. This probably is a consequence of the fact that most models are optimized for identifying positive and negative sentiment, which does not fully align with the needs of text analysis where neutrality is the dominant category. Hence, there is a need to develop models that are more precise in distinguishing neutral tones.

Beyond sentiment analysis and the basic classification of articles (as positive, neutral, or negative), future research should incorporate more detailed emotion recognition analysis and objectivity analysis, which would allow for a better understanding of the tone of articles. To further validate these models, it would also be beneficial to conduct tests on a different set of articles, exploring variations in topics and languages to assess their robustness and adaptability.

This study can be used by companies to efficiently monitor and evaluate their brand perception on such popular platforms as Wikipedia, enabling quicker responses to reputation challenges. The comparison of multiple sentiment analysis methods provides businesses with guidance on choosing effective tools suited to their specific analytical needs. By applying transformer-based models like DistilBERT or RoBERTa, companies can achieve deeper insights due to their superior accuracy in detecting nuanced sentiments. Additionally, employing the proposed aggregation methods helps organizations reliably interpret overall sentiment trends (also in relation to competitors), facilitating better-informed strategic decisions.

It is important to acknowledge that the limitations in this study results from the capabilities of the models used. Furthermore, the scope of this research could be expanded to include additional languages, as the current dataset was limited to English for the purposes of this study. It should also be noted that the data may not provide a comprehensive representation of the market, as not all companies are documented in Wikipedia articles. Therefore, the findings may not fully reflect the state of specific industries. These elements could be considered valuable contributions to future work.

References

1. Abiola, O., Abayomi-Alli, A., Tale, O.A., Misra, S., Abayomi-Alli, O.: Sentiment analysis of COVID-19 tweets from selected hashtags in nigeria using vader and text blob analyser. J. Electr. Syst. Inf. Technol. **10**(1), 5 (2023)

2. Aljedaani, W., et al.: Sentiment analysis on twitter data integrating textblob and deep learning models: The case of us airline industry. Knowl.-Based Syst. **255**, 109780 (2022). https://doi.org/10.1016/j.knosys.2022.109780, https://www.sciencedirect.com/science/article/pii/S0950705122009017
3. Azwarni, A., Packier Mohammad, N.S.: Evaluating textblob, lexicon, support vector machine, naive bayes, and chatgpt approaches for sentiment analysis nasdaq listed companies. J. Theor. Appl. Inf. Technol. **102**(13), 5357–5366 (2024)
4. Barbieri, F., Camacho-Collados, J., Neves, L., Espinosa-Anke, L.: Tweeteval: unified benchmark and comparative evaluation for tweet classification. arXiv preprint arXiv:2010.12421 (2020)
5. Bose, R., Aithal, P., Roy, S.: Sentiment analysis on the basis of tweeter comments of application of drugs by customary language toolkit and textblob opinions of distinct countries. Int. J. **8**, 3684–3696 (2020)
6. Chou, S.: Social media sentiment analysis on multilingual dataset (2024)
7. Danny, A.E., Bashir, S., Ibrahim, M.: Enhancing user experience through sentiment analysis for Katsina state transport agency: a textblob approach. FUDMA J. Sci. **7**(6), 117 – 122 (2023). https://doi.org/10.33003/fjs-2023-0706-2057, https://fjs.fudutsinma.edu.ng/index.php/fjs/article/view/2057
8. data.lewoniewski.info: Supplementary materials for this research (2025). https://data.lewoniewski.info/wikisentiment/
9. Diyasa, I.G.S.M., et al.: Twitter sentiment analysis as an evaluation and service base on python textblob. In: IOP Conference Series: Materials Science and Engineering, vol. 1125, p. 012034. IOP Publishing (2021)
10. Fakharchian, S.: Designing a forecasting assistant of the bitcoin price based on deep learning using market sentiment analysis and multiple feature extraction. Soft Comput. **27**(24), 18803–18827 (2023)
11. Hermansyah, R., Sarno, R.: Sentiment analysis about product and service evaluation of pt telekomunikasi indonesia tbk from tweets using textblob, naive bayes & k-nn method. In: 2020 International Seminar on Application for Technology of Information and Communication (iSemantic), pp. 511–516 (2020). https://doi.org/10.1109/iSemantic50169.2020.9234238
12. Hevner, A.R., March, S.T., Park, J., Ram, S.: Design science in information systems research. MIS q. 75–105 (2004)
13. Hube, C.: Bias in wikipedia. In: Proceedings of the 26th International Conference on World Wide Web Companion (2017). https://doi.org/10.1145/3041021.3053375
14. Illia, F., Eugenia, M.P., Rutba, S.A.: Sentiment analysis on pedulilindungi application using textblob and vader library. In: Proceedings of The International Conference on Data Science and Official Statistics, vol. 2021, pp. 278–288 (2021)
15. Isnan, M., Elwirehardja, G.N., Pardamean, B.: Sentiment analysis for tiktok review using vader sentiment and svm model. Procedia Comput. Sci. **227**, 168–175 (2023)
16. Jahin, M.A., Shovon, M.S.H., Mridha, M.F., Islam, M.R., Watanobe, Y.: A hybrid transformer and attention based recurrent neural network for robust and interpretable sentiment analysis of tweets. Sci. Rep. **14**(1) (2024). https://doi.org/10.1038/s41598-024-76079-5
17. Jojoa, M., Eftekhar, P., Nowrouzi-Kia, B., Garcia-Zapirain, B.: Natural language processing analysis applied to COVID-19 open-text opinions using a distilbert model for sentiment categorization. AI Soc. **39**(3), 883–890 (2024)
18. Joshy, A., Sundar, S.: Analyzing the performance of sentiment analysis using bert, distilbert, and roberta. In: 2022 IEEE International Power and Renewable Energy Conference (IPRECON), pp. 1–6 (2022). https://doi.org/10.1109/IPRECON55716.2022.10059542

19. Khan, R., Rustam, F., Kanwal, K., Mehmood, A., Choi, G.S.: Us based COVID-19 tweets sentiment analysis using textblob and supervised machine learning algorithms. In: 2021 International Conference on Artificial Intelligence (ICAI), pp. 1–8 (2021). https://doi.org/10.1109/ICAI52203.2021.9445207
20. Kulev, I., et al.: Extraction of medication names from twitter using augmentation and an ensemble of language models. arXiv preprint arXiv:2111.06664 (2021)
21. Lewoniewski, W., Węcel, K., Abramowicz, W.: Companies in multilingual wikipedia: articles quality and important sources of information. In: Special Sessions in the Advances in Information Systems and Technologies Track of the Conference on Computer Science and Intelligence Systems, pp. 48–67. Springer (2022)
22. Lewoniewski, W., Węcel, K., Abramowicz, W.: Identifying reliable sources of information about companies in multilingual wikipedia. In: 2022 17th Conference on Computer Science and Intelligence Systems (FedCSIS), pp. 705–714. IEEE (2022)
23. Madanagopal, K., Caverlee, J.: Towards ongoing detection of linguistic bias on wikipedia. In: Companion Proceedings of the Web Conference 2021 (2021). https://doi.org/10.1145/3442442.3452353
24. Matei, S.A., Dobrescu, C.: Wikipedia's "neutral point of view": settling conflict through ambiguity. Inf. Soc. **27**(1), 40–51 (2011)
25. Pérez, J.M., et al.: Pysentimiento: a python toolkit for opinion mining and social NLP tasks (2023)
26. Rateb, M.N., Alansary, S., Elzouka, M.K., Galal, M.: Sentiment analysis in times of crisis: evaluating svm, cnn-lstm, and psyntimiento. preprint (version 1) (2024). https://doi.org/10.21203/rs.3.rs-3994513/v1
27. Roessing, T., Einwiller, S.: Portrayals of large corporations in the English and German version of wikipedia-exploring similarities and differences. Corp. Reput. Rev. **19**, 108–126 (2016)
28. Sayeed, M.S., Mohan, V., Muthu, K.S.: Bert: a review of applications in sentiment analysis. HighTech Innov. J. **4**(2), 453–462 (2023)
29. Singgalen, Y.A.: Implementation of perspective, vader, and textblob in toxicity and sentiment analysis of food and tourism. J. Inf. Syst. Res. **5**, 1292–1302 (2024)
30. Singgalen, Y.A.: Implementation of the glove in topic analysis based on vader and textblob sentiment classification. Build. Inform. Technol. Sci. (BITS) **5**(4), 649–658 (2024)
31. Wambui, G.D., Waititu, G.A., Wanjoya, A.: The power of the pruned exact linear time (pelt) test in multiple changepoint detection (2015)

AI-Driven Patent Claim Construction: Opportunity or the Dawn of Inventive Cannibalism?

Jan Černý(✉)

Department of Information Technology, Prague University of Economics and Business, Nám. W. Churchilla 1938/4, Prague 3, Czechia
cerj07@vse.cz

Abstract. The integration of Artificial Intelligence (AI) in intellectual property management has sparked significant debate regarding its impact on patentability and authorship. This study first reviews the contemporary legal landscape surrounding inventions and copyright and then examines the role of AI-driven models, particularly through prompt engineering techniques, in generating novel patent claims. By leveraging large language models (LLMs) such as ChatGPT, Gemini 2.0, DeepSeek R1, Llama 3.1, and Claude Sonnet 3.5, the research explores how AI can enhance the inventive process, supporting human creativity while raising concerns about *inventive cannibalism*—the potential fragmentation and appropriation of innovation.

Keywords: intellectual property · artificial intelligence · patent · patent law · copyright · innovation · novelty · large language models · LLMs · patent searches

1 Introduction

Since OpenAI's first model became publicly available, numerous expert and legal debates have centred on the paradigm shift in evaluating intellectual property, particularly concerning authorship on one hand and industrial intellectual property—primarily patents—on the other.

On one side, it has become evident that individual intellectual property rights do not have clearly defined boundaries when artificial intelligence is involved in creative processes. On the other, intellectual property protection for businesses tends to focus on the acceptance of AI's role, particularly in inventive activities. There is broad consensus on the issue of authorship—fully AI-generated literary works, for example, cannot be considered copyrightable. However, once human involvement is introduced into the creative process, with AI serving as an assistive tool, the situation becomes more complex. The intersection of artificial intelligence (AI) and industrial intellectual property protection presents complex challenges and opportunities within patent law frameworks globally with a critically needed examination whether existing patent systems can adequately accommodate AI-generated inventions and how AI technologies are transforming patent

K. Węcel (Ed.): BIS 2025, LNBIP 554, pp. 116–124, 2025.
https://doi.org/10.1007/978-3-031-94193-1_9

application processes [1–4]. The fundamental tension lies between traditional conceptions of inventorship predicated on human creativity and the growing capabilities of AI systems to generate novel innovations independently of direct human intervention. Marsoof [1] provides a comprehensive analysis of key judicial decisions in the DABUS[1] series of cases across multiple jurisdictions including the United States, United Kingdom, and Australia. These landmark cases consistently rejected recognizing AI as an inventor under current patent law frameworks, highlighting the anthropocentric orientation of existing legal definitions of inventorship. The similar statement came from the European Patent Office and USPTO [5, 6].

Conversely, patent authorities do not explicitly exclude the integration of AI solutions in inventive activities. In scenarios where a human inventor employs AI systems to advance their novel solutions, they simultaneously establish authorship of such innovations [7–11].The critical determinant lies in the provision of unique instructions to the AI system for generating specific outputs. This instructional specificity creates the necessary nexus between human cognitive input and AI-mediated technical outcomes, thereby satisfying the legal requirements for inventorship while accommodating technological augmentation of the creative process. The human inventor's intellectual contribution manifests through the deliberate parameterization and direction of the AI system toward particular problem domains, rather than being displaced by autonomous computational processes.

1.1 Research Contribution

A core aspect of this research is the systematic application of LLMs to patent claim analysis and innovation forecasting. The methodology involves prompting an LLM with structured patent claims and providing specific instructions to analyze novelty. The LLM is then required to identify novel elements within existing claims and use its broader knowledge base to predict potential future solutions that build upon these claims. This process forces the model to synthesize prior knowledge, identify gaps, and propose forward-looking innovations, making novelty assessment a crucial factor in the evaluation.

To ensure the validity of AI-generated novel claims, the newly identified solutions will undergo verification using traditional patent search approaches. This includes cross-referencing the generated claims with existing patents in established database, the Patentscope of the WIPO. By combining AI-driven prediction with rigorous human verification methods, this study seeks to evaluate the feasibility of AI-assisted patent claim generation while maintaining alignment with existing intellectual property frameworks.

Lastly, the research addresses potential risks for R&D companies regarding their rivalry and the possible misuse of AI for parasitic activities—specifically in the development of cannibalistic innovations.

[1] DABUS (Device for the Autonomous Bootstrapping of Unified Sentience) is an artificial intelligence system developed by Dr. Stephen Thaler, designed to autonomously generate novel ideas and inventions without human intervention [12].

2 Methodology

The methodology of this study focuses on evaluating the ability of large language models (LLMs) to assess the novelty of existing patent claims and predict potential future developments. To achieve this, a structured approach is employed, combining AI-driven analysis with traditional patent verification methods. The study selects patent applications from the cosmetics industry, specifically focusing on patent claims published in official patent gazettes. These applications are chosen based on their relevance, diversity in technological advancements, and recent publication dates to ensure alignment with current innovation trends. The core of the study involves prompting an LLM to analyze these patent claims systematically with the following prompt:

> **Step 1:** Novelty Analysis
> Carefully examine the following patent claims:
> [PATENT CLAIMS]
> Identify Unique and Distinguishing Aspects
> Compare the claimed solution to conventional formulations, processes, or technologies in [FIELD].
> Highlight specific elements (e.g., composition, method, apparatus features) that stand out as novel.
> Reference Prior Art
> Indicate any known, similar technologies or patents that could be considered part of the prior art. If possible, reference patent databases (e.g., USPTO, EPO, WIPO) or key scientific literature to bolster your analysis.
> Explain how the existing claims differ in scope or inventive concept from these references.
> Conclusion on Novelty
> Summarize why the claims are (or are not) novel, providing a rationale based on your comparisons and prior art findings.
>
> **Step 2:** Prediction of Future Developments
> Analyze Current Trends in [FIELD]
> Examine emerging materials, formulations, processes, or system integrations that are on the horizon.
> Consider recent scientific breakthroughs, industry research, and patent filings that signal upcoming directions.
> Propose Future Extensions or Improvements
> Suggest at least two plausible, non-obvious technological steps or enhancements that could extend or improve upon the existing patent claims.
> Provide a brief rationale for each proposed development (e.g., how it addresses limitations in the current patent, satisfies unmet market needs, or leverages advanced materials).

Step 3: Generation of New Patent Claims
Draft New Claims Based on Predicted Developments
Formulate at least three new patent claims that incorporate the future developments identified in Step 2.
Structure your claims in a legally precise format typical of official patent filings, specifying whether they are independent or dependent claims.
Ensure Patentability Criteria
Novelty: Clearly distinguish each new claim from known prior art or existing patent claims.
Inventive Step: Demonstrate why the proposed solutions are not obvious to those skilled in the art.
Industrial Applicability: Briefly outline how each claim can be applied in a real-world context, emphasizing its utility.

Provide Justifications
For each claim, include a short explanation highlighting its innovative aspects and how it addresses existing technical challenges or gaps.

Use Clear Claim Language
For example, start with phrases such as "A method of…" or "A composition comprising…" to maintain consistency with standard patent drafting conventions.

2.1 AI-Generated Claims Evaluation

For the testing purposes of this study, we have selected five large language models: ChatGPT o1, Gemini Experimental Advanced 2.0., DeepSeek R1, Llama 3.1 and Claude Sonnet 3.5. These models were chosen based on their advanced natural language processing capabilities, their ability to analyze complex textual information, and their relevance in AI-assisted research applications. The primary objective is to evaluate how each model performs in assessing the novelty of existing patent claims, predicting future inventive developments, and generating new patent claims that adhere to established patenting principles.

For fundamental verification, we have identified key terms describing the novel claims generated by different AI models and constructed a precise search syntax tailored for the Patentscope database to conduct a novelty search. This search is designed to ensure that the AI-generated claims meet the fundamental patentability criteria, particularly novelty and non-obviousness. For the novelty search we used the search strategy involves a combination of Boolean operators, field-specific queries (e.g., title, claims, abstract, descriptions), in all cases focused on precision.

To compare the patent claims from different language models, TF-IDF vectorization and cosine similarity were applied as standard NLP techniques for textual analysis. The claims were first extracted and structured to retain key technical aspects. TF-IDF assigned weights to terms based on their importance relative to the dataset, highlighting unique innovations. Using cosine similarity, the conceptual overlap between models was quantified, where a score of 1 indicates high similarity and 0 represents no overlap. The results were presented in a similarity matrix, enabling a structured comparison of shared and distinct innovations across models.

3 Results

As a demonstration we have selected the claims of the patent application Heating-Capable Seating Unit With Composite Stone Components published 9 Jan 2025 (no. 20250009135). We conducted prompt engineering based on the claims in the given patent application. The analyzed claims introduce innovations in smart heated seating, combining AI-driven automation, sustainable energy solutions, and sensor-based adaptability (see Table 1). ChatGPT focuses on wireless power transfer, occupancy-based heating, and solar panel integration for energy efficiency. Gemini 2.0 introduces thermochromic materials for visual heat indicators and biometric sensors for adaptive heating. Llama 3.1 emphasizes eco-friendly modular seating, AI-optimized energy management, and wireless charging for user devices. DeepSeek R1 contributes photovoltaic energy storage, moisture-responsive heating, and a hydrophobic nanoparticle coating for water resistance. Claude Sonnet 3.5 advances AI-driven thermal optimization, wireless power sharing between units, and predictive maintenance monitoring.

A novelty search returned zero results, suggesting that the claims may be novel and distinct from existing patented technologies. However, to confirm the possible novelty, a cross-search across patent databases and extensive non-patent literature must be further conducted to rule out any implicit prior art and ensure the uniqueness of the proposed innovations.

Most similar pairs (see Table 2) were ChatGPT & Llama 3.1 (0.197764) both focus on smart seating technology, incorporating automated heating control and sensor-based energy efficiency and Llama 3.1 & Claude Sonnet 3.5 (0.233824) Both models incorporate AI-driven automation.

4 Implications

The AI-assisted method for analyzing patent claims presents both opportunities and risks for R&D organizations, patent offices, and the broader innovation ecosystem.

4.1 Opportunities

1. Immediate applicability for R&D organizations – Any further research and development organization can immediately adopt this method to assess the novelty of their solutions. The approach is accessible, cost-effective, and adaptable across different industries, making it an efficient tool for validating inventions before filing patents.
2. AI-assisted, not fully AI-generated – The method serves as an AI-assisted solution rather than a fully AI-generated one, ensuring that human expertise remains central in assessing novelty. This human oversight introduces an opportunity for no-issue patent filings, as the approach remains within traditional patentability frameworks, mitigating concerns over AI-generated patents.
3. Enhanced hypothesis verification – The approach significantly reduces the time needed to verify the potential novelty of a claim, allowing inventors and organizations to rapidly iterate on their ideas and refine their intellectual property strategies.

Table 1. AI-Generated Claims Based on Prompt Engineering Techniques with Results from a Traditional Novelty Search

Model	Field	Challenged Innovation	No AI-generated Novel Claims	Novelty Described in Keywords	Novelty Search Results (in Claims)
ChatGPT o1	Furniture	20250009135	3	Wireless power transfer, inductive charging, energy storage, heating elements, smart control interface, solar panel integration, off-grid energy, occupancy-based heating automation, weather-adaptive heating adjustment	0
Gemini EA 2.0	Furniture	20250009135	5	Thermochromic material, temperature-responsive visual indicators, biometric-adaptive heating, heart rate sensor, skin temperature sensor, flexible solar panels, curved surface integration	0
Llama 3.1	Furniture	20250009135	3	Eco-friendly materials, modular seating, drainage cavity system, smart sensors, AI-driven energy optimization, wireless charging, automated heating, lighting adjustment	0
DeepSeek R1	Furniture	20250009135	3	Photovoltaic layer, battery storage, charge controller, moisture-responsive heating, hydrophobic nanoparticle coating, microchannel drainage system, precipitation detection	0
Claude Sonnet 3.5	Furniture	20250009135	3	AI-driven thermal management, wireless communication, neural network-based heat optimization, wireless power distribution, load balancing, predictive maintenance, remote control heating	0

4. Competitive and technology intelligence tool – This method can be effectively integrated into competitive intelligence (CI) and technology intelligence (TI) workflows, helping companies track emerging trends, monitor competitors' innovations, and refine their R&D priorities.

Table 2. Model Claim Similarity Based on Vectorized Text Representation and Cosine Similarity Techniques

Model	ChatGPT	Gemini 2.0	Llama 3.1	DeepSeek R1	Claude Sonnet 3.5
ChatGPT	1.0	0.12803921968662463	0.19776415722881502	0.1424976004831935	0.15239844397303148
Gemini 2.0	0.12803921968662463	1.0	0.1374601999980244	0.10353025724550723	0.026286827817527992
Llama 3.1	0.19776415722881502	0.1374601999980244	1.0	0.14216945551483282	0.23382439218117876
DeepSeek R1	0.1424976004831935	0.10353025724550723	0.14216945551483282	1.0	0.059167750581292686
Claude Sonnet 3.5	0.15239844397303148	0.026286827817527992	0.23382439218117876	0.059167750581292686	1.0

4.2 Risks

1. Inventive cannibalism – AI-driven inventive cannibalism can be defined as the deliberate utilization of artificial intelligence technologies by anyone other than the inventor or rights holder of a specific invention or innovation to extract tacit ideas, analyze them, and synthesize novel technological solutions that may be considered non-obvious. The process may unintentionally encourage the appropriation of ideas from granted patents and non-granted, but recently published patent applications, leading to strategic fragmentation of innovation and increasing the risk of overlapping claims. This issue should be formally addressed in upcoming debates on legal reforms within global patent frameworks, particularly focusing on preventing unethical exploitation of AI-generated claims.
2. Overloaded Patent Offices – The ease of idea validation may contribute to an exponential increase in patent filings, potentially overwhelming patent offices. This situation could be exploited as part of an unethical yet legally permissible patent strategy, wherein organizations flood the system with applications to create artificial barriers for competitors. However, this scenario might also accelerate the adoption of AI-driven evaluation processes for patent applications.
3. Increased complexity in patent searches – The widespread use of AI-driven patent descriptions may make prior art searches more challenging, as AI-generated technical language could create overly complex or ambiguous filings, making it harder for patent examiners and researchers to identify relevant existing technologies.

5 Limitations

The AI-predicted novelty method and the generation of claims based on it require further development across a broader range of claims. While the novelty search conducted on Patentscope, including its non-patent literature module, provided extensive coverage, cross-searching additional databases (e.g., Espacenet, USPTO, EPO, CNIPA) remains necessary to confirm the absence of prior art across different jurisdictions. Although the method offers a cost-effective validation tool for SMEs, contextual window limitations in AI models may restrict the depth of analysis for complex claims, necessitating human oversight to capture intricate technical nuances. Future research should focus on expanding the scope, refining search methodologies, and addressing contextual limitations to enhance the reliability and cross-domain applicability of this approach.

6 Conclusion

Ongoing debates on AI's role in the invention process underscore a clear trend toward AI-driven inventive activities. While patent offices continue to reject patents listing AI as the sole inventor—not due to a lack of inventability but for legal and policy reasons—they increasingly accept AI-assisted innovations, acknowledging their role in enhancing human creativity. This paper explores two key perspectives in this evolving landscape. First, it demonstrates how large language models, when applied with structured prompt engineering techniques, can effectively assist in identifying novel solutions and optimizing patentability assessments. Second, it raises concerns about AI-assisted claims, particularly regarding inventive cannibalism, overloaded patent offices, and the complexity AI-generated patent descriptions introduce to prior art searches. As AI becomes a fundamental tool in R&D and intellectual property management, its benefits and risks must be carefully balanced to ensure responsible, ethical, and effective integration into the global innovation ecosystem.

References

1. Marsoof, A.: Artificial intelligence and inventorship. In: Chakraborty, M., Chakrabarty, S.P., Penteado, A., Balas, V.E. (eds.) Proceedings of 5th International Ethical Hacking Conference. eHaCON 2024. LNNS, vol. 1148, pp. 39–46. Springer, Singapore (2025). https://doi.org/10.1007/978-981-97-8457-8_4
2. European Patent Office: Guidelines for examination in the European Patent Office (2024). https://www.epo.org/en/legal/guidelines-epc/2024/index.html
3. World Intellectual Property Organization: WIPO Conversation On Intellectual Property (IP) And Artificial Intelligence (AI) (2021). https://www.wipo.int/edocs/mdocs/mdocs/en/wipo_ip_ai_3_ge_20/wipo_ip_ai_3_ge_20_inf_5.docx
4. United States Patent and Trademark Office: Request for Comments: Artificial Intelligence and Inventorship (2023). https://www.federalregister.gov/documents/2023/02/14/2023-03066/request-for-comments-regarding-artificial-intelligence-and-inventorship
5. European Patent Office: AI inventorship: EPO publishes grounds for rejecting DABUS patent applications. European Patent Office (2020). https://www.epo.org/en/news-events/news/epo-publishes-grounds-its-decision-refuse-two-patent-applications-naming-machine
6. United States Patent and Trademark Office: Decision on Petition (2019). https://www.uspto.gov/sites/default/files/documents/16524350_22apr2020.pdf
7. United States Patent and Trademark Office: USPTO issues inventorship guidance and examples for AI-assisted inventions (2024). https://www.uspto.gov/subscription-center/2024/uspto-issues-inventorship-guidance-and-examples-ai-assisted-inventions
8. Japan Patent Office: Recent Trends in AI-related Inventions (2024). https://www.jpo.go.jp/e/system/patent/gaiyo/ai/ai_shutsugan_chosa.html
9. United Kingdom Intellectual Property Office: Guidelines for examining patent applications relating to artificial intelligence (AI) (2025). https://www.gov.uk/government/publications/examining-patent-applications-relating-to-artificial-intelligence-ai-inventions/guidelines-for-examining-patent-applications-relating-to-artificial-intelligence-ai--2
10. Bhushan, T., Tahiru, H.A.: Protecting AI-generated inventions under the UK's IP regulatory regime. J. Intellect. Prop. Rights **30**(1), 86–100 (2025). https://doi.org/10.56042/jipr.v30i1.7973

11. Japan Patent Office: Examination guidelines for patent and utility model (2024). https://www.jpo.go.jp/e/system/laws/rule/guideline/patent/index.html
12. Thaler, S.L.: Vast topological learning and sentient AGI. J. Artif. Intell. Conscious **8**(1), 81–111 (2021). https://doi.org/10.1142/S2705078521500053

Political Narratives and Misinformation During the COVID-19 Pandemic: A Comparative Analysis of Polish Political Parties on Twitter

Ewelina Księżniak(✉) and Marcin Sawiński

Poznań University of Business and Economics, 61-875 Poznań, Poland
ewelina.ksiezniak@ue.poznan.pl

Abstract. The COVID-19 pandemic not only posed a global public health crisis but also fueled the rapid spread of misinformation, particularly on social media. Existing studies suggest that political ideology influences susceptibility to fake news, with right-leaning individuals generally displaying higher acceptance of misinformation. However, little research has quantitatively compared how misinformation and skepticism were distributed among different political fractions in Poland. In this study, we analyze pandemic-related narratives among supporters of four major Polish political parties: Law and Justice (PiS), Civic Platform (PO), Confederation, and The Left (Lewica). Using a novel dataset of politically affiliated tweets, we employ BERTopic for topic modeling and GPT-4o for large-scale stance classification to investigate the prevalence and distribution of misinformation. Our findings reveal significant differences in how parties engaged with pandemic-related discourse.

Keywords: Misinformation · COVID-19 · Twitter · Topic Modeling · GPT-4o · BERTopic · Political Discourse

1 Introduction

The COVID-19 pandemic has not only posed a significant public health crisis but also intensified the spread of misinformation. Numerous studies have examined factors contributing to susceptibility to fake news, including age, personality traits, media literacy, and cognitive style (e.g., analytical thinking and the ability to override heuristics). However, an important dimension of susceptibility relates to political partisanship. Research has shown that political ideology influences perceptions of COVID-19 threats and the ability to discern between real and fake news. Conservatives, in particular, have been found to perceive the virus as less threatening, show greater skepticism toward media coverage, and perform worse in distinguishing real from fake COVID-19 headlines [1].

These ideological differences in susceptibility have prompted scholars to explore potential psychological and informational mechanisms. Although there is

K. Węcel (Ed.): BIS 2025, LNBIP 554, pp. 125–133, 2025.
https://doi.org/10.1007/978-3-031-94193-1_10

no scholarly consensus on the root causes of this asymmetry, several contributing factors have been identified. One explanation is that false claims often align more closely with right-wing political narratives, and due to biases such as motivated reasoning and confirmation bias, conservative audiences may be less inclined to reject them [2,12]. Another explanation highlights psychological traits more commonly associated with conservatism, such as a stronger need for cognitive closure and heightened intergroup loyalty. These dispositions-particularly under conditions of polarization-may increase the motivation to defend the ingroup and strategically endorse misinformation [4]. Moreover, algorithmically driven echo chambers reinforce these dynamics by exposing users, especially strong partisans, to ideologically consistent content while filtering out opposing viewpoints [14,15].

While ideological asymmetries in susceptibility to COVID-19 misinformation have been extensively studied in the United States [12,13], fewer investigations have focused on the Polish context. Existing research has shown that certain political actors, such as the far-right Confederation party (Konfederacja) and figures like Grzegorz Braun, actively promoted anti-vaccine rhetoric [3]. Braun, for example, framed pandemic policies as mechanisms of totalitarian control, drawing on nationalist and historical imagery [3]. Social media further amplified such narratives, as Polish Confederation deputies used their platforms to disseminate vaccine skepticism [5]. Despite these findings, there remains a lack of large-scale quantitative analyses comparing COVID-19-related topics and political stances across multiple parties, including mainstream actors such as Law and Justice (PiS) and Civic Platform (PO).

This study aims to address this gap by analyzing pandemic-related topics and stances among voters and sympathizers of four major Polish political parties: Law and Justice (PiS), Civic Platform (PO), Confederation, and The Left. Unlike prior studies that focused on binary divisions between left- and right-wing movements, we investigate how different factions engaged with COVID-19 discourse, particularly regarding attitudes toward government actions and vaccinations. To this end, we constructed a novel dataset of tweets from politically affiliated users, applying BERTopic for topic modeling and GPT-4o for large-scale stance classification. We decided to use GPT-4o due to its recent achievements in annotation tasks, including stance classification, where it has demonstrated high consistency and alignment with expert judgments [16]. However, given documented risks of misclassification, prompt sensitivity, and ideological bias in politically charged contexts, we complement its use with manual validation to ensure robustness.

The main contributions of this study are as follows:

- We introduce a new dataset of tweets from supporters of the four largest political parties in Poland (PiS, PO, Confederation, The Left), enabling a more detailed analysis of COVID-19-related political discourse.
- To the best of our knowledge, this is the first study to quantitatively compare pandemic-related topics and stances across multiple Polish parties using Twitter data.

– We demonstrate the applicability of BERTopic and GPT-4o in identifying and classifying political narratives, showcasing the utility of large-scale language models in computational political analysis.

2 Dataset

The dataset was created by scraping Twitter data using the academic API within the period January 2020 - January 2021. In the initial stage, tweets were filtered based on hashtags associated with specific political parties in Poland, including Law and Justice (PiS), Civic Platform (PO), Confederation (Konfederacja), and the Left (Lewica). The filtering process included both general party-related hashtags and those actively used by the respective parties during the 2020 electoral campaign. For PiS, the selection included hashtags such as *#Duda2020*, *#ObronimyPolskęPlus (#WeWillDefendPolandPlus)*, *#MimoWszystkoDuda (#DudaDespiteEverything)*; for PO, *#SilniRazem (#StrongTogether)*, *#Trzaskowski2020*; for Confederation, *#Konfederacja (#Confederation)*, *#WakacjeZKonfederacją (#HolidaysWithConfederation)*, *#DrużynaKonfederacji (#ConfederationTeam)*; and for the Left, *#LewicawMediach (#TheLeftInTheMedia)*, *#LewicaRazem (#TheLeftTogether)*, *#NowaLewica (#NewLeft)*, *#KonwencjaLewicy (#TheLeftConvention)*. Additionally, tweets containing explicit self-identifications such as *I vote for PiS* were also extracted. User accounts with more than 10,000 total posts were excluded from the dataset, as they were likely to be bots or highly automated users.

Subsequently, a manual annotation process was conducted on a sample of 300 user accounts that had posted tweets matching the above criteria. These users were categorized based on their stance toward the respective parties on a three-level scale: supportive, opposing, or neutral. To scale up the annotation process, we employed the GPT-4o model [8] via the OpenAI API to annotate the remaining tweets containing the identified hashtags and keywords. The accuracy of the automated annotations was validated by comparing GPT-based annotations with the manually labeled dataset, achieving an accuracy of 80%. As a result, we constructed a dataset comprising 3 376 277 tweets, with 423 users classified as PiS supporters, 447 as PO supporters, 443 as Confederation supporters, and 379 as Left supporters

Moreover, to obtain a broader perspective on COVID-19 narratives on Polish Twitter-beyond users who share explicitly political content-an additional 5,000 randomly selected users were included in the dataset. This approach allowed for a more comprehensive analysis of general trends in pandemic-related discourse. As a result, 2 195 710 additional tweets were collected.

Next, tweets were filtered based on key terms related to the pandemic, such as *COVID-19*, *vaccination*, *coronavirus*, *pandemic* etc. This filtering process resulted in a dataset of 85 215 tweets specifically discussing COVID-19, of which 6,515 belonged to users with an assigned political affiliation.

3 Topics Extraction

To extract dominant topics from COVID-19 dataset, we employed topic modeling technique BERTopic. BERTopic uses transformer-based models to generate embeddings, reduces their dimensionality with UMAP (Uniform Manifold Approximation and Projection), clusters them into topics using HDBSCAN (Hierarchical Density-Based Spatial Clustering of Applications with Noise), and finally assigns each document a unique topic identifier [7]. The allegro/herbert-base-cased model [6] was used to generate tweet embeddings. We tested a number of hyperparameters, such as minimum topic sizes (100–300), UMAP parameters (number of neighbors: 5–15; minimum distance: 0.1–0.5), and HDBSCAN cluster sizes (100–300), in an effort to achieve high coherence while maintaining reasonable topic sizes. To assess topic consistency, coherence scores were calculated for every iteration. Ultimately, tweets related to the COVID-19 pandemic were divided into 66 topics.

To test the hypothesis that voters of different political parties in Poland exhibit variations in the prevalence of specific COVID-19 narratives, we analyzed the distribution of topics across political groups. Specifically, we calculated the proportion of tweets classified into each detected topic relative to the total number of tweets authored by users affiliated with a given political party (PiS, PO, Confederation, Left). This allowed us to quantify differences in narrative engagement across political groups.

To assess the statistical significance of these differences, we employed Pearson's chi-squared test of independence [9]. Pearson's chi-squared test determines whether there is a significant association between categorical variables (in this case, political affiliation and the relative frequency of topics within each group). The test relies on the following assumptions: 1) Observations are independent, 2) Expected frequencies in each cell of the contingency table are sufficiently large (typically 5) [9].

These assumptions were verified and met in our analysis. The hypotheses for the chi-squared test are formulated as follows:

- **Null hypothesis** (H_0): There is no association between political affiliation and the relative frequency of topics; in other words, the distribution of topics is independent of political affiliation.
- **Alternative hypothesis** (H_A): There is a significant association between political affiliation and the relative frequency of topics, indicating that topic distributions differ across political groups [9].

The results of Pearson's chi-squared test of independence indicate a statistically significant association between political affiliation and the relative frequency of topics. The test yielded a chi-squared statistic of **1118.96** with a **p-value $< 2.59 \times 10^{-128}$**, which is far below the conventional significance threshold (e.g., **0.05**). Given this extremely low p-value, we reject the null hypothesis (H_0) and conclude that topic distributions differ significantly across political groups, suggesting that political affiliation influences the discourse on COVID-19.

To identify the topics that most distinctly differentiate political discourse among various parties, we analyzed the standardized residuals from the chi-squared test of independence. Standardized residuals measure the deviation between observed and expected topic frequencies within each political group, highlighting significant differences in topic prominence and were computed with the formula:

$$z_{ij} = \frac{O_{ij} - E_{ij}}{\sqrt{E_{ij}}}$$

where z_{ij} represents the standardized residual for party i and topic j, O_{ij} is the observed frequency, E_{ij} is the expected frequency, and $\sqrt{E_{ij}}$ is the standard deviation of the expected value [10]. By ranking topics based on the mean absolute standardized residuals, we identified the topics with the greatest disparities across political affiliations.

To further examine the thematic divisions in political discourse, we conducted an in-depth analysis of the five topics that exhibited the greatest differentiation among political parties, as determined by standardized residuals. This analysis was based on the unique words assigned to each topic by BERTopic and a manual expert evaluation. These topics are summarized as follows: **Topic 6 - Economic & Social Impact of the Pandemic:** Discussions on the broader impact of the pandemic on employment, the economy, and societal changes, including lockdowns and economic restrictions. Notably, the phrase *"w czasie pandemii"* ("during the pandemic") frequently appears among the keywords; **Topic 4 - Vaccinations & Combating Misinformation:** Discussions on vaccinations, including efforts to counter misinformation about the pandemic and vaccines. Occasionally, rhetoric related to vaccinations in the context of the Women's Strike was also present; **Topic 1 - Government Response & Policies:** Discussions on how the government managed the pandemic, the policies enforced, and their effectiveness. The keyword *"PiS"* frequently appears, indicating a strong association with discussions about government actions; **Topic 7 - Legislation & Political Debates:** Conversations focusing on legal aspects, political conflicts, and legislative measures related to COVID-19; **Topic 3 - Public Health & Medical Aspects:** Discussions from a medical perspective, including vaccines, hospital capacity, and medical recommendations. The keyword *"lekarze"* ("doctors") is commonly found in this topic.

In Table 1, we present the standardized residuals for these topics across the four political affiliations: the Left (Lewica), Confederation (Konfederacja), Civic Platform (PO), and Law and Justice (PiS). Standardized residuals greater than zero indicate that a given topic was more frequently discussed by a specific party than expected, while negative residuals indicate less frequent discussion.

The analysis of standardized residuals revealed differences in how frequently these topics were discussed across political groups: **Topic 6**: The Left discussed this topic significantly more frequently (10.53) compared to the other parties, while Civic Platform (PO) discussed it much less than expected (-7.97); **Topic 4**: This topic appeared more frequently in the discourse of Civic Platform (7.11), whereas Confederation engaged with it far less often (-6.19); **Topic 1**: Law and

Table 1. Standardized residuals for the top 5 most differentiating topics across political affiliations.

Political Affiliation	Topic 6	Topic 4	Topic 1	Topic 7	Topic 3
Left (Lewica)	10.53	−1.61	−2.73	−5.30	−0.73
Confederation	0.95	−6.19	3.17	1.42	−2.81
PO	−7.97	7.11	−2.12	2.19	3.67
PiS	−0.77	−0.21	3.96	0.38	−1.96

Justice (PiS) mentioned this topic more frequently than expected (3.96), while Civic Platform discussed it less frequently (−2.12); **Topic 7**: Civic Platform brought up this topic more frequently (2.19), while the Left (Lewica) discussed it significantly less often (−5.30); **Topic 3**: Civic Platform mentioned this topic most often (3.67), whereas it was less frequently discussed by Law and Justice (−1.96).

4 Stance Classification

The distinction between a topic and a stance lies in their scope and specificity. A topic represents a broad subject of discussion, such as the economic impact of the pandemic or vaccination efforts. In contrast, a stance reflects a particular attitude expressed toward specific aspects of the discussion. Our objective was to examine whether, within the identified topics, political parties exhibit different stances, revealing variations in their perspectives and messaging strategies.

We focused on classifying tweets into four categories based on their stance toward two key dimensions: trust in government actions and trust in the scientific-medical community. The categories were as follows:

- **Government Criticism**: Statements expressing criticism of government actions.
- **Government Support**: Statements expressing support for government actions.
- **Pandemic and Vaccine Skepticism**: Statements reflecting skepticism about the existence of the pandemic or the scientific-medical community.
- **Pro-Medical Science**: Statements advocating medical science.
- **Other**: Tweets that did not fit into the above categories.

To perform this classification, we utilized the GPT-4o model [8] via the OpenAI API, instructing it to assign each tweet to one of five stance categories based on its content: government criticism, government support, pandemic and vaccine skepticism, pro-medical science, or other.

The accuracy of the automatic annotations was validated by comparing them against a manually annotated sample of 200 tweets, with the GPT-4o model achieving an accuracy of 83%. In total, we annotated using GPT4o model [8] 6,515 tweets from topics that most distinctly differentiate between political parties.

5 Stance Analysis

The objective of this analysis was to examine the dominant types of stances expressed by different political parties. To achieve this, we calculated the percentage share of tweets classified into each category relative to the total number of tweets from each party within the five most distinguishing topics. We focused on four key categories: Pandemic and Vaccine Skepticism, Pro-Medical Science, Government Criticism, and Government Support. The plots Fig. 1a, 1b 1c, 1d illustrate these trends.

- In the *Pandemic and Vaccine Skepticism* category, the highest share of tweets is observed for **Confederation**, while the lowest is for **The Left**. **PiS** and **PO** are in between, with no significant differences between them.
- In the *Pro-Medical Science* category, **The Left** (11.86%) and **PO** (11.22%) have the highest shares. The proportions are lower for **Confederation** (6.47%) and **PiS** (5.13%).
- In the *Government Criticism* category, the largest share is for **PO** (37.18%) and **The Left** (34.38%). The shares are significantly lower for **Confederation** (17.46%) and **PiS** (14.74%).
- In the *Government Support* category, **PiS** has the highest share (23.72%), while for the other parties, the share is marginal (below 1%).

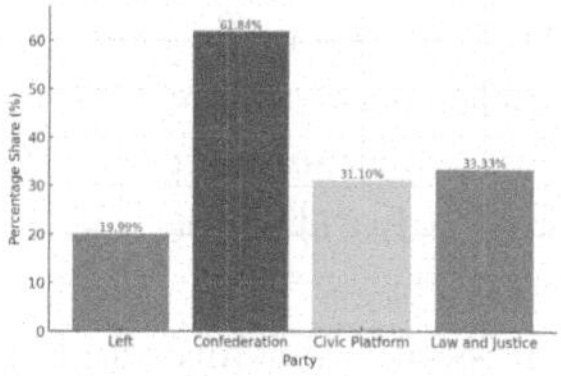

(a) *Pandemic and Vaccine Skepticism*

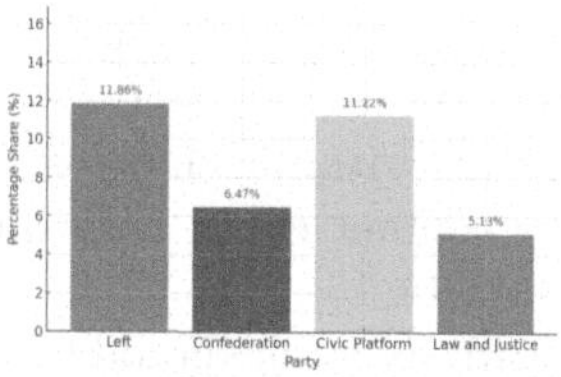

(b) *Pro-Medical Science*

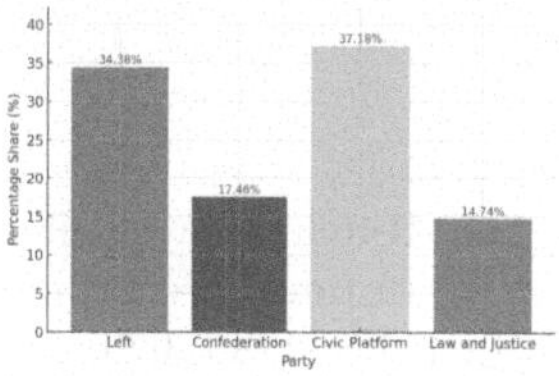

(c) *Government Criticism*

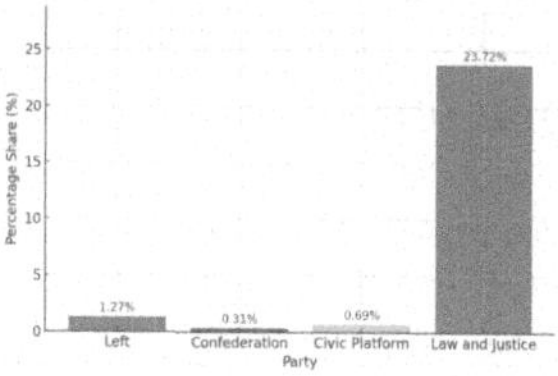

(d) *Government Support*

Fig. 1. Percentage of tweets across different political parties for each category.

6 Conclusions

This study provides a comprehensive analysis of political discourse and misinformation during the COVID-19 pandemic on Polish Twitter. By employing

BERTopic for topic modeling and GPT-4o for stance classification, we identified significant differences in how supporters of major Polish political parties-Law and Justice (PiS), Civic Platform (PO), Confederation, and The Left (Lewica)-engaged with pandemic-related discourse.

Our findings highlight that political affiliation played a crucial role in shaping attitudes towards COVID-19 policies, vaccines, and government actions. Confederation supporters exhibited the highest levels of pandemic and vaccine skepticism, whereas PO and The Left were more likely to endorse pro-medical science narratives. PiS supporters, in contrast, showed the highest level of government support, while opposition parties (PO and The Left) were the most vocal in criticizing government measures. The results underscore the polarization of political discourse on social media, where misinformation and skepticism were disproportionately present in right-wing narratives, aligning with findings from other international studies [12,13]. This suggests that political ideology remains an important factor influencing susceptibility to misinformation and trust in scientific institutions.

Given these insights, several policy implications emerge. First, interventions aimed at increasing media literacy-especially among populations most vulnerable to misinformation-have shown promise in reducing susceptibility to fake news [17]. Second, evidence suggests that misinformation is more effectively countered when corrective messages are framed in ways that align with the audience's values [18], such as emphasizing group loyalty or freedom. Third, building institutional trust is essential, as distrust in scientific and governmental institutions has been linked to higher acceptance of conspiracy narratives. Finally, algorithmic changes on social media platforms-such as promoting cross-cutting content-could help mitigate the effects of echo chambers and selective exposure [14,15].

Future research should examine the effectiveness of these interventions across political contexts and cultures, and evaluate how large language models may assist or hinder efforts to detect and counter misinformation.

References

1. Calvillo, D.P., Ross, B.J., Garcia, R., Smelter, T.J., Rutchick, A.M.: Political ideology predicts perceptions of the threat of COVID-19 (and susceptibility to fake news about it). Soc. Psychol. Personal. Sci. **11**(8), 1119–1128 (2020)
2. Baptista, J.P., Gradim, A.: Who believes in fake news? Identification of political (a)symmetries. Soc. Sci. **11**(10), 460 (2022)
3. Kosman, M.: Sanitary segregation imposed by big brother-a multimodal critical discourse analysis of Grzegorz Braun's radical anti-vaccine rhetoric. J. Contemp. Eur. Stud. 1–23 (2024)
4. Pechmann, C., Zhu, R.: Polarization increases strategic sharing of misinformation to promote ingroup dominance. J. Mark. **88**(3), 123–140 (2024). https://doi.org/10.1177/00222429231220827
5. Skrzypek, M.: Polish deputies' support for the anti-vaccination movement in social media: the case of confederation leaders. In: Digital Communication and Populism in Times of Covid-19: Cases, Strategies, Examples, pp. 107–117. Springer (2023)

6. Mroczkowski, R., Rybak, P., Wróblewska, A., Gawlik, I.: HerBERT: efficiently pretrained transformer-based language model for polish. In: Proceedings of the 8th Workshop on Balto-Slavic Natural Language Processing, Kyiv, Ukraine, pp. 1–10. Association for Computational Linguistics (2021). https://www.aclweb.org/anthology/2021.bsnlp-1.1
7. Grootendorst, M.: BERTopic: Neural Topic Modeling with a Class-Based TF-IDF Procedure. arXiv preprint arXiv:2203.05794 (2022)
8. OpenAI: GPT-4o: OpenAI's Omnimodal AI Model (2024). https://openai.com/research/gpt-4o
9. Pearson, K.: On the criterion that a given system of deviations from the probable in the case of a correlated system of variables is such that it can be reasonably supposed to have arisen from random sampling. Philos. Mag. Ser. 5 **50**(302), 157–175 (1900). https://doi.org/10.1080/14786440009463897
10. Sharpe, D.: Your chi-square test is statistically significant: now what? Pract. Assess. Res. Eval. **20**(8), n8 (2015)
11. van Hulst, M., Metze, T., Dewulf, A., De Vries, J., Van Bommel, S., Van Ostaijen, M.: Discourse, framing, and narrative: three ways of doing critical, interpretive policy analysis. Crit. Policy Stud. 1–23 (2024)
12. Garrett, R.K., Bond, R.M.: Conservatives' susceptibility to political misperceptions. Sci. Adv. **7**(23), eabf1234 (2021)
13. Nikolov, D., Flammini, A., Menczer, F.: Right and left, partisanship predicts vulnerability to misinformation. arXiv preprint arXiv:2010.01462 (2020)
14. Guess, A.M., Nyhan, B., Reifler, J.: Selective Exposure to Misinformation: Evidence from the Consumption of Fake News During the 2016 U.S. Presidential Campaign. European Research Council Working Paper (2018). https://doi.org/10.2139/ssrn.2968341
15. Eady, G., et al.: How social media shapes polarization: evidence from the facebook and instagram research partnership. Science **380**(6650), 711–717 (2023). https://doi.org/10.1126/science.abo3420
16. Gilardi, F., Gessler, T., Kubli, M.: ChatGPT Outperforms Humans in Coding Political Texts. arXiv: 2303.15056 (2023). https://arxiv.org/abs/2303.15056
17. Guess, A., Lerner, M., Lyons, B., Montgomery, J., Nyhan, B., Reifler, J.: A digital media literacy intervention increases discernment between mainstream and false news in the united states and India. Proc. Natl. Acad. Sci. **117**(27), 15536–15545 (2020). https://doi.org/10.1073/pnas.1920498117
18. van der Linden, S., Roozenbeek, J., Compton, J.: Inoculating against fake news about COVID-19. Front. Psychol. **11**, 566790 (2020). https://doi.org/10.3389/fpsyg.2020.566790

Process Mining

Can Users Trust Process Mining?

Irina Tentina(✉), Felix Mannhardt, Francesca Zerbato, and Boudewijn van Dongen

Eindhoven University of Technology, Eindhoven, The Netherlands
{i.tentina,f.mannhardt,f.zerbato,b.f.v.dongen}@tue.nl

Abstract. Process mining enables organizations to gain insights into their business processes using event logs from IT systems and applications. Some organizations manage to achieve valuable results from process mining, while others still struggle with its adoption or even end their process mining initiatives. Many factors influencing the adoption of process mining have been studied recently, such as lack of management support or poor data quality. However, little attention has been paid to business users' trust in process mining analytics. In this paper, we explore how business users assess their overall trust and confidence in process mining implemented within their organization. We report on the results obtained from interviews with 18 process mining users from various industries. We present common trust-building approaches followed by business users and challenges they face when validating the quality of process mining output. We then discuss practical implications for practitioners that can help build trust into process mining.

Keywords: process mining · trust · data quality · business users

1 Introduction

With the increasing use of Information Systems (IS) for the day-to-day execution of business processes, Process Mining (PM) solutions are gaining more attention in practice. PM enables organizations to gain insights into their business processes using event logs from IT systems and applications [1]. Event logs consist of data records of completed process executions. According to Gartner, the PM software market increased by 39.5% to $871.6 million in 2023 [17]. This growth in solutions seems indicative of the demand for PM within the industry. Some organizations have already managed to achieve valuable results using PM, while others still struggle with PM adoption [22], or even end their PM initiatives [3].

Ultimately, the value-adding proposition of PM relies on its adoption by PM consumers. A typical PM project follows different phases and involves several user roles [4]. For example, first, data engineers and domain experts interact to identify and extract the raw data from the source system, e.g., a set of tables, and transform it into an event log [1]. Then, depending on the PM software available and the use case investigated, PM analysts set up a series of analyses

K. Węcel (Ed.): BIS 2025, LNBIP 554, pp. 137–151, 2025.
https://doi.org/10.1007/978-3-031-94193-1_11

and visualizations that are used by end business users, e.g., interactive dashboards to explore the process and identify actionable insights. In this paper, we call the original source data transformed into the event log as PM *input* and the set of analyses and visualizations as PM *output*. We refer to data engineers and analysts as PM output *producers*, and to end business users are PM output *consumers*. By PM adoption, we refer to the effective use of PM *output* showing process visualizations along with the process performance insights by individual PM *consumers*, i.e., we focus on the individual level [2].

There are many different factors affecting adoption by PM consumers that have been studied by previous research [8,15,22] such as lack of management support, poor data quality, or challenging (real-time) system integration. Among others, the *lack of trust in PM insights* was one of the shortlisted challenges in a recent Delphi study by Martin et al. [15], while the lack of trust in the quality of the input data [26] was also reported as one of the challenges perceived by process mining analysts, i.e., PM producers. Although previous research studied the work practices of process analysts [13,24], it is not yet clear what role trust plays in the adoption of PM by individual PM *consumers* and what may cause their lack of trust in PM input and output.

Our work addresses this gap by empirically studying the trust in PM input and output by individual PM consumers. We designed and conducted an interview study with PM consumers, inquiring about trust in the context of real-life PM use cases they have worked with. Since mistrust can be related to data quality validation or lack thereof [26], we also investigate the role of such validation in relation to trust. In particular, we address the following research questions:

RQ1: Do PM consumers follow specific approaches to build trust in their PM input and output within their organizations?

RQ2: Do PM consumers experience any challenges during their PM output quality validation to support their trust building?

Our study reveals that PM consumers follow different trust-building approaches, and encounter challenges when validating data quality validation to support their trust in PM output. We then discuss the interview results in light of existing literature and provide recommendations for PM producers that may have practical implications on consumers' PM adoption and trust in PM output. To our knowledge, this is one of the first empirical studies that investigates trust in PM from the perspective of end users.

This paper is structured as follows. In Sect. 2, we introduce background concepts. Next in Sect. 3, we provide details on our research method. In Sect. 4 we summarize our main findings and observations. In Sect. 5 we discuss the results, share practical implications for PM producers, as well as limitations of our study. Finally, we conclude the paper and outline future work in Sect. 6.

2 Background and Related Work

Trust. Research on human trust in IS has been shown to be positively related to the users adoption of IS [18,19]. In this paper, we use the definition of *trust* in technology by McKnight et al. [19] as the 'belief that a specific technology has the attributes necessary to perform as expected in a given situation in which negative consequences are possible'. More specifically, we define trust in PM as the level of confidence and willingness of an individual user to depend on PM output and use a PM solution. According to this definition, trust relations involve two parties: the *trustor* and the party to be trusted, the *trustee*. In this paper study, the trustee is the end business user (PM consumer) and the trustees are the PM input and PM output. Numerous studies explored trust for specific types of IS. For instance, there is work on trust in management IS (MIS). Hertel et al. conceptualize the trust of individuals in MIS and introduce a new model of trust in management IS [16]. To our knowledge, there is limited work addressing trust in PM. Some studies propose to utilize PM as a means to improve the trustworthiness of information systems and technologies. For example, PM can be used to detect untrustworthy behaviors [14] or to promote trustworthy artificial intelligence [20]. However, these studies do not consider the trust of individual PM users.

Quality. Trusting data does not always equate high data quality, but quality is a crucial driver of trust. According to KPMG research [12], 60% of organizations report that they are *not very confident* in their data and analytics (D&A) insights, indicating that trust in D&A remains a significant challenge. KPMG identified four main pillars of trust [12]: quality, effectiveness, integrity, and resilience, with quality being the most frequently cited by respondents. Previous research on data quality in PM, seems to confirm an impact of data quality on trust. For instance, Hofstede et al. [7,10] report that data and information quality challenges are one of the factors influencing user confidence in process mining results. In [10], authors discuss that improved quality measures may lead to increased trust by customers. Results of the recent Delphi study on data quality [7,10] show that there is a need to build trust in the truthfulness of process data to gain the confidence of business users in process mining results. Concerning the management of data quality in PM, existing research has focused on developing tools and methods to identify data quality issues [6,21]. The framework introduced in [6] is also validated with PM experts [5]. While these works propose solutions to issues with data quality, they mainly focus on PM input. Instead, in this paper, we explore trust in PM considering the quality of both PM input and output, focusing on the following well-known quality dimensions [10] such as data freshness, accuracy, completeness and reliability. As PM output we consider visualization and analysis artifacts derived from the PM input. Despite the possible high quality of PM input e.g., event log, during the transformation and set up tasks the output can be still affected.

PM Consumers. Data processing tasks performed by PM *producers* are the most time-consuming. According to the XES Survey, 61%–80% of the effort of share is typically spent on data pre-processing [23]. The fact that data processing activities are "too laborious" can become one of the terminators for PM initiatives [3] and has shown to be challenging for PM producers [25]. However, one of the goals of those processing activities is to achieve accurate and reliable PM outputs that can be trusted by *consumers*. In this paper, we focus on *consumers* since we expect them to be less involved in the technical development—and therefore less knowledgeable about the processing that has been made on the input data—but still essential to understanding the final output and responsible for identifying and realizing the value out of PM use.

3 Research Method

In this section, we describe the study design and execution in Sect. 3.1 and the procedure we followed to collect and analyze the data in Sect. 3.2.

3.1 Study Design

To address our research questions, we followed a qualitative approach and designed a semi-structured interview study. The main reason for choosing semi-structured interviews as a data collection method was to gather as much input as possible through open questions that allowed participants to explain and detail their answers.

Since our aim was to comprehensively understand the trust of PM consumers in PM input and output, we focused on investigating real-life work settings of PM consumers, i.e., end business users involved with PM in their organizations. To this end, we requested our participants to select one *use case* that had already been implemented within their organization and to refer to this concrete use case when answering the questions. Due to the exploratory nature of our study, we did not have a preference for particular use cases and, thus, we asked our participants to focus on a use case they "consumed" frequently. To recruit interviewees, we followed convenience sampling [9] and invited people in our professional network. Since we aimed to target PM consumers across a variety of PM experience levels and application domains, we formulated the following three participation criteria. Participants must (1) have worked in industry and have had a business user role in PM; (2) have been involved in PM consumer activities; (3) have been involved in at least one active PM user case implemented within their organization.

We designed our interview protocol with questions covering two main themes that aligned with our RQs: (1) the user's overall *trust* in PM and (2) their challenges with respect to validating the *quality* of PM's outputs. Concerning the first theme, we aimed to understand (i) how much users trust the source system *input data* that was used by PM producers to develop the event log for the selected use case, and (ii) how much trust and confidence users have in the final *PM output* for the selected use case. For the second theme, we focused on the four PM output quality dimensions introduced in Sect. 2, i.e., *freshness, accuracy, completeness and reliability* and asked about the validation of data quality as a means to build trust and the challenges related to it.

For the first theme, we designed both closed and open questions. Closed questions consisted of statements concerning trust that participants had to rate

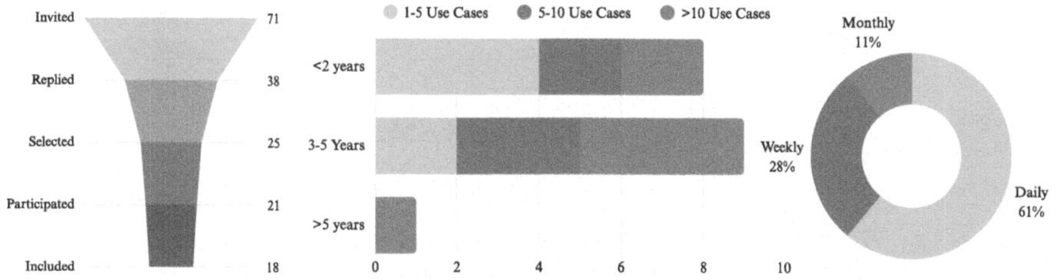

(a) Number of participants from invitation to inclusion in the results (b) Number of participants (incl. in the results) per their PM experience in years and number of PM use cases (c) Percentage of participants (incl. in the results) per the frequency of their PM use

Fig. 1. Funnel of participants from invitation until inclusion into the study and an overview on the PM experience of included participants.

from *strongly disagree* to *strongly agree* thinking about their selected use case. Examples of such statements are: (1) "I feel that I can trust data quality overall and data that was used for building PM analysis.", (2) "I feel that results of PM analysis are accurate and reliable." Instead, open questions aimed at gaining further insights into their assessment of the statement. For the second theme, we designed both open and closed questions, asking participants whether they perform validation for each of the above-mentioned quality dimensions to support their trust and whether they encounter challenges in doing so.

3.2 Data Collection and Analysis

Data Collection. The first author collected the data between July and September 2024. The interviews were conducted in English as Microsoft Teams meetings, which we recorded following the participants' consent. Before each interview we presented the objective of our study and collected demographic information from the participants. Each interview lasted approximately 60 min on average.

We sent invitations to 71 potential participants via LinkedIn or e-mail. Our invitation introduced our research goals and participation criteria. We received replies from 38 invitees. For replied invitees, we conducted an initial screening based on the participation criteria introduced in Sect. 3.1. If our criteria were met, we selected the participant to proceed with the interview. We rejected 13 candidates as twelve had a role that cannot be considered as a consumer and one had no active use case to discuss. Four out of 25 selected candidates were not able to participate due to personal reasons or never responded to the follow-up communication. Figure 1 gives an overview of our invitees and participants.

We performed interviews with 21 participants. Three participants could not provide answers to all the questions and, thus, we excluded them from our data. The remaining 18 participants have an average of 2.5 years of PM experience with at least two PM projects or use cases they have worked on. They use different PM solutions, such as Celonis, Microsoft Process Mining, QPR Process Analyzer and Fluxicon Disco. Our participants are experienced in different processes such as Human Resources, Quality Management, Purchase-to-Pay, Order-to-Cash, IT

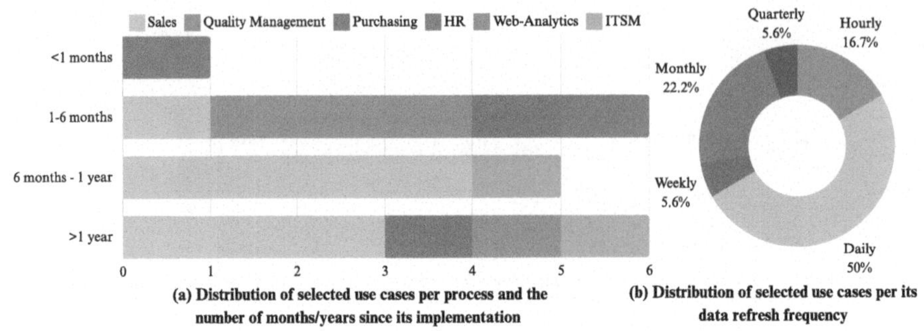

Fig. 2. Selected use cases per process, lifetime, and data refresh mode.

Service Management, Web-Analytics. Five participants perform their PM activities globally, the rest only in one of the regions in Europe, Northern America, South Africa and Asia. The majority of the selected use cases have been in productive use for at least one to six months and require daily data refresh. Figure 2 shows characteristics of the selected use cases.

Data Analysis. We transcribed the video recordings. For analyzing the interview data, we followed a qualitative coding approach [9]. Due to the interview structure, we had two high-level themes in the collected data: answers related to the overall trust in PM input and output and answers related to the questions for the quality of PM output. Answers related to the overall trust theme captured reasons for trust or mistrust in PM input and output. The second theme includes multiple sub-themes related to the different dimensions of PM output quality, i.e., freshness, accuracy, completeness and reliability. The main coding structure emerged around those main themes and sub-themes. We followed an inductive approach for coding responses to the open-ended questions.

The coding was performed in multiple rounds. First, the first author individually studied all the results and analyzed each participant individually using "in-vivo" coding. By doing this, the author used direct quotes from participants as codes and captured main *reasons* and *challenges* for the trust or mistrust in PM output by users for the overall trust theme. For the second theme, main activities performed by users and their challenges with respect to the PM output quality were identified and coded. Then the first author continued with the descriptive coding. In parallel, the second author studied the interview data to get an understanding of the content. Separately, the two authors developed initial descriptive codes, which they then shared, discussed to find agreement and refined. This refinement happened in several rounds and led to the main results.

Our supplementary material with the interview questions, participants' details and the coding scheme are at https://doi.org/10.5281/zenodo.14755285

4 Results and Findings

We present the trust-building approaches emerging from the interviews that address RQ1 in Sect. 4.1 and challenges with the quality validations of PM output experienced by consumers for RQ2 in Sect. 4.2.

4.1 Trust-Building Approaches of PM Consumers

Our participants had to rate their trust in PM input, i.e., data that was used for building their PM analysis, and the PM output, i.e., the results of their analysis, on a scale from strongly disagree to strongly agree. From hereon, we refer to the participants as consumers. Overall, only 22% of consumers (4/18) highly trust in **both** their input and output, while others selected lower ratings. We observed that most (3/4) consumers with high trust level have PM experience of five and more use cases, as well as more strategic than operational roles.

Table 1. Number of consumers per trust approach grouped by rating and statement. Multiple approaches could be assigned to one consumer. One consumer could be assigned to the category only for input or output, or both.

	Input (Statement 1)			Output (Statement 2)			#Consumers
	High	Medium	Low	High	Medium	Low	
OT1 Relying on personal previous experience	3	**4**	0	3	**5**	0	8
OT2 Relying on colleagues experience	**3**	1	0	**3**	2	0	5
OT3 Trusting by default	**3**	0	0	**2**	0	0	5
OT4 Using PM output regularly without major issues	2	2	0	**3**	**3**	0	6
OT5 Being aware of known issues	0	**3**	1	0	**5**	0	6
OT6 Accepting risk of possible issues	0	1	1	0	1	1	2
OT7 Having limited trust due to validation complexity	0	**4**	2	0	**4**	2	6

Based on what we discovered during the follow-up questions, we grouped consumers' explanations into seven **o**verall **t**rust-building approaches (OT#). In Table 1 we provide an overview of all approaches in which we aggregate the level of trust reported in the closed questions into high (strongly agree), medium (agree), and low (neither agree or disagree, or lower). Consumers could provide multiple explanations, therefore multiple approaches could be assigned to one consumer. One consumer could be assigned to the different approaches for the input and output. In this section, we describe each approach. We start by introducing approaches that emerged from the interviews with consumers who reported a high or medium level of trust (OT1-3). OT4 and OT6 were mentioned equally by consumers with medium and high/low level of trust. We continue with approaches that are more common for the medium level of trust (OT6) category. Finally, we explain the approach followed by medium and low trust consumers (OT7). We use U# to refer to the consumer ID in the supplementary materials.

The first approach, reported by 8/18 consumers, is **Relying on personal previous experience and/or validations (OT1)**. Some consumers reported being involved in the implementation of the selected use case or other use cases, *"and for the results, I was involved in other developments"* (u3) from the very beginning. This helped them gain trust directly from their personal experience: *"the reason that I am involved into the validation and customization"* (u9). Other consumers instead reported that they were newly introduced to existing use cases and have not been involved into the initial validations. They emphasized that, once they started to work on the use case, they had to perform their own checks to be able to trust the outcome: *"I was not a part of the initial project. I had to perform a lot of my own explorations and validations."* (u6). Consumers with this approach have high or medium trust based on their personal validation experience: *"I trust because of the number of times that I checked cases"* (u6).

The second approach is **Relying on colleagues experience and/or validations (OT2)**. In contrast to the previous approach, some consumers (5/18) indicated that they did not perform any or only a limited number of validations themselves. Some of them find it time-consuming and *"would not like to spend that much time on it"* (u17). Nevertheless, they trust that input and/or output are accurate, because they trust their colleagues who were either involved into the initial validation of the selected use case or worked on the same use case. For example, u1 reported that *"I trust my colleagues performed a sufficient number of tests and quality checks."*

The next approach related to a high level of trust is **Trusting by default (OT3)**. 5/18 consumers rated their trust rather high, but could not find any strong arguments for why they do so. U4 shared that it is *"one of my job responsibilities is also about convincing others to trust the data"*. U5 referred to their trust in the PM solution and explained that they do not develop customized PM output but rather use the standard PM solution functionality: *"I rely on [name of PM solution], because we do not develop a lot of things on top"*.

Higher frequency of use by more than one consumer was described as an important aspect to gain trust. Consumers (6/18) stated that they gained trust over time: *"At the beginning it was different and now it is stable."* (u15) and with more regular use. We clustered those consumers' explanations into the approach **Using PM output regularly without major issues (OT4)**. This type of trust is more common for operational consumers who use PM outcome on a daily (or even hourly) basis: *"This is my daily job and I am seeing stuff that if it would not be in the dashboard, it would be missed"* (u18). These consumers have not observed any major issues since the initial use case validation: *"So far, we have not find any missing cases. We are just relying on the accuracy of the data since its initial validation"* (u10).

The fifth approach is **Being aware of known issues (OT5)**. Consumers (6/18) shared that they may not fully trust all the results shown as PM output and rate their trust level as limited but sufficient. However, as soon as they are aware of issues and problems, they can use the PM outcome. Among issues users mentioned are problems that affect their trust in both PM input and output. U3 reported: *"We see that the raw data is not correct, in ERP system itself. We can see that fields [in the source system] are not properly used"*. U13 described their trust as limited, but sufficient, because *"We did not go into specific analyses as we know that the data is not accurate."*. However, since the limitations of the data are known, PM was still in use for the use case. As another example, u17 rated the trust in PM input as high, but reported low trust in PM output due to issues with the logic in the dashboard: *"We are busy fine-tuning with our [PM experts] the definition, the criteria, trying to understand why it is not working and we are not yet there. After 6 months, we are still not 100% convinced. It takes some time."* One user could not rate the trust in PM output high due to the PM solution: *"Because there are still some bugs in the tool from [Vendor] and due to the bugs, you cannot rely on the data"* (u14). To summarize, among known issues, users named different aspects, such as raw data quality, bugs in the PM solution, and the difficulty to define the logic to get a correct output.

Fig. 3. The percentage of PM consumers who perform quality validation in their work.

Sometimes consumers were not aware of any issue but chose to trust based on **Accepting risk of possible issues (OT6)** (2/18). They could not point to any specific issue in their PM input or output. These consumers rated their trust as limited but sufficient with the main explanation that *"we can never come to 100%"* (u5) acknowledging that even if the latest validation was successful "there could be sometimes false-alarms" (u1).

The main approach associated with consumers reporting medium and low trust is **Having limited trust due to validation complexity (OT7)**. (6/18) As we have learned from the presented approaches, the validation performed by the consumers themselves or their colleagues seems to be an important factor for building trust. As a result, consumers who found it difficult to perform such a validation assessed their trust level lower than others. U14 trusts the data and the source, i.e., PM input, but still struggles with the validation of PM output: *"I trust that it is coming from the right source. However, I struggle with understanding what those numbers/values mean"*. Consumers highlighted that the difficulty exists due to the complexity of the PM output (u9) or customizations in the source system that are not always considered in the PM output (u8). Next, we elaborate more on the consumer's challenges related to the validation process in Sect. 4.2.

4.2 Challenges Related to PM Output Quality Validation

To address RQ2, we investigated how consumers perform validation checks to support their trust and which challenges are related to the latter. We summarized the results from our interviews in Fig. 3.

(C1) Lack of Validation Strategy: 94% of the consumers were involved in PM output validations during the initial implementation of their selected use case or still perform some validation checks regularly. Only few consumers (2/18) described a standard practice for their *accuracy* and *completeness* checks. U5 mentioned that they use the checklist provided by the PM vendor as the minimum basis to perform validation. U13 described a standard practice of performing user acceptance testing (UAT) once the development phase is finished. The rest could not point us to a concrete *validation strategy* within their organization and shared only some examples of what they usually validate.

(C2) Need of Regular Accuracy Validation: Despite a significant number of consumers rating their overall trust as sufficient or higher (Sect. 4.1), almost half of the consumers continues *accuracy* or *completeness* checks on a regular basis in production. For example, u7 performs daily validation to ensure that PM output can be trusted: *"In the morning, when I open the dashboard I choose a few cases as samples to check them in the source ERP system"*. On the contrary, u11 who reported limited but sufficient trust, shared *"I perform those checks only if there is a really good reason. Otherwise, you loose a lot of time."*

(C3) Difficulty with Completeness Checks: The questions related to completeness were the most difficult for consumers to answer. Basic *case-level* completeness checks seem to be performed often (u3, u4, u5). For example, the comparison of the total number of cases with the source system or any other reporting available in the organization was mentioned by many consumers, e.g., *"we have another solution where we should have more or less the same scope as in PM solution. I expect that results should be similar."* (u3). Some consumers (11/18) replied that it is hard or even "impossible to verify"(u6) the completeness from the activities perspective and they do perform no or limited checks.

(C4) Lack of Sampling Approach: The most common *accuracy* validation method mentioned is sample testing. We found that the particular sampling approach is a frequent concern. Most consumers reported that they choose samples based on a "feeling" or "randomly". Only one consumer shared exemplary logic for the sample selection: *"We have a look at some trends over time to see if we observe any interesting results, like any outliers, or if we have some missing values"*(u5). 39% of the consumers agreed that they would benefit from a "smarter" way of sampling.

(C5) Complexity of PM Output: It can be challenging to reconcile the values for the accuracy checks, if there is a lot of data filtering applied in the PM solution,e.g., there are values or groups of values filtered out in PM input and/or output by PM producers. Consumers may *"get confused, why do I see this number and how did we scope it. I know we excluded some exceptions, but*

I am coming back to our PM experts to double check what are those hidden filters"(u4).

(C6) Lack of Data Freshness Awareness: In most of the discussed use cases data is refreshed daily. Some consumers would like to have a more frequent schedule, however, PM producers were not able to satisfy this request yet. U12 stressed that a daily data refresh can slow down adoption because, if during the day another consumer has already followed up on the specific case and made changes to the source system, this change only becomes visible in the PM solution the next day. 56% of consumers reported that their dashboards show the date and time of the latest data refresh. 22% do not have it immediately visible, but they know how to check it, 17% cannot check it themselves and trust it is refreshed according to the schedule. One of the consumers cannot see the date and, thus, "cannot be always sure if the data is freshest". Almost all consumers agreed that awareness of data freshness would improve their confidence in the PM output.

(C7) Data Manipulations vs Reliability: We asked whether consumers are confident to share their PM output without any additional data manipulations with colleagues, stakeholders or external people. Most users (16/18) were confident to share them as is for the selected use case. U9 regularly send the use case outcome to the customer *"we are doing it, and we are also facing the risk, that we are sharing the information that is incorrect"*. Some consumers stressed that they find it crucial to make sure that their recipients of the PM outcome have been introduced to PM and are aware of the use case, to be able to interpret it. U5 reported that it is not possible to directly share PM output with other stakeholders and some manipulations have to be performed first. This was attributed to the complexity of the output, which made it necessary to translate the results into an easier to understand format for other stakeholders. U14 shared: *"I do not feel comfortable [sharing] this use case because of the bugs."*

Concluding, most consumers (17/18) performed some quality validation. We summarized seven challenges related to the consumers' validation process. We follow up on reported challenges and discuss them in Sect. 5.

5 Discussion

This section discusses the results and limitations of our study and draws implications for practitioners.

As highlighted in the introduction, more and more organizations are already using PM on a regular basis and have implemented PM solutions. As a result, a growing number of business users consume PM output in their daily work. However, there is limited understanding of PM users' confidence in PM output as well as of their involvement in the validation process and the challenges that this may pose. Our interviews reveal that most of our participants trust their PM results. However, depending on the degree to which our participants trust PM output and the trust-building approaches they follow, we could identify four types of trust: (1) **learned trust** based on the extensive and continuous PM output use and validations performed by consumers themselves (OT1, OT4); (2) **conditional trust** based on the awareness of the existing issues with the data or the difficulties to perform complete and accurate validation (OT5, OT7); (3) **naive trust** based on trust in colleagues and their experience or accepting the risk that some issues or inaccuracies may occur (OT2, OT6); (4) **blind trust** by the default without any strong argumentation (OT3).

While explaining trust-building approaches (cf. RQ1), our participants referred to data quality validations or known quality issues. Awareness of data quality seems to be a critical factor for consumers' trust. For example, to gain learned or conditional trust, participants perform quality validations. However, based on our results for RQ2, for most participants the validation can be challenging and there is no standard strategy in place to guide them (C1). Existing literature confirms that data quality management, including data validation, can be challenging [10]. However, recent work by Goel et al. [7], takes a step forward and presents a framework that explains how to govern process data holistically. The proposed framework could help developing standard PM input and PM output quality validation strategies on the organizational level [2]. In such a scenario, PM consumers would be less involved in continuous manual validation (C2), which, in turn, would limit its negative influence on their trust level.

We suggest that guidance to validate data quality should also be tailored to individual PM consumers. Our results show that individual PM consumers could benefit from more structured validation methods. For example, there is a risk with the current sampling approaches (C3) to select non-representative subsets of cases and miss potential quality issues. Several approaches have already been proposed to support the sampling process [11]. However, there was no focus on the PM output and individual PM consumers. Our work suggests that further research in this direction is required. For instance, research on how to perform PM output sampling such that provides more interpretable results for the validation by PM consumers.

We also observed that difficulties arising while performing validation checks (C2) and the complexity of PM (C4, C6) may lead to lower trust levels or misguided trust, because participants follow a naive or blind trust approach. Similar challenges have also been highlighted by [10]. The importance of the guidance for the stakeholders have been emphasized, for example, in the form of methodological support or with improved visual analytics. Current research only addresses this for PM input, while PM consumers would require guidance for the PM output validation.

In summary, we found that although approaches exist to address some of the challenges discussed, they lack focus on the individual PM consumers and their implementation in PM practice is still limited.

Practical Implications. Several practical implications for practitioners arise from the interview results that can help PM producers foster consumers trust and prevent mistrust in PM inputs and outputs when developing and implementing PM use cases. First, the lack of strategy (C1) and structured approach to perform

PM output validation (C3) may be challenging for PM consumers. Therefore, it is advisable to develop sustainable PM quality strategy with standard quality testing procedures for PM use cases. Second, the sampling procedure for the quality assessment can be improved and potentially automated in PM solutions. Third, our findings on trust-building approaches indicate that consumers gain trust by performing their own validations. To facilitate this process, templates can be developed to provide traceability and transparency on the previously performed validations and its results. These templates can be shared with a new consumers as evidence and help reduce the time and manual effort that consumers spend on performing their own validations. Moreover, guidance for PM producers on the standard visualization principles for PM output, e.g., including time of the latest data refresh mentioned in C5, could improve consumers' trust.

Limitations. When interpreting the results of our study, some limitations should be taken into account. As our findings emerged from the data gathered during interviews, our results are subject to validity threats typical of interview studies, such as respondent bias, researcher bias and generalization. To mitigate these risks we: (1) pilot-tested our interview and guaranteed the anonymity of the participants, (2) coded and analyzed the data with multiple authors, and (3) designed our study to achieve diversity in the level of PM experience of our participants and the frequency of PM solution use. At last, it is important to acknowledge that trust can change and develop over time for the same consumer [19]. Therefore, we cannot guarantee that the same consumer may give a different answer for the same use case at a different point in time or for a different use case. However, by focusing on the most frequently used use case, we ensured that consumers engaged with a use case they know well and, as such, have developed a good "feeling" for their trust over repeated interactions.

6 Conclusion

In this work, we investigated the approaches PM consumers follow to build trust in PM inputs and outputs and whether they experience challenges during quality validation of the PM output to support their trust building. By analyzing the interviews, we identified seven trust-building approaches and seven challenges for quality validation of PM output. Our results show that trust in PM input and output varies among consumers and is influenced by their approach to validation and their awareness of data quality issues. We conclude that future research and practical efforts should focus on developing new methods and better visualization guidance to support individual PM consumers in validating the PM input and output used in their analyses.

References

1. van der Aalst, W.M.P.: Process Mining - Data Science in Action, 2nd edn. Springer (2016)
2. vom Brocke, J., Jans, M., Mendling, J., Reijers, H.A.: A five-level framework for research on process mining. Bus. Inf. Syst. Eng. **63**(5), 483–490 (2021)
3. Dani, V.S., Leopold, H., van der Werf, J.M.E.M., Beerepoot, I., Reijers, H.A.: From loss of interest to denial: a study on the terminators of process mining initiatives. In: CAiSE. LNCS, vol. 14663, pp. 371–386. Springer (2024)
4. van Eck, M.L., Lu, X., Leemans, S.J.J., van der Aalst, W.M.P.: PM^2: a process mining project methodology. In: CAiSE. LNCS, vol. 9097, pp. 297–313. Springer (2015)
5. Eden, R., et al.: Revealing the root causes of digital health data quality issues: a qualitative investigation of the Odigos framework. In: HICSS, pp. 3035–3044. ScholarSpace (2023)
6. Emamjome, F., Andrews, R., ter Hofstede, A.H.M., Reijers, H.A.: Alohomora: unlocking data quality causes through event log context. In: ECIS (2020)
7. Goel, K., Martin, N., ter Hofstede, A.: Demystifying data governance for process mining: insights from a delphi study. Inf. Manag. **61**(5), 103973 (2024)
8. Grisold, T., Mendling, J., Otto, M., vom Brocke, J.: Adoption, use and management of process mining in practice. Bus. Process. Manag. J. **27**(2), 369–387 (2021)
9. Hennink, M., Hutter, I., Bailey, A.: Qualitative Research Methods. SAGE Publications (2020)
10. ter Hofstede, A.H.M., et al.: Process-data quality: the true frontier of process mining. ACM J. Data Inf. Qual. **15**(3), 29:1–29:21 (2023)
11. Kabierski, M., Nguyen, H.L., Grunske, L., Weidlich, M.: Sampling what matters: relevance-guided sampling of event logs. In: ICPM, pp. 64–71. IEEE (2021)
12. KPMG: Building trust in analytics. Technical report, KPMG (2019). https://assets.kpmg.com/content/dam/kpmg/ie/pdf/2019/03/ie-building-trust-in-analytics-data.pdf
13. Kubrak, K., Milani, F., Nolte, A.: A visual approach to support process analysts in working with process improvement opportunities. Bus. Process. Manag. J. **29**(8), 101–132 (2023)
14. López, J., Maag, S., Saint-Pierre, C., Bustos-Jiménez, J., Cavalli, A.R.: Process mining for trust monitoring. In: AINA Workshops, pp. 605–610. IEEE (2015)
15. Martin, N., et al.: Opportunities and challenges for process mining in organizations: results of a delphi study. Bus. Inf. Syst. Eng. **63**(5), 511–527 (2021)
16. Meeßen, S.M., Thielsch, M.T., Hertel, G.: Trust in management information systems (MIS). Zeitschrift für Arbeits-und Organisationspsychologie A&O (2019)
17. Mehta, V., Kerremans, M., Tornbohm, C., Chibber, H.: Market share analysis: process mining software, worldwide. Gartner Report (2023)
18. Müller, L.S., Hertel, G.: Trusting information systems in everyday work events–effects on cognitive resources, performance, and well-being. Ergonomics 1–18 (2023)
19. Müller, L.S., Nohe, C., Reiners, S., Becker, J., Hertel, G.: Adopting information systems at work: a longitudinal examination of trust dynamics, antecedents, and outcomes. Behav. Inf. Technol. **43**(6), 1096–1128 (2024)
20. Pery, A., Rafiei, M., Simon, M., van der Aalst, W.M.P.: Trustworthy artificial intelligence and process mining: challenges and opportunities. In: ICPM Workshops. LNBIP, vol. 433, pp. 395–407. Springer (2021)

21. Suriadi, S., Andrews, R., ter Hofstede, A., Wynn, M.T.: Event log imperfection patterns for process mining: towards a systematic approach to cleaning event logs. Inf. Syst. **64**, 132–150 (2017)
22. Syed, R., Leemans, S., Eden, R., Buijs, J.: Process mining adoption. In: Fahland, D., Ghidini, C., Becker, J., Dumas, M. (eds.) BPM 2020. LNBIP, vol. 392, pp. 229–245. Springer, Cham (2020). https://doi.org/10.1007/978-3-030-58638-6_14
23. Wynn, M.T., et al.: Rethinking the input for process mining: insights from the XES survey and workshop. In: ICPM Workshops. LNBIP, vol. 433, pp. 3–16. Springer (2021)
24. Zerbato, F., Soffer, P., Weber, B.: Process mining practices: evidence from interviews. In: BPM. LNCS, vol. 13420, pp. 268–285. Springer (2022)
25. Zimmermann, L., Zerbato, F., Weber, B.: Process mining challenges perceived by analysts: an interview study. In: BPMDS, pp. 3–17. LNBIP, Springer (2022)
26. Zimmermann, L., Zerbato, F., Weber, B.: What makes life for process mining analysts difficult? A reflection of challenges. Softw. Syst. Model. 1–29 (2023)

Do not Stay Behind: Embracing Process Mining in the Italian Public Sector

Serena Racis(✉) and Alessandro Spano

University of Cagliari, Cagliari, Italy
s.racis97@gmail.com, spano@unica.it

Abstract. The public sector urgently needs to digitalize its infrastructures and services to better satisfy citizens' needs. Process Mining (PM) is a process-oriented tool that allows organizations to objectively and timely analyse their processes, resulting in improved process efficiency and higher organizational performances. Nevertheless, PM successful implementation depends on the acceptance and support of key stakeholders, including civil servants in the public sector context. This study is among the first attempts to adopt Information Systems acceptance models to quantitatively evaluate civil servants' perceptions of PM prior to its adoption to understand its implementation viability. Through a survey conducted among Italian public organizations, we found that civil servants with basic theoretical and practical knowledge about PM show a positive attitude towards PM use, perceiving it as useful, easy to use and job-fitting.

Keyword: Process Mining · Information Systems · Public Sector

1 Introduction

Globally, the public sector is undergoing unforeseen and complex challenges that require changing the way it manages its processes and delivers public services to satisfactorily meet citizens' needs [1]. The use of Information and Communication Technologies (ICT) is fundamental to promote processes' digitalization and organizational structures' transformation to deliver better public services [2] and achieve public sector goals such as increased efficiency, cost and waste reduction, higher transparency, accountability, and citizen centricity [3]: public organizations that properly adopt ICT and digital technologies achieve these goals fast and effectively and are able to deliver higher quality services to citizens and businesses [2].

However, the use of ICT and process digitalization are not sufficient alone to generate value [4]: what is essential is to go through a full digital transformation, which implies organizational, relational, and cultural changes that affect both the technical and the social systems [3]. While many public organizations have already undergone digitalization processes that primarily involved technical changes, broader cultural and organizational changes are still on the roadmap [5]. Understanding how to introduce new technologies considering how they are perceived by organizational members is crucial

K. Węcel (Ed.): BIS 2025, LNBIP 554, pp. 152–164, 2025.
https://doi.org/10.1007/978-3-031-94193-1_12

to successfully adopting and implementing them [6], as both individuals' characteristics and their perceptions of innovation's characteristics can influence the innovation's implementation process [7].

This study aims to analyse ICT acceptance in the public sector from the perspective of civil servants, focusing on an emerging technology such as Process Mining (PM). Although PM is just one of the many technologies public organizations can adopt to achieve public sector goals, it is particularly suitable to help them overcome many challenges, especially if used in combination with other promising technologies [8] and business process management methods. For example, while public organizations are urged to digitalize their processes and make them more efficient [5], simply doing it without a proper process analysis that provides full process transparency will only amplify process inefficiencies [8]. PM provides exactly this transparency: by extracting log-data directly from organizations' Information Systems (IS), PM allows to reconstruct the real processes' execution, detect bottlenecks and inefficiencies [9], and identify potential and priorities for process improvement, optimization, digitalization, and automation [9, 10], which is decisive to go through digital transformation.

Despite PM potential to contribute to digitalizing public sector processes and make them more efficient through objective and prompt analyses, its adoption and acceptance is little explored [11–13]: we found only few studies focusing on users' acceptance of PM [11, 14–17].

Based on IS research, we conducted an exploratory study in the Italian public sector to investigate civil servants' attitude towards PM use, prior to its eventual adoption to understand if it can be successfully implemented in the public sector. A key success factor for public sector organizations is in fact the identification of barriers and challenges before starting change initiatives [3]. Investigating civil servants' general attitude towards PM use is thus fundamental to discover in advance factors fostering or inhibiting its introduction and avoid implementation failures.

Our findings reveal that civil servants with basic PM understanding show a positive attitude towards its use, as they perceive it as useful to improve their overall job performance, easy to use and job-fitting. Socio-demographic, job-related and organizational variables should also be considered by public managers to support an effective PM implementation in their organizations, as women, younger individuals, those with higher organizational positions, those with more years of job experience and those working in smaller organizations show a more favourable attitude towards PM use.

2 Theoretical Background

Numerous studies have devoted significant attention to individuals' acceptance, adoption, and use of new technologies [18–21], particularly in the private sector. In the public sector, this topic is increasingly being studied [6, 22, 23]. However, research mostly investigates a citizen's or a business perspective towards innovation and technology acceptance [24], while literature on civil servants' ICT acceptance [25–28] is missing [29].

Nevertheless, the success of ICT in public organizations strongly depends on civil servants' acceptance and intention to use it [30]. Civil servants face many barriers when adopting new technologies, ranging from lacking competencies and skills [31] to security and privacy concerns [30]. At the same time, public organizations must deal with limited funding [32] and budgetary resources, institutional and legal constraints, political pressures, multiple stakeholders' relationships [32–34], inadequate Information Technology (IT) infrastructures [35], and inadequate organizational structure and culture [31]. All these challenges contribute to the high failure rate of public sector organizations' transformation initiatives [6], which could be lessened by optimally changing and managing their internal and external processes [36].

PM is an indispensable tool for data-driven and process-oriented organizations willing to optimize their processes and generate value. By extracting and analysing process-related event data readily available in organizations' IS, PM helps organizations understand how their processes are performed, where bottlenecks, inefficiencies, and deviations are, and how to intervene to improve them [37]. Moreover, by providing objective process insights in a faster and more reliable way with respect to traditional process analysis based on assumptions and workers' interviews [38], PM can also contribute to enhancing process transparency and knowledge-sharing among individuals [39].

Situated between data science and process science disciplines [9], PM allows to perform backward-looking analyses based on past data, online analyses based on real-time data, and forward-looking analyses [40]. Starting from event logs, which are sequences of recorded events related to a specific case and an activity [41], PM basic capabilities enable to (automatically) discover the real processes' execution (*Process Discovery*), compare an ideal or existent process model with an event log of the same process to check for deviations (*Conformance Checking*), and improve and enrich the existing process model with the results of the performed analyses (*Process Enhancement*) [41]. Another PM capability that could be particularly beneficial to the public sector to better assign tasks to civil servants and facilitate performance evaluation is the *social network/organizational mining* [41], which allows to get a deeper understanding of how human resources involved in a process work and interact with each other [42].

3 Methods

As the public sector is characterised by a strong resistance to change [43], and PM is expected to bring substantial technical, operational, and organizational changes [38, 39], PM acceptance may be harder in the public than in the private sector. We conducted an exploratory study to understand whether PM would be generally favourably accepted by civil servants and which variables affect this choice, by testing several hypotheses based on existing IS acceptance models through a questionnaire administered between February and June 2023 to a sample of civil servants with different roles, responsibilities and tasks (managers, officials, office holders, employees) and from different segments of the Italian public sector (local and central administrations, regional and national agencies, health organizations, universities, and others) to assess their attitude towards PM use. We focused on the Italian public sector as Italy strongly committed to digitally transform and simplify public administrations' processes and foster the country's competitiveness through investments in innovative technologies, infrastructures, digital services, and civil servants and citizens' digital skills [44].

To consider both PM's perceived characteristics and attitudinal variables we extended and modified the original Technology Acceptance Model (TAM) by adopting constructs from the Diffusion of Innovation Theory and the Social Cognitive Theory. Specifically, perceived usefulness, perceived ease of use, perceived compatibility, personal outcome expectations, and anxiety were hypothesised as influential on the civil servants' attitude towards PM use. We also included six control variables (gender, age, education, job position, years of experience in current job position, and organization size) to consider those factors that could influence the attitude towards PM use in the public sector, as differences in access and use of ICT exist depending on demographic, cultural, and socio-economic factors [45]. Figure 1 shows the conceptual framework guiding this research.

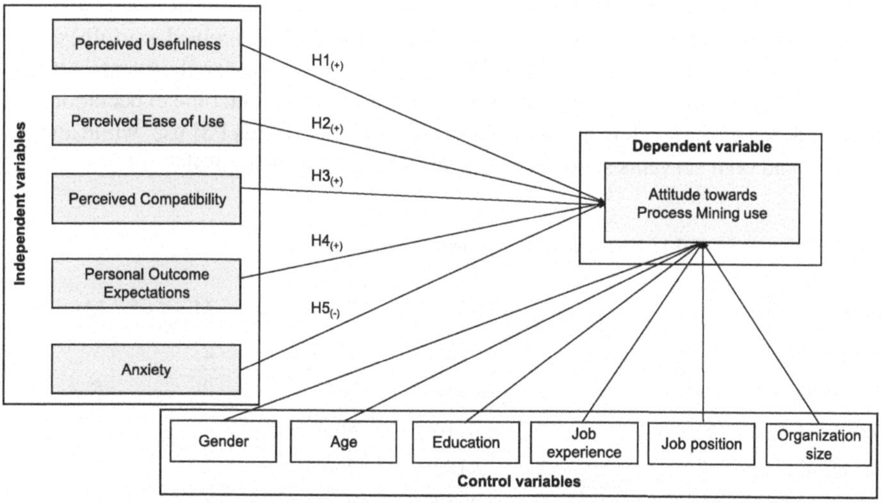

Fig. 1. Conceptual framework for Attitude towards PM use in the public sector

One novelty of this study is the investigation of civil servants' attitude towards PM use prior to its (eventual) adoption: this implies that public organizations involved in this study had not introduced PM yet. To obtain reliable answers, we adopted a purposive sampling technique [46]: we selected civil servants with different roles and from different segments of the Italian public sector attending theoretical and practical courses given by academics and consultants on process analysis and PM. Despite not using PM in their job activities, participants were knowledgeable about PM, which ensured more reliable results with respect to distributing surveys to individuals completely unfamiliar with PM use and techniques. In fact, attitudes activated from memory are more likely to influence perceptions, information processes, and behaviours with respect to those purposefully constructed in response to a question [47].

4 Data Analysis and Results

We administered a questionnaire to a total of 294 civil servants, obtaining a net response rate of 41% (122 complete answers). The majority of the civil servants is of female sex (68.03%), highly educated (87.71%), older than 50 years old (53.28%) and with more than 20 years of working experience (41.8%). Mainly, they work as officers (61.48%) and in large organizations (53.28%).

The descriptive analyses reveal a positive attitude towards PM use: 95.09% of civil servants favourably perceive PM introduction in the organization they work for (32.79% moderately agree, 62.30% strongly agree), while less than 5% are neutral or slightly disagree. On average, civil servants moderately agree with the suggested items, except for those related to outcome expectations and anxiety (Table 1). However, the question for anxiety was negative in meaning: a moderate level of disagreement means that civil servants do not feel apprehensive about using a PM tool. Perceived usefulness and perceived compatibility are the most decisive variables to enhance civil servants' attitude towards PM use, followed by perceived ease of use. Personal outcome expectations are not considered of great relevance for influencing attitude towards PM use, while anxiety is crucial and civil servants seem to be not apprehensive of using PM.

Table 1. Level of agreement for the model's variables.

Model variables	Mean	Std. dev.	Min	Max
Attitude (y)	4.5655	.6164	2	5
Perceived usefulness (x)	4.1885	.7855	2	5
Perceived ease of use (x)	3.8114	.8656	1	5
Perceived compatibility (x)	4.4180	.7367	2	5
Personal outcome expectations (x)	3.3442	.9339	1	5
Anxiety (x)	2.1229	1.064	1	5

We performed two ordered probit regression models: one to investigate the influence of the independent variables on the attitude towards PM use (Model 1); the other to control for socio-demographic, job-related, and organizational variables (Model 2). As we assumed (and results later confirmed) perceptions would differ depending on the different job position, we clustered the standard errors for the job position (role) to have better estimates. Table 2 reports the results of the two ordered probit models side by side. Multicollinearity among variables was checked by running two linear regression models, whose variance inflation factor was equal to 1.48 (Model 1) and 1.62 (Model 2), thus far below the upper threshold of 10 [48].

Model 1 shows that perceived usefulness, perceived ease of use, and perceived compatibility are positively correlated with the attitude towards PM use, while anxiety is negatively correlated. On the other hand, personal outcome expectations show no correlation with attitude towards PM use. This suggests that the more civil servants perceive PM as useful to improve their job performances, job fitting, and easy to use, and the less they feel negative feelings towards its use, the more they will have positive attitudes towards it. Controlling for socio-demographic, job-related, and organizational variables (Model 2), results are consistent, despite perceived ease of use's coefficient reducing significance.

Table 2. Model estimation for attitude towards PM use.

	Model 1 Attitude towards PM use	Model 2 Attitude towards PM use
Perceived usefulness	0.767***	0.791***
	(0.0984)	(0.162)
Perceived ease of use	0.230***	0.204**
	(0.0729)	(0.0928)
Perceived compatibility	0.488**	0.509**
	(0.243)	(0.200)
Personal outcome expectations	0.172	0.212
	(0.143)	(0.167)
Anxiety	-0.190***	-0.225***
	(0.0390)	(0.0858)
Gender		-0.500***
		(0.129)
Age		-0.284**
		(0.130)
Education		0.0373
		(0.119)
Years of job experience		0.165*
		(0.0902)
Employee*		0

(*continued*)

Table 2. (*continued*)

		(.)
Officer		0.360***
		(0.0748)
Office holder		1.636***
		(0.222)
Manager		0.972***
		(0.141)
Other		-0.0135
		(0.242)
Organization size		-0.0629*
		(0.0372)
N	122	122
Pseudo R^2	0.30	0.36
AIC	149.7	138.0
BIC	160.9	149.3

Standard errors in parentheses
* $p < 0.10$, ** $p < 0.05$, *** $p < 0.01$
Notes: When all job position's dummies are equal to zero, Employee is the considered observation.

In Model 1, the relationships perceived usefulness-attitude and perceived ease of use-attitude from the original TAM are found to be significant, which is consistent with much work in the IS literature [40, 50, 51]. Also perceived compatibility and anxiety explain a significant proportion of variance in attitude towards PM use, while personal outcome expectations are not influential. In Model 2 results are confirmed: perceived usefulness, perceived ease of use and perceived compatibility continue to positively influence attitude towards PM use, anxiety continues to have a negative effect, and personal outcome expectations continue to have no influence. Table 3 reports a summary of the findings and the related hypotheses, which, except for H4, are all supported.

Control variables all are related to attitude towards PM use, except for the level of education. This may be due to civil servants receiving the same formation on process analyses and PM, starting from the same level of knowledge (they knew nothing about PM).

Table 3. Summary of findings and related hypotheses.

Variable	Hypothesis	Relationship with Attitude	Supported?
Perceived usefulness	H1	+	Yes
Perceived ease of use	H2	+	Yes
Perceived compatibility	H3	+	Yes
Personal outcome expectations	H4	+	No
Anxiety	H5	–	Yes

Regarding gender differences on technology acceptance, it seems that men show a less positive attitude towards PM use with respect to women, which suggests that PM may have some peculiar features that make women more willing to accept PM with respect to men. For example, previous findings highlighted that women are more process-oriented than men [52]: as PM is intrinsically process-oriented, this may explain why women are more favourable than men to introduce PM in the organizations they work.

Consistent with previous literature on IS [27, 45, 53], age negatively influences attitude towards PM use, while being workers with higher organizational positions than employees and the years of work experience in the current job position positively influence it. The latter result may be explained by the fact that as job experience increases, individuals become more knowledgeable about organizations' processes and skilled with their tasks, so they are better able to acknowledge how innovation and new technologies, such as PM, can improve their job performances [54]. In addition, consistent with the Schumpeterian hypothesis [55], we found that civil servants working in smaller organizations show a more favourable attitude towards PM use, which may be due to the higher levels of flexibility and collaboration that allow an easier introduction of IT innovation [56].

5 Discussion and Implications

The Italian public sector urgently needs to digitalize its infrastructures and services to better satisfy citizens' needs and adhere to EU rules, standards, and goals. PM solutions can be key enablers to digitally transform and simplify public administration processes and generate public value through improved service quality for internal and external users. However, these achievements strongly depend on civil servants' usage of PM, which depends on their attitudes towards its use. This exploratory study is a pioneering effort that contributes to IS and PM theory advancements in the public sector, by providing a first understanding of those factors grounded in the IS theory that influence civil servants' attitude towards PM use prior to its adoption to understand its implementation viability. Identifying them and their relationships is fundamental to understand in advance whether PM could be successfully implemented in the public sector and avoid wrong investments.

This paper reveals that civil servants with basic theoretical and practical knowledge about PM positively perceive its introduction in the organizations they work for. Specifically, civil servants will more favourably adopt PM if they perceive it as useful to improve their job activities and performances, easy to use, job-fitting, and they do not feel apprehensive about using this tool. We also found that women, younger individuals, those with higher organizational positions, and those with more years of work experience more positively perceive PM introduction, which suggests that specific PM training programs based on gender, age, job position, and job experience could increase positive attitudes towards PM use.

6 Conclusions

Our findings provide policymakers and public managers with valuable insights to consider PM introduction in public organizations to effectively ride the wave of digital transformation and not stay behind, as civil servants, if adequately and previously informed about PM usage, and later supported with PM tailored trainings, would be more ready and willing to use it, which avoid implementation failures and public money's waste.

However, some limitations, especially related to the exploratory nature and the research approach adopted, must be reported. The most critical limitations concern the instrument development. While adopted constructs and items derive from existing IS theory and validated scales [18, 49, 57–59], we adapted them to fit the research context and subjectively selected only items considered relevant to the examined setting. Moreover, based on evidence suggesting using single-item measures when small samples are used and weak effect sizes are expected [60, 61], such as in exploratory studies [61–63], we chose one single item for each construct. However, this implies that we could not statistically analyse our model's constructs' reliability and convergent and discriminant validity, as in the case of multi-item measures. Therefore, our findings should be considered as initial undertakings to be deeper investigated and refined by further research.

Another limitation concerns the data sample, which may be too small to generalize our findings to the entire Italian public sector population, as it only involves Italian civil servants familiar with PM. Additionally, cultural aspects may have influenced civil servants' answers, limiting the applicability of the findings to other contexts. Nevertheless, we selected civil servants with different organizational positions, from different segments of the Italian public sector, and from organizations of different size to have the most heterogeneous sample as possible.

Though we are aware of these limitations, this exploratory study provides a first overview of civil servants' attitude towards PM use, which is crucial to evaluate its future effective introduction in the public sector. The proposed framework represents a first effort to quantitatively analyse PM adoption in the public sector by adopting IS lenses and can be a starting point to better refine it through more advanced statistical methodologies or qualitative interviews to gain deeper insights into the unique factors influencing PM acceptance in the public sector. Other areas for further research include conducting longitudinal studies to track changes in attitudes towards PM use over time and assess its long-term impact.

Disclosure of Interests. The authors have no competing interests to declare that are relevant to the content of this article.

References

1. De Vries, H., Tummers, L., Bekkers, V.: The diffusion and adoption of public sector innovations: a meta-synthesis of the literature. Perspect. Public Manag. Gov. **1**(3), 159–176 (2018)
2. Anthopoulos, L., Reddick, C.G., Giannakidou, I., Mavridis, N.: Why e-government projects fail? An analysis of the Healthcare. gov website. Gov. Inf. q. **33**(1), 161–173 (2016)
3. Weerakkody, V., Janssen, M., Dwivedi, Y.K.: Transformational change and business process reengineering (BPR): lessons from the British and Dutch public sector. Gov. Inf. Q. **28**(3), 320–328 (2011)
4. Nograšek, J., Vintar, M.: E-government and organisational transformation of government: black box revisited? Gov. Inf. Q. **31**(1), 108–118 (2014)
5. Tangi, L., Janssen, M., Benedetti, M., Noci, G.: Digital government transformation: a structural equation modelling analysis of driving and impeding factors. Int. J. Inf. Manag. **60**, 102356 (2021)
6. Gil-Garcia, J.R., Flores-Zúñiga, M.Á.: Towards a comprehensive understanding of digital government success: integrating implementation and adoption factors. Gov. Inf. Q. **37**(4), 101518 (2020)
7. Kim, J.S., Chung, G.H.: Implementing innovations within organizations: a systematic review and research agenda. Innovation **19**(3), 372–399 (2017)
8. Reinkemeyer, L.: Process Mining, RPA, BPM, and DTO. Process Mining in Action: Principles, Use Cases and Outlook, 41–44 (2020)
9. van der Aalst, W.M.P.: Process mining: data science in action. Springer (2016)
10. Geyer-Klingeberg, J., Nakladal, J., Baldauf, F., Veit, F.: Process mining and robotic process automation: a perfect match. In: BPM (Dissertation/Demos/Industry), pp. 124–131 (2018, July)
11. Grisold, T., Mendling, J., Otto, M., vom Brocke, J.: Adoption, use and management of process mining in practice. Bus. Process. Manag. J. **27**(2), 369–387 (2021)
12. Rott, J., Böhm, M.: Mine the right process–towards a method for selecting a suitable use case for process mining adoption (2022)
13. vom Brocke, J., Jans, M., Mendling, J., Reijers, H.A.: A five-level framework for research on process mining. Bus. Inf. Syst. Eng. **63**(5), 483–490 (2021)
14. Graafmans, T., Turetken, O., Poppelaars, H., Fahland, D.: Process mining for six sigma: a guideline and tool support. Bus. Inf. Syst. Eng. **63**, 277–300 (2021)
15. Lashkevich, K., Milani, F., Danylyshyn, N.: Analysis templates for identifying improvement opportunities with process mining (2023)
16. Maneschijn, D.G.J.C., Bemthuis, R.H., Arachchige, J.J., Bukhsh, F.A., Iacob, M.E.: Balancing simplicity and complexity in modeling mined business processes: a user perspective. In: International Conference on Enterprise Information Systems, pp. 3–21. Springer Nature, Cham, Switzerland (2022, April)
17. Wynn, M.T., et al.: ProcessProfiler3D: a visualisation framework for log-based process performance comparison. Decis. Support. Syst. **100**, 93–108 (2017)
18. Davis, F.D.: Perceived usefulness, perceived ease of use, and user acceptance of information technology. MIS q. 319–340 (1989)
19. Rogers, E.M.: Diffusion of Innovations (4th ed.). The Free Press, New York (1995)
20. Venkatesh, V., Davis, F.D.: A theoretical extension of the technology acceptance model: four longitudinal field studies. Manag. Sci. **46**(2), 186–204 (2000)

21. Venkatesh, V., Morris, M.G., Davis, G.B., Davis, F.D.: User acceptance of information technology: toward a unified view. MIS q. 425–478 (2003)
22. Carter, L., Bélanger, F.: The utilization of e-government services: citizen trust, innovation and acceptance factors. Inf. Syst. J. **15**(1), 5–25 (2005)
23. Titah, R., Barki, H.: E-government adoption and acceptance: a literature review. Int. J. Electron. Gov. Res. (IJEGR) **2**(3), 23–57 (2006)
24. Rehouma, M.B., Hofmann, S.: Government employees' adoption of information technology: a literature review. In: Proceedings of the 19th Annual International Conference on Digital Government Research: Governance in the Data Age, pp. 1–10 (2018, May)
25. Dečman, M.: Understanding technology acceptance of government information systems from employees' perspective. Int. J. Electron. Gov. Res. **11**(4), 69–88 (2015)
26. Oschinsky, F.M., Stelter, A., Niehaves, B.: Cognitive biases in the digital age–How resolving the status quo bias enables public-sector employees to overcome restraint. Gov. Inf. Q. **38**(4), 101611 (2021)
27. Wickramasinghe, V., Wickramasekara, J.: Putting public services into enterprise system-predicting employees' acceptance of transformational government technology in an expanded technology acceptance model. Public Organ. Rev. **22**(2), 345–365 (2022)
28. Wirtz, B.W., Piehler, R.: EGovernment applications and public personnel acceptance: an empirical analysis of the public servant perspective. Int. J. Public Adm. **39**(3), 238–247 (2016)
29. Batara, E., Nurmandi, A., Warsito, T., Pribadi, U.: Are government employees adopting local e-government transformation? The need for having the right attitude, facilitating conditions and performance expectations. Transform. Gov. People Process Policy **11**(4), 612–638 (2017)
30. Gupta, K.P., Bhaskar, P., Singh, S.: Prioritization of factors influencing employee adoption of e-government using the analytic hierarchy process. J. Syst. Inf. Technol. **19**(1/2), 116–137 (2017)
31. Altameem, T., Zairi, M., Alshawi, S.: Critical success factors of e-government: a proposed model for e-government implementation. In: 2006 Innovations in Information Technology, pp. 1–5. IEEE, November 2006
32. Damanpour, F., Schneider, M.: Characteristics of innovation and innovation adoption in public organizations: assessing the role of managers. J. Public Adm. Res. Theory **19**(3), 495–522 (2009)
33. Jonathan, G.M.: Digital transformation in the public sector: Identifying critical success factors. In: European, Mediterranean, and Middle Eastern Conference on Information Systems (pp. 223–235). Springer, Cham (2020)
34. Moser-Plautz, B., Schmidthuber, L.: Digital government transformation as an organizational response to the COVID-19 pandemic. Gov. Inf. Q. **40**(3), 101815 (2023)
35. Mergel, I., Gong, Y., Bertot, J.: Agile government: systematic literature review and future research. Gov. Inf. Q. **35**(2), 291–298 (2018)
36. Kim, H., Pan, G., Pan, S.: Managing IT-enabled transformation in the public sector: a case study on e-government in South Korea. Gov. Inf. Q. **24**, 338–352 (2007)
37. van der Aalst, W.M.: Process mining: a 360-degree overview. In: Process Mining Handbook (pp. 3-34). Springer International Publishing, Cham (2022)
38. Reinkemeyer, L.: Status and future of process mining: from process discovery to process execution. In: Process Mining Handbook (pp. 405–415). Springer International Publishing, Cham (2022)
39. Eggers, J., Hein, A., Böhm, M., Krcmar, H.: No longer out of sight, no longer out of mind? How organizations engage with process mining-induced transparency to achieve increased process awareness. Bus. Inf. Syst. Eng. **63**(5), 491–510 (2021)

40. Kerremans, M., Srivastava, T., Choudhary, F.: Market Guide for Process Mining, Gartner (2021). https://it.scribd.com/document/567567818/Market-Guide-for-Process-Mining-Gartner. Accessed 30 Oct 2023
41. van der Aalst, W., et al.: Process mining manifesto. In: Business Process Management Workshops: BPM 2011 International Workshops, Clermont-Ferrand, France, August 29, 2011, Revised Selected Papers, Part I, vol. 9, pp. 169–194. Springer, Berlin, Heidelberg (2012)
42. Eggers, J., Hein, A.: Turning big data into value: a literature review on business value realization from process mining. In: ECIS (2020, June)
43. Bannister, F.: Dismantling the silos: extracting new value from IT investments in public administration. Inf. Syst. J. **11**(1), 65–84 (2001)
44. Di Mascio, F., Natalini, A., Profeti, S.: Administrative reforms in the Italian national recovery and resilience plan: a selective approach to bridge the capacity gap. Contemp. Ital. Polit. **14**(4), 487–501 (2022)
45. Czaja, S.J., et al.: Factors predicting the use of technology: findings from the center for research and education on aging and technology enhancement (CREATE). Psychol. Aging **21**(2), 333 (2006)
46. Etikan, I., Musa, S.A., Alkassim, R.S.: Comparison of convenience sampling and purposive sampling. Am. J. Theor. Appl. Stat. **5**(1), 1–4 (2016)
47. Fazio, R.H.: On the power and functionality of attitudes: the role of attitude. Attitude Struct. Funct. **153**, 179 (1989)
48. Diamantopoulos, A., Siguaw, J.A.: Formative versus reflective indicators in organizational measure development: a comparison and empirical illustration. Br. J. Manag. **17**(4), 263–282 (2006)
49. Davis, F.D., Bagozzi, R.P., Warshaw, P.R.: User acceptance of computer technology: a comparison of two theoretical models. Manag. Sci. **35**(8), 982–1003 (1989)
50. Moore, G.C., Benbasat, I.: Development of an instrument to measure the perceptions of adopting an information technology innovation. Inf. Syst. Res. **2**(3), 192–222 (1991)
51. Salahshour Rad, M., Nilashi, M., Mohamed Dahlan, H.: Information technology adoption: a review of the literature and classification. Univ. Access Inf. Soc. **17**, 361–390 (2018)
52. Venkatesh, V.: Determinants of perceived ease of use: integrating control, intrinsic motivation, & emotion into the technology acceptance model. Inf. Syst. Res. **11**(4), 342–365 (2000)
53. Rogers, W.A., Fisk, A.D., Mead, S.E., Walker, N., Cabrera, E.F.: Training older adults to use automatic teller machines. Hum. Factors **38**(3), 425–433 (1996)
54. Dahabiyeh, L., Mowafi, O.: Challenges of using RPA in auditing: a socio-technical systems approach. Intell. Syst. Account. Financ. Manag. (2023)
55. Utterback, J.M.: Innovation in industry and the diffusion of technology. Science **183**(4125), 620–626 (1974)
56. Lee, G., Xia, W.: Organizational size and IT innovation adoption: a meta-analysis. Inf. Manag. **43**(8), 975–985 (2006)
57. Compeau, D., Higgins, C.A., Huff, S.: Social cognitive theory and individual reactions to computing technology: a longitudinal study. MIS q. 145–158 (1999)
58. Melone, N.P.: A theoretical assessment of the user-satisfaction construct in information systems research. Manag. Sci. **36**(1), 76–91 (1990)
59. Taylor, S., Todd, P.A.: Understanding information technology usage: a test of competing models. Inf. Syst. Res. **6**(2), 144–176 (1995)
60. Diamantopoulos, A., Sarstedt, M., Fuchs, C., Wilczynski, P., Kaiser, S.: Guidelines for choosing between multi-item and single-item scales for construct measurement: a predictive validity perspective. J. Acad. Mark. Sci. **40**, 434–449 (2012)
61. Stebbins, R.A.: Exploratory Research in the Social Sciences, vol. 48. Sage, Thousand Oaks (2001)

62. Rossiter, J.R.: The C-OAR-SE procedure for scale development in marketing. Int. J. Res. Mark. **19**(4), 305–335 (2002)
63. Stanton, J.M., Sinar, E.F., Balzer, W.K., Smith, P.C.: Issues and strategies for reducing the length of self-report scales. Pers. Psychol. **55**(1), 167–194 (2002)

Exploring the Role of Visualization in Process Mining: A Systematic Literature Review

Florian Eckhard(✉), Holger Wittges, and Stefanie Rinderle-Ma

School of Computation, Information and Technology, Technical University of Munich, Boltzmannstr. 3, 85748 Garching, Germany
{florian.eckhard,holger.wittges,stefanie.rinderle-ma}@tum.de

Abstract. Visualization is an integral part of process mining. It can be used to visualize the process model or show information about process bottlenecks. This review gives an overview of the current usage of visualization for process mining. Additionally, we derive open research fields based on the current literature. We conduct a systematic literature review to reach this overview, including 33 papers. In addition, we define a taxonomy to categorize research within the field of visualization for process mining. Our results show a significant interest in process discovery and process monitoring, but fewer papers for process reengineering. The most used visualization type is the Directly Follows Graph. However, this is mainly used for control flow and not for other process mining perspectives. Instead, other types of visualization are used to cover the remaining perspectives. For future work, we propose stronger cooperation between the field of process mining and information visualization to combine the benefits of both areas and see open research in upcoming approaches like object-centric process mining.

Keywords: Process Mining · Visualization techniques · Literature Review

1 Introduction

Visualization is an integral part of process mining. It can be used, for example, to visualize the process model or show information about process bottlenecks. Nevertheless, it is often neglected in practice. At the same time, more efforts are made to bring information visualization and process mining closer together. Process mining bridges the gap between machine learning and data mining on one hand and process modeling and analysis on the other. Its primary objective is to uncover, monitor, and enhance real-world processes by extracting valuable insights from event logs obtained from systems. An event log is a list of activities recorded in a system during their execution. Each element is typically assigned to a case and has an activity name, a timestamp, and optional further properties. An elemental part of process mining is visualizing the mined models or generated

K. Węcel (Ed.): BIS 2025, LNBIP 554, pp. 165–178, 2025.
https://doi.org/10.1007/978-3-031-94193-1_13

results [1,2]. Information visualization is defined as a technology using human visual abilities to extract meanings from visual information. Combining color, size, position, or orientation of objects changes in visualizations can lead to faster and better analytic results and improve usability for process analysts [9,14]. To give an overview of visualization in process mining, we conduct a systematic literature review answer the following research questions:

- RQ1: What is the current state of research in visualizing process mining results?
- RQ2: Which visualization types are most commonly used in process mining research for which process mining type and perspective?
- RQ3: What are open research areas in visualizing process mining results?

To answer these questions, we define a taxonomy to categorize the review results. Based on the taxonomy, we create a concept matrix to assign each paper to its relevant categories. In addition, we identify open research questions and areas for future study based on the existing literature and the gaps in current research. To the best of our knowledge, this is the first work that provides a survey in the area of process mining and visualization with a combined view of process mining types and perspectives. In previous works like Sirgmets et al. [36] from 2018, an overview of two areas is given: 1) process mining techniques using visualization and 2) visualization types used for process mining. In contrast to our approach, the results are shown separately and not in a combined taxonomy, as well as without process mining perspectives. Another survey focusing on combining the research field information visualization and process mining is presented by Yeshchenko and Mendling [43]. While this work gives a comprehensive overview of the visualization of event sequence data, no mapping to process mining types and perspectives is provided. Additionally, this research focuses on six significant journals of information visualization and process mining, which could exclude relevant visualizations published at conferences.

This paper is separated into seven Sections. After the introduction, we briefly overview some relevant process mining concepts and define terms used in this review. In the third Section, we introduce our visualization taxonomy, which will be used to classify papers and create our concept matrix. Section 4 presents the literature review design, followed by the findings in Sect. 5. Finally, our discussion follows before the conclusion Section, and an outlook for future research is given.

2 Background

In this Section, we want to give an overview of important process mining topics and define related terms. The three main process mining types are process discovery, conformance checking and process reengineering. In van der Aalst [4], it was extended by a fourth type, operational support. In the context of this work, we define operational support as equivalent to process monitoring and will hereafter use only the term process monitoring. Process discovery is a technique to identify processes based on an event log without the need to provide information in advance. As a result, a process model, for example a petri net, is generated. During conformance checking, a comparison is made between the existing process model and an event log of the same process. This makes it possible to check whether the log follows the expected process and vice versa. The third type, process reengineering, extends or improves an existing process model based on the event log. As a result, an improved model is generated. Process monitoring aims to display recommendations or warnings about a process, preferably in real-time, to improve work with the process. Unlike process reengineering, the model is not changed [2,4].

Besides the process mining types, a process can be viewed from different perspectives. Four perspectives are defined in the process mining manifesto [1]. First is the control flow, often seen as the basis of the other perspectives [4]. This perspective visualizes the order in which the activities are executed. The second is the organizational perspective, including the information on which actors are part of the process and how they are connected. The time perspective covers all aspects of the timing and frequency of events. Lastly, the case perspective, sometimes called data perspective, includes the properties of objects used in the process [1]. Another form of process mining is the comparison of two process models, also known as comparative process mining. The goal is to recognize the difference between two processes. This is used, for example, in combination with process cubes that distinguish processes by various domains, e.g., regions. The processes from the different areas can then be compared to discover regional differences within a company [29].

3 Visualization Taxonomy

In this Section we introduce our taxonomy. The aim of this taxonomy is to categorize the works more easily according to their characteristics and to be able to find and compare similar works. An overview of the taxonomy is presented in Fig. 1. The taxonomy is based on the process mining types and perspectives presented in Sect. 2.

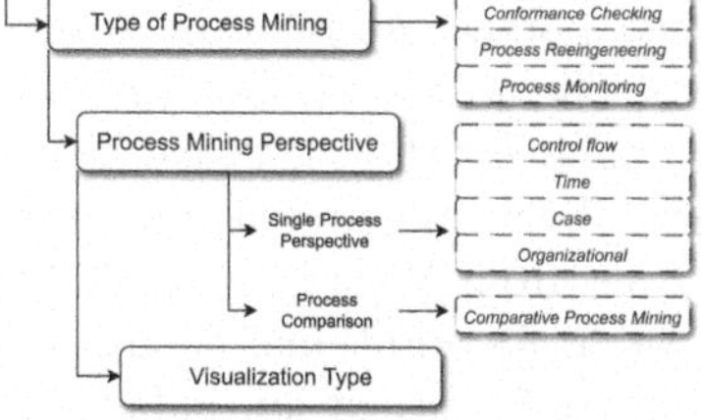

Fig. 1. Taxonomy for process mining visualizations

At the first level, a distinction is made between the research type of applied research and theoretical research. The applied research category includes all papers applying process mining and visualization to address business or social challenges or develop a tool for an existing visualization concept. In contrast, the theoretical research category includes all papers that develop new concepts of how visualization can be used for process mining or apply existing visualization for a theoretical process mining topic. In this work, we will focus on theoretical research, but we also want to emphasize the relevance of applied research in process mining and visualization. Following the research type classification, studies are further categorized by process mining type. The taxonomy includes all four process mining types as presented in Sect. 2. At the third level, a distinction is made between the process mining perspectives and a further differentiation between single process perspectives and process comparison. All results in the area of process comparison are assigned to the comparative process mining perspective. For the single process perspective, the categories from the process mining manifesto [1] are adopted in this categorization. In the context of this paper, the assessment of whether a perspective is included in a work is not based on which perspectives are addressed within a paper but exclusively on whether a perspective is taken into account in the visualization. The last level considers the visualization type. In addition to widely used business process visualizations such as Directly Follows Graph (DFG), petri nets or BPMN, new visualization approaches are also relevant. There is no active exclusion of visualization types, but newly developed or individual visualizations are assigned to a custom category instead of being added as their own category.

4 Literature Review Design

This Section outlines the methodological approach used for this literature review. As a guideline, this literature review follows a concept-centric approach presented by Webster and Watson [39]. Figure 2 presents an overview of our methodological process, which consists of four main steps. The process begins with identifying potentially relevant articles from Scopus, Web of Science, IEEE Xplore, ACM Digital Library and AIS eLibrary. The search string[1] was intentionally formulated without too many constraints to ensure that a broad overview of visualization topics in process mining is covered. We considered all papers published between January 2014 and December 2024. In total, 519 papers were found, of which 189 were identified as potentially relevant after removing the duplicates and checking the abstracts. The abstracts were screened by two reviewers independently of each other. Based on the identified papers, a full-text screening was conducted to categorize the results and assess their relevance to the review. The Concept Matrix, derived from Webster and Watson [39], is based on the taxonomy presented in Sect. 3. As additional exclusion criteria, all entries that are not in English or are not conference or journal articles have been withdrawn.

[1] The syntax between all databases is slightly different, but the search terms were chosen identically for all databases.

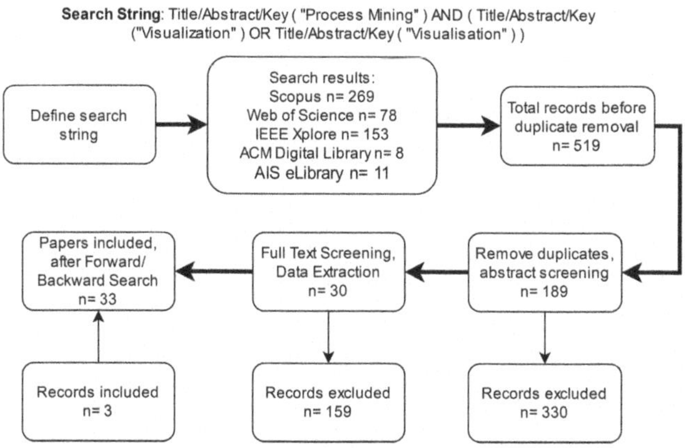

Fig. 2. Literature review process

For concepts that have been published in several formats (for example, journal, conference or workshop) by the same author, we have only included the journal version as long as no relevant content is lost. After the full-text screening, 30 papers remain relevant. A forward and backward search was conducted to find relevant articles that were not part of the search string. In this step, three papers were included. In total, 33 papers are included in this literature review.

5 Findings

In the following Section, the findings of our literature review are presented. The concept matrix is shown in Table 1. The table is structured such that the rows represent different visualization types, whereas the columns illustrate the process mining types, single-process perspectives, and comparative process mining. The papers in the table are assigned to the respective concepts based on their visualization type. It is important to mention that for papers with several visualization types, an assignment to a perspective only occurs if the specific type also visualizes it. In the following sub-sections, relevant developments in the area of process mining visualization based on the concept matrix are discussed.

5.1 Directly-Follows Graph

The most commonly used visualization is DFG. This is plausible, as most process mining tools (e.g., ProM, Disco, Celonis) support this visualization. Surprisingly, petri nets and BPMN, which are also widely used process modeling notations, are minor or not represented. Despite the widespread use of DFG, its use in most papers is limited to the control flow [12,18,19,23,25,27,37,40,42]. Although DFG's are widely used, their adoption can be viewed critically. For example, van

Table 1. Concept Matrix: Theoretical Research

Visualization type	Type of Process Mining				Process Mining perspective				Comparative Process Mining	sum
	Process Discovery	Conformance Checking	Process Reeingeneering	Process Monitoring	Control flow	Time perspective	Case perspective	Organizational perspective		
DFG	[12, 23, 33, 37, 40]	[25, 42]	[19]	[18, 21, 26, 27]	∀	[21, 26]	[21, 33]	-	[25]	12
BPMN	-	-	-	-	-	-		-	-	0
Petri Net	[34]	-	-	[41]	∀	[41]	-	-	-	2
Gantt chart	[35]	[24, 31]	-	[24]	[24, 35]	[24, 31, 35]	[24]	-	-	3
Drift Map/Chart	-	[42]	-	-	-	[42]	-	-	-	1
Bar chart	-	-	[19]	[16, 18]	-	[18, 19]	[18, 19]	[16]	-	3
Dot plot	-	-	-	[18]	-	[18]	[18]	-	-	1
Sankey	-	-	-	[38]	[38]	-	-	-	-	1
Heatmap	-	-	-	[16]	[16]	[16]	-	[16]	-	1
Declare	[8]	-	-	-	[8]	-	-	-	-	1
Chevron diagram	[35]	-	-	-	[35]	[35]	-	-	-	1
Graph	-	[7, 15]	-	-	[7]	[7]	-	-	[7, 15]	2
Custom	[17, 20, 22, 32, 35]	[11, 28, 32]	-	[6, 13, 21]	$\forall x(x \neq [27])$	[6, 13, 21, 28, 35]	[6, 21, 28]	[17]	[11, 28, 32]	10
sum	**12**	**9**	**1**	**10**	**29**	**14**	**7**	**2**	**6**	

der Aalst [5] discusses the disadvantages of DFG's. The main problem is that the simplification used to present DFG results could be misleading and result in false analytical results. To overcome these limitations, the papers Pfahlsberger et al. [27] and Wetzel et al. [40] present DFG extensions. Pfahlsberger et al. [27] proposes eight different path semantics and possible path combinations to add information about the process behavior. Two examples are shown in Fig. 3a. The arrow with a solid gray line and an angular course connecting nodes 2 to 1 represents an allowed backjump path. On the opposite, the arrow with a solid red line (nodes 4 to 3) and a curvilinear course shows a prohibited backjump path. The allowed and prohibited path combinations, in this example refinement and rework, can be used for further analysis of process flows. For example, prohibited process flows can be analyzed from time, cost, and quality perspectives to identify negative process impacts. In contrast, Wetzel et al. [40] focuses on so-called history dependencies. Figure 3b shows an example of the visualization. A history dependency based on this example means that activity D can't be executed if activity A is missing. Therefore, the trace $\langle B, C, D \rangle$ is not valid. A standard DFG can't visualize this dependency and can lead to false analytical results. Layout format optimization is another research direction to improve graph-based representations like DFG. In Mennens et al. [23] and Sonke et al. [37], two approaches are presented. Mennens et al. [23] presents an algorithm to generate a stable layout for the same process. While current industry-leading process mining tools could generate different visualizations for the same process, the presented algorithm transforms visualizations into a similar or same layout. The authors argue that this supports the analyst's mental model and generates better visualizations in total, based on the results from a user study. While the visualization in Mennens et al. [23] follows a top-down approach, Sonke et al. [37] present alternative layouts besides top-down approaches. The nodes are arranged both vertically and horizontally in order to make better use of the aspect ratio.

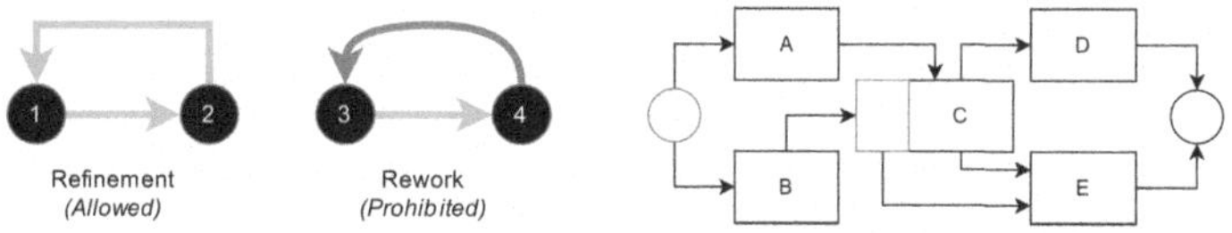

(a) Path semantics [27] (b) History dependencies [40]

Fig. 3. DFG extensions

While the presented approaches show promising approaches for reducing DFG shortcomings, they are currently in an early stage, and no production-ready implementation in widely used process mining tools exists.

5.2 Time and Case Perspective

Besides the control flow, the time perspective is the most researched perspective, with 14 distinct papers, followed by the case perspective with seven papers. The significant coverage of the time perspective is plausible since timestamps are a central component of the event log [1]. This distinguishes the time perspective from the organizational and case perspective, whose information must first be added to the log or process model. We found two focus areas regarding time perspective: 1) new approaches to visualize time-specific information, 2) adding time information for conformance checking. For the first focus area, Denisov et al. [13] present performance spectra to visualize all activities and cases individually over time, Schuster et al. [35] introduce chevron diagrams and a combination of points and vertical bars to visualize partially ordered event data, and Nagy and Werner-Stark [24] visualize processes as a gantt chart like diagram to improve time-related analysis. The second focus expands traditional conformance-checking methods, focusing on control-flow changes, to integrate time-related drift detection. For such changes, two visulisations based on gantt charts and drift maps/charts are proposed [31,42]. Both visualizations support process analysts in identifying temporal changes in their processes. As seen in all approaches, time is a challenging but also a relevant topic in adopting process mining to real world use cases. While new visualization approaches [13,35,42] show promising results, they require further research to enhance user-friendliness and practical tool support.

In contrast to the time perspective, we can't find focus areas for the case perspective. Especially new visualization approaches for the case perspective are missing. Instead, the case perspective is often used as a filter or attribute [18], visualized by established visualization types or text-based information beside the process diagram [6,19,21,24]. We only identified the work from Pini et al. [28] presenting an approach that integrates case information directly into process visualization. Unfortunately, due to missing data, the evaluation of a real-world dataset was not possible. While using established visualizations is not negative by default; the stronger integration of the case perspective into the visualization

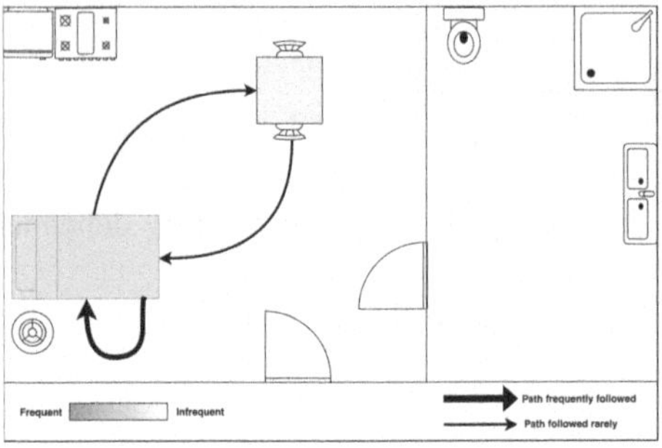

Fig. 4. Geo-aware visualization: Tiramisù approach [6]

could bring advantages. There could be interrelations to the emergence of object-centered process mining, which integrates process objects into the process model.

5.3 Geo-Aware Visualization

Standard process mining visualizations do not provide geographic context. While the first approaches for geo-awareness were presented, for example, by de Leoni et al. [21] in 2014, new approaches to extend this concept arose in 2024. Alman et al. [6], which extends previous process map approaches with additional dimensions, and Corea and Delfmann [12] present a new 3D geo-aware approach based on DFG.

The Tiramisù framework from Alman et al. [6] defines three layers. Firstly, the backdrop serves as the background for the visualization and provides a common context, represented by the room with furniture in Fig. 4. Secondly, the process representation shows the process flow via nodes and edges, shown by arrows and furniture in Fig. 4. The geo-awareness is represented by the nodes with fixed positions in the backdrop so that context-relevant information can be displayed. Finally, the Dimensions Layer augments the backdrop or provides additional information for the end user, for example, by coloring the furniture based on activity frequency. There is no limit to how many Dimension Layers are provided, but it should be possible to activate these dynamically to maintain clarity. The 3D-based geo-aware approach by Corea and Delfmann [12] uses a DFG combined with a 3D world as the basis. For example, the process is linked to a virtual hospital, whereby the process is displayed geo-aware in the hospital rooms. Both approaches open the possibility of creating visualizations optimized for a business domain and, therefore, increasing the value of the visualization. The main problem with both approaches is that a significant amount of domain knowledge is required to design suitable visualizations. Additionally, a high technical effort is needed to create an appropriate 3D environment based on the process map. As a future research field, it may be interesting to see how the generated 3D environments can be displayed using virtual or augmented reality.

5.4 Comparative Process Mining

Besides the main process mining types and perspectives, we also focus on comparative process mining. Although most of the work dates from 2015/2016, the topic is highly relevant from the authors' point of view because processes need to be adapted more and more dynamically, and visualizations of the differences can be highly relevant. In the identified papers, there is one focus on control-flow comparison [11,15,25,28,32] and another focus on time comparison [7,28]. In Gall et al. [15], a combination of color and symbolism is used to visualize the difference between two processes, showing the control flow of both processes together. This is similar to Cordes and Vogelgesang [11] using color and line patterning and Saito [32] using a 3D city metaphor. In contrast, Pini et al. [28] define a leading process variant as the basis and visualize if an activity is executed earlier or later in the compared variants, without showing the exact difference. While previous approaches are evaluated on small event logs, Neubauer et al. [25] focus on comparing complex processes after trace clustering. The analysis of complex event logs is a relevant research field, which we will discuss in Sect. 6. As for the control flow, there are approaches to compare time information in one single diagram [7,28] and multiple separated diagrams [28]. Bolt et al. [7] present an annotated transition system that uses colorization to indicate, for each activity, in which process a metric is statistically significantly higher. Currently, two metrics are supported: occurrence and elapsed time. In Pini et al. [28], both options are presented, whereas the single model visualization does not indicate whether the value is higher or lower but merely shows that the metrics differ. While multiple works were published in 2015/2016 we found limited research in the following years. This is surprising because multiple questions like comparing multiple process variants (more than three), enhancing process attribute comparisons, and comparing complex processes remain without further investigation.

5.5 Under-Researched Areas

The currently most under-researched areas are process reengineering (1 paper), conformance checking (excluding process comparison, 3 papers) and organizational perspective (2 papers). For process reengineering [19] and the organizational perspective [16,17] first approaches exist. While for the organizational perspective, first implementations based on ProM and custom development exist, there are no developed prototypes for process reengineering. In Kubrak et al. [19] the focus is on developing wireframe mock-ups to guide the development of usable tools. While this is an important step, developing usable process reengineering visualizations is still an open question. The third underresearched area is conformance checking, although nine works are assigned to the category. Without comparative process mining, only three papers remain for conformance checking [24,31,42]. Two papers [31,42] focus on identifying temporal changes, and Nagy and Werner-Stark [24] is the only identified work also visualizing control flow changes. The conformance checking results align with previous work from Rehse et al. [30].

Table 2. Future research areas

Future research areas
Investigate visualizations for process reengineering, conformance checking and case perspective
Adopt current or investigate new visualizations for object-centric process mining
Develop a framework to better integrate user evaluations into process mining visualization development processes
Development of a maturity model to quantify how effectively a perspective is covered in a visualization
Develop guidelines on how to choose the right visualization type for a specific process mining use case

5.6 Tool Usage

We also look at the question with which tools the found approaches are implemented. 10 papers used ProM[2]. The second most used approach, with 8 papers, is individual development for visualization. Individual development is defined as using a programming language with its associated frameworks (e.g., Python and Matplotlib) to visualize results. Three papers combine a custom development with the usage of ProM or PM4Py. PM4Py[3] is the third most used tool (4 papers). The remaining papers do not implement a visualization (5 papers), don't name the tool used (2 papers), or use other tools like Cortado or Noreja (1–2 papers each). For all papers, a trend toward open-source and easy-to-extend systems is visible.

6 Discussion

This Section presents the key findings of our research and discusses their implications for future studies. To answer the research questions outlined in Sect. 1, we create a taxonomy for classifying process mining and visualization papers and present a concept matrix mapping the results to these categories. The results indicate a strong interest in combining visualization and process mining. The most underresearched process mining type is currently process reengineering. We could not find a paper explaining why this is the case. A possible explanation could be that process reengineering partially relies on process discovery and conformance checking, and current research focuses on these types first. Another reason could be that research focuses on process monitoring, where the analytic results can also be used for process reengineering. Nevertheless, we believe focusing on specific process reengineering visualizations would be helpful. Regarding the perspectives, we identified control flow as the most researched area. This is

[2] https://promtools.org/.
[3] https://github.com/pm4py/pm4py-core.

plausible because the control flow is seen as the basis of the other perspectives [4]. Regarding the remaining three perspectives, we see increasing interest in the time perspective. Another important finding is that proposed visualization approaches are predominantly evaluated using small process models. Although first approaches [25] explore methods for visualizing complex processes or their abstractions, there is limited support for visualization techniques specifically designed for large and complex process models.

To answer the second research question, we mapped the process mining types and perspectives to the visualization types used. The most used visualization type through all researched categories is DFG. The focus on DFG or node-based visualization aligns with the previous work from Sirgmets et al. [36]. The visualization focus on DFG remains for research as well as most process mining tools, despite the remarks about their disadvantages [5]. Besides DFG, we found multiple works defining new visualization approaches. Classic visualization techniques such as bar charts, heatmaps, and graphs are also used. These mainly show information about the time and case perspective and are used beside a control flow visualization. Only in a few cases, such as chevron diagrams or gantt charts, the control flow is combined with another perspective in a single diagram. No explanation was found for the low usage of BPMN and petri nets. One possibility is that the existing focus on DFG in research and industry has promoted further research in this direction. Another reason could be that the comparatively easier representation as DFGs was considered adequate for most current process mining requirements.

The further integration of process mining and information visualization is an important aspect and is part of the third research question about current challenges. As suggested by Yeshchenko and Mendling [43] and Pini et al. [28], we also want to emphasize the need for stronger cooperation between these two fields. From deeper cooperation, all process mining types and perspectives could benefit. Our results show that the main research focus lies on case-centric process mining and only limited results for new approaches like object-centric process mining [3]. As a relatively new process mining approach, visualization is in its infancy. The increased complexity resulting from adding objects also increases the requirements for suitable visualizations. Adopting approaches from the information visualization domain could be advantageous here. Besides other, an essential aspect of information visualization is evaluating the proposed approach in a realistic testing environment [10]. 17 of all reviewed papers do not specify a user evaluation. The better integration of evaluations into the development of process mining visualizations is a relevant field of research for the future. To summarize the results from RQ3, an overview of open research directions is shown in Table 2.

While this study offers valuable insights into the synergy of process mining and visualization, it is crucial to recognize limitations that should be considered when interpreting the findings. The first limitation relates to our search string. Although our search string is very openly formulated, it is possible that we did not capture papers that used synonyms for visualization. The second limitation is the focus on scientific publications, excluding the features of commercial/open-source software tools. It cannot be guaranteed that functions identified as missing in this paper are present in existing software solutions. The last limitation refers to considering the different process mining levels (case, variant, process). For this review, we don't make any such differentiation but look only at the different perspectives presented.

7 Conclusion

Our research has two primary research contributions. First, we developed a taxonomy to classify papers in the field of visualization and process mining. This makes it easier to categorize literature and identify gaps and developments in different research areas. It can also be used in future reviews to track the development of selected fields. Secondly, we conduct a literature review to give an overview of the current research state of process mining and visualization. This review provides insights into under-researched areas and an overview of the used visualizations in process mining. Based on the results, we reveal open research areas for future researchers. Besides further research from a technical standpoint, quantifying the coverage of a process mining perspective would be a future area of improvement for our taxonomy. For example, the time perspective assignment currently does not capture how good the visualizations of, for example, bottlenecks are. Our next step is to investigate ways to combine information visualization and process mining more effectively to address the open gaps discussed in this review.

References

1. van der Aalst, W., et al.: Process mining manifesto (2021). https://www.tf-pm.org/resources/manifesto. Accessed 25 Oct 2024
2. van der Aalst, W.M.P.: Process Mining, 2nd edn. Springer, Heidelberg (2016)
3. van der Aalst, W.M.P.: Object-Centric Process Mining: Dealing with Divergence and Convergence in Event Data, pp. 3–25. Springer (2019). https://doi.org/10.1007/978-3-030-30446-1_1
4. van der Aalst, W.M.: Process discovery from event data: relating models and logs through abstractions. WIREs Data Mining Knowl. Discov. **8**(3) (2018). https://doi.org/10.1002/widm.1244
5. van der Aalst, W.M.: A practitioner's guide to process mining: limitations of the directly-follows graph. Procedia Comput. Sci. **164**, 321–328 (2019)
6. Alman, A., Arleo, A., Beerepoot, I., Burattin, A., Di Ciccio, C., Resinas, M.: Tiramisù: making sense of multi-faceted process information through time and space. J. Intell. Inf. Syst. (2024)
7. Bolt, A., de Leoni, M., van der Aalst, W.M.P.: A Visual Approach to Spot Statistically-Significant Differences in Event Logs Based on Process Metrics, pp. 151–166. Springer (2016)
8. Burattin, A., Cimitile, M., Maggi, F.M.: Lights, Camera, Action! Business Process Movies for Online Process Discovery, pp. 408–419. Springer (2015). https://doi.org/10.1007/978-3-319-15895-2_34

9. Card, S.K., Mackinlay, J.D., Shneiderman, B.: Readings in information visualization. The @Morgan Kaufmann Series in Interactive Technologies. Morgan Kaufmann, San Francisco (2007)
10. Carpendale, S.: Evaluating Information Visualizations, pp. 19–45. Springer, Heidelberg (2008). https://doi.org/10.1007/978-3-540-70956-5_2
11. Cordes, C., Vogelgesang, T., Appelrath, H.J.: A Generic Approach for Calculating and Visualizing Differences Between Process Models in Multidimensional Process Mining, pp. 383–394. Springer (2015)
12. Corea, C., Delfmann, P.: Geo-aware process mining, vol. 3758, pp. 101–105 (2024)
13. Denisov, V., Fahland, D., van der Aalst, W.M.P.: Unbiased, Fine-Grained Description of Processes Performance from Event Data, pp. 139–157. Springer (2018). https://doi.org/10.1007/978-3-319-98648-7_9
14. Fekete, J.D., Plaisant, C.: Excentric labeling: dynamic neighborhood labeling for data visualization. In: Proceedings of the SIGCHI Conference on Human Factors in Computing Systems, pp. 512–519 (1999)
15. Gall, M., Wallner, G., Kriglstein, S., Rinderle-Ma, S.: A Study of Different Visualizations for Visualizing Differences in Process Models, pp. 99–108. Springer (2015). https://doi.org/10.1007/978-3-319-25747-1_10
16. Hoogmoed, A., Vidgof, M., Djurica, D., Rubensson, C., Mendling, J.: Visual Representation of Resource Analysis Insights for Process Mining, pp. 117–128. Springer (2024). https://doi.org/10.1007/978-3-031-61007-3_10
17. Jalali, A.: Supporting Social Network Analysis Using Chord Diagram in Process Mining, pp. 16–32. Springer (2016)
18. Kaouni, A., Theodoropoulou, G., Bousdekis, A., Voulodimos, A., Miaoulis, G.: Visual Analytics in Process Mining for Supporting Business Process Improvement. IOS Press (2021). https://doi.org/10.3233/faia210089
19. Kubrak, K., Milani, F., Nolte, A.: A visual approach to support process analysts in working with process improvement opportunities. Bus. Process. Manag. J. **29**(8), 101–132 (2023)
20. Lamghari, Z.: Process mining: a new approach for simplifying the process model control flow visualization. Transdiscipl. J. Eng. Sci. **13** (2022). https://doi.org/10.22545/2022/00193
21. de Leoni, M., Suriadi, S., ter Hofstede, A., van der Aalst, W.: Turning event logs into process movies: animating what has really happened. Softw. Syst. Model. **15**(3), 707–732 (2014)
22. Manoj Kumar, M.V., Prashanth, B.S., Sneha, H.R., Thomas, L., Annappa, B., Murthy, Y.V.S.: Process Logo: An Approach for Control-Flow Visualization of Information System Process in Process Mining, pp. 481–492. Springer, Singapore (2021). https://doi.org/10.1007/978-981-16-1338-8_40
23. Mennens, R.J., Scheepens, R., Westenberg, M.A.: A stable graph layout algorithm for processes. Comput. Graph. Forum **38**(3), 725–737 (2019)
24. Nagy, Z., Werner-Stark, A.: Visualization methods to support real-time process monitoring. In: CEUR Workshop Proceedings, vol. 3628, pp. 1–14 (2023)
25. Neubauer, T.R., Pamponet Sobrinho, G., Fantinato, M., Peres, S.M.: Visualization for enabling human-in-the-loop in trace clustering-based process mining tasks. In: 2021 IEEE International Conference on Big Data (Big Data), vol. 53, pp. 3548–3556. IEEE (2021). https://doi.org/10.1109/bigdata52589.2021.9671985
26. Nguyen, H., Dumas, M., ter Hofstede, A.H.M., La Rosa, M., Maggi, F.M.: Business Process Performance Mining with Staged Process Flows, pp. 167–185. Springer (2016)

27. Pfahlsberger, L., Rubensson, C., Knoblich, S., Vidgof, M., Mendling, J.: Multi-perspective path semantics in process mining based on causal process knowledge. In: CEUR Workshop Proceedings, vol. 3645 (2023)
28. Pini, A., Brown, R., Wynn, M.T.: Process Visualization Techniques for Multi-perspective Process Comparisons, pp. 183–197. Springer (2015). https://doi.org/10.1007/978-3-319-19509-4_14
29. Process and Data Science Group -RWTH Aachen University: Comparing multiple processes (2024). https://www.processmining.org/comparative-process-mining.html. Accessed 09 Nov 2024
30. Rehse, J.R., Pufahl, L., Grohs, M., Klein, L.M.: Process mining meets visual analytics: the case of conformance checking. In: Proceedings of the Annual Hawaii International Conference on System Sciences, vol. 2023-January, pp. 5452–5461 (2023)
31. Richter, F., Seidl, T.: Looking into the tesseract: time-drifts in event streams using series of evolving rolling averages of completion times. Inf. Syst. **84**, 265–282 (2019). https://doi.org/10.1016/j.is.2018.11.003
32. Saito, S.: Digital twincity: a holistic approach towards comparative analysis of business processes. In: Proceedings of the 25th International Conference on Model Driven Engineering Languages and Systems: Companion Proceedings, MODELS 2022, vol. 214, pp. 17–21. ACM (2022)
33. Salas-Urbano, M., Capitán-Agudo, C., Cabanillas, C., Resinas, M.: Designing a User Interface to Explore Collections of Directly-Follows Graphs for Process Mining Analysis, pp. 35–47. Springer (2024)
34. de San Pedro, J., Cortadella, J.: Mining structured petri nets for the visualization of process behavior. In: Proceedings of the 31st Annual ACM Symposium on Applied Computing, SAC 2016, vol. 80, pp. 839–846. ACM (2016)
35. Schuster, D., Zerbato, F., van Zelst, S.J., van der Aalst, W.M.: Defining and visualizing process execution variants from partially ordered event data. Inf. Sci. **657**, 119958 (2024). https://doi.org/10.1016/j.ins.2023.119958
36. Sirgmets, M., Milani, F., Nolte, A., Pungas, T.: Designing Process Diagrams–A Framework for Making Design Choices When Visualizing Process Mining Outputs, pp. 463–480. Springer (2018)
37. Sonke, W., Verbeek, K., Meulemans, W., Verbeek, E., Speckmann, B.: Optimal algorithms for compact linear layouts. In: 2018 IEEE Pacific Visualization Symposium (PacificVis), pp. 1–10. IEEE (2018)
38. Völzer, H., Zerbato, F., Sulzer, T., Weber, B.: A fresh approach to analyze process outcomes. In: 2023 5th International Conference on Process Mining (ICPM), pp. 97–104. IEEE (2023). https://doi.org/10.1109/icpm60904.2023.10271968
39. Webster, J., Watson, R.T.: Analyzing the past to prepare for the future: writing a literature review. MIS Q. **26**(2), xiii–xxiii (2002)
40. Wetzel, M., Koschmider, A., Wilke, T.: Visually Representing History Dependencies in Event Logs, pp. 5–16. Springer (2021)
41. Wynn, M., et al.: Processprofiler3d: a visualisation framework for log-based process performance comparison. Decis. Support Syst. **100**, 93–108 (2017)
42. Yeshchenko, A., Di Ciccio, C., Mendling, J., Polyvyanyy, A.: Visual drift detection for event sequence data of business processes. IEEE Trans. Visual Comput. Graphics **28**(8), 3050–3068 (2022)
43. Yeshchenko, A., Mendling, J.: A survey of approaches for event sequence analysis and visualization. Inf. Syst. **120**, 102283 (2024)

A Data-Driven Framework for Retail Inventory Optimization: Integrating Object-Centric Process Mining and Mathematical Models

Dina Kretzschmann[1(✉)], Alessandro Berti[1], and Wil M. P. van der Aalst[1,2]

[1] Process and Data Science (PADS), RWTH Aachen University, Aachen, Germany
{dina.kretzschmann,a.berti,wvdaalst}@pads.rwth-aachen.de
[2] Celonis, Munich, Germany

Abstract. Efficient inventory management is essential for retail companies, significantly impacting customer satisfaction and cost efficiency. However, managing related processes is highly complex due to the dynamic interaction of multiple object types along the supply chain (e.g. suppliers, customers, materials) and the reliance on diverse and fragmented information systems. Consequently, retail companies face challenges with inefficient inventory structures, including understock (loss of sales) and overstock (high capital commitment). This paper proposes an object-centric process mining approach to analyze process-related causes of inefficient inventory management. A novel Object-Centric Data Model (OCDM) is introduced, capturing key entities and interactions across inventory management-related processes. Unlike traditional models, the OCDM is enriched with established inventory metrics, enabling multi-perspective analyses that link process interactions to metrics such as *understock, healthy stock*, and *overstock*. This approach uncovers inventory management inefficiencies and their process-related root causes, delivering actionable insights. A case study at a leading European pet retailer demonstrates the approach's practical value in identifying bottlenecks and suggesting improvement actions, showcasing the potential for improved inventory control in retail settings.

Keywords: Inventory Management · Object-Centric Data Model · Process Mining

1 Introduction

Retail companies operate in a dynamic environment where efficient inventory management is crucial for balancing customer satisfaction (through product availability) and profitability (through cost minimization). Inventory management functions as the temporal bridge between customer orders and supplier replenishment [4,34]. Core processes affecting inventory management are therefore Order-to-Cash (O2C) for sales and Purchase-to-Pay (P2P) for purchasing,

K. Węcel (Ed.): BIS 2025, LNBIP 554, pp. 179–193, 2025.
https://doi.org/10.1007/978-3-031-94193-1_14

both supported by diverse information systems, including Enterprise Resource Planning (ERP) systems and demand forecasting tools (see Fig. 1).

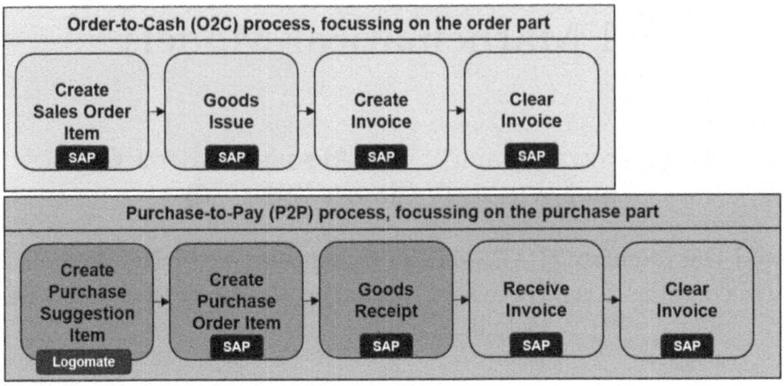

Fig. 1. Inventory management serves as the link between customer demand, managed through Order-to-Cash (O2C) processes, and purchasing activities, governed by Purchase-to-Pay (P2P) processes.

Problem Statement and Research Motivation: Efficient inventory management is critical for retail success but often hampered by the interplay of O2C and P2P processes. Diverse systems and suboptimal process executions can lead to understock (lost sales) or overstock (tied-up capital and storage costs) [4,23,34]. Traditional optimization models and ETL methodologies mainly provide static analysis, while advanced data analytics solutions primarily focus on improving demand forecasting models, lacking insights into the underlying process-related causes [32].

Process mining can reveal these causes by examining process flows [1,32]. Yet, standard methods struggle to capture the complex, multi-object nature of inventory management (e.g. suppliers, customers, materials) addresses this gap by incorporating these intertwined relationships [2]. By analyzing dynamic interactions and dependencies, OCPM helps identify root causes of understock and overstock, enabling a deeper understanding of process-driven inefficiencies.

This paper builds upon and extends the work presented in [17], where overstock issues in a P2P process were analyzed using object-centric process mining in collaboration with a leading European pet retailer. While the previous study focused on identifying overstock patterns and proposing targeted improvements within P2P processes, this research broadens the scope to integrate both O2C and P2P processes, introducing a novel object-centric data model enriched with established inventory management metrics to address both understock and overstock inefficiencies comprehensively.

Research Questions: This paper aims to address the limitations of existing approaches by investigating how the enrichment of state-of-the-art object-centric process mining techniques [2] with established inventory management exact

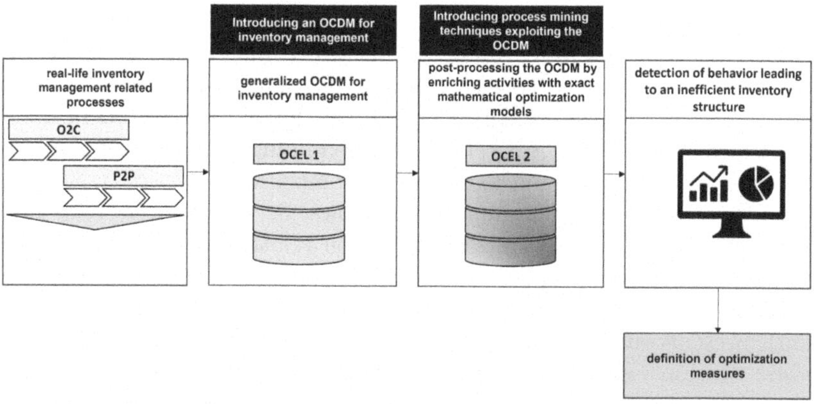

Fig. 2. Outline of the contributions of the paper.

mathematical optimization models [34] can be leveraged to optimize inventory management and gain a deeper understanding of process-related drivers for an inefficient inventory structure and define optimization measures. Specifically, we address the following research questions:

RQ1 How can we effectively model the complex interplay of objects and events within the O2C and P2P processes to facilitate a holistic, object-aware understanding of inventory management?

RQ2 How can we enrich the object-centric perspective with established inventory management exact mathematical optimization models (e.g. Min-Max method) to enable data-driven analysis of process inefficiencies leading to understock and overstock situations?

RQ3 What actionable insights can be derived from the application of OCPM to a real-life inventory management scenario, and how can these insights be used to improve existing practices?

Contributions:

C1 *OCDM for Inventory Management*: An Object-Centric Data Model (OCDM) that integrates data from diverse sources (e.g., ERP, forecasting) into a single object-centric event log (OCEL A, Fig. 2), enabling comprehensive inventory process analysis.

C2 *OCDM Enrichment with Inventory Metrics*: A method to enrich the OCDM with established inventory management exact mathematical optimization models (e.g., Min-Max method), producing OCEL B (Fig. 2) for deeper assessments of inventory efficiency and targeted optimization actions.

C3 *Real-Life Case Study*: A case study at a European pet retailer demonstrating how the OCDM and enrichment approach uncover process inefficiencies in demand forecasting and supplier management, informing concrete improvements.

The introduced Object-Centric Data Model (OCDM) (**C1**) and the enrichment with relevant inventory management exact mathematical optimization models (**C2**) are generalizable and reproducible across different scenarios and systems in the retail environment. This reduces the effort for retailers to define and implement their own object-centric process mining analyses, as the approach can be applied flexibly without extensive customization.

The rest of the paper is organized as follows. Section 2 covers related work on inventory management optimization. Section 3 outlines the metrics needed to understand the rest of the paper. Section 4 describes the proposed object-centric data model, the computation of metrics, and the post-processing operations. Section 5 presents an implementation based on simulated data for illustrative purposes. Section 6 details a case study conducted in a pet retailer company. Finally, Sect. 7 concludes the paper by discussing limitations and future work.

2 Related Work

Inventory management optimization techniques and tools have evolved significantly, yet existing approaches often fail to holistically connect data-driven decisions with the underlying processes that shape stock levels. While exact mathematical optimization models provide clear guidelines for optimal inventory structures, they often rely on idealized assumptions and static parameters that rarely hold in dynamic, real-world contexts [23,34]. In contrast, various approaches-from business process management techniques to lean management methods—offer qualitative insights into inefficiencies. ETL methodologies and advanced data analytics approaches concentrate on analyzing inventory structures and improving forecasting models. However, these approaches share a common limitation: the capability to systematically connect these insights to the precise causes manifest in event data and process executions.

Exact mathematical optimization models (e.g., ABC analysis [5,23,27,33], Economic Order Quantity (EOQ) [13,14,28,29], and the Min-Max method [5,12,29,31]) are well-established for classifying inventory items, minimizing costs, and determining reorder thresholds. These models, however, often hinge on stable demand and constant lead times, assumptions that cannot capture the full complexity of modern supply chains [23,34]. The resulting gaps limit the explanatory power needed to diagnose why certain deviations-such as persistent understock or overstock—arise in practice.

In parallel, traditional business management methods (e.g., process mapping, fishbone diagrams) and lean management approaches (e.g., Just-In-Time) highlight the value of understanding process-related factors [7,14,15,18,26]. However, these qualitative methods do not inherently exploit event data at scale, nor do they link inventory metrics directly to underlying process variations. Similarly, ETL methodologies centralize and standardize data for improved decision-making [11,21], while advanced data analytics, AI, and ML approaches focus on demand forecasting models and anomaly detection [20,24,25,30]. Though these tools improve accuracy and speed, they often operate as black boxes and remain detached from the actual process flows and their interdependencies, leaving practitioners without a clear path from insight to actionable process-oriented interventions.

Traditional process mining techniques have been applied to logistics and inventory management [3,32], offering a complementary view by revealing how processes unfold and where inefficiencies occur. Yet these methods struggle to handle the multi-object nature of retail operations. The complexity lies in the interplay of various objects—suppliers, customers, materials—interacting through events spread across disparate systems. Recently, object-centric process mining (OCPM) has emerged as a powerful paradigm that incorporates multi-object event data [2]. While OCPM-based case studies for inventory management exist [17], they are not yet fully integrated with established inventory optimization models, nor do they systematically enrich process-level analyses with mathematical metrics.

This gap underscores why we have chosen our approach. By enriching OCPM with exact mathematical optimization models, we combine two complementary perspectives: the proven theoretical rigor of inventory metrics and the dynamic event-centric, object-aware lens of OCPM. The resulting synergy enables both to identify where a company's inventory structure diverges from optimal conditions and explain how these deviations originate from underlying process behaviors. In other words, we do not merely detect understock or overstock situations—we illuminate their root causes embedded in the company's operational processes, paving the way for targeted interventions and sustainable improvements.

3 Preliminaries

3.1 Inventory-Related Metrics and KPIs

Achieving an optimal balance between delivery service level, inventory costs, and capital commitment is crucial. Higher inventory ensures availability but increases costs, so finding an efficient, *normal* stock level is key [10,23,34].

We apply the classical formulas of the *Min-Max* method [5,12,29,31], to simplify our approach. These methods use four main metrics calculated for each material (m):

Economic Order Quantity (EOQ) [28] minimizes total ordering and holding costs:

$$\mathrm{EOQ}_m = \sqrt{\frac{2D_m S}{H}},$$

where D_m is demand, S the fixed cost per order, and H the holding cost.

Safety Stock (SS) [35] provides a buffer against variability:

$$\mathrm{SS}_m = z \times \sigma_m \times \sqrt{l_m},$$

with z (service level), σ_m (standard deviation of demand), and l_m (lead time).

Table	Description
Sales Order Documents	Sales Document Number
	Document Creation Date
	Customer Number
	Document Date in Document
	Sales Document Type
	Order Type
	Order Reason
Sales Order Items	Sales Document Number
	Item Number
	Material Number
	Plant
	Order Quantity
	Net Price
Materials	Material Number
	Material Type
	Industry Sector
	Material Group
	Valuation Class
	Gross Weight
	Net Weight
	Weight Unit
	Volume
	Volume Unit
	Transport Group
Purchase Requisitions	Client
	Purchase Document Number
	Item Number of Purchasing Document
	Purchase Requisition Number
	Purchase Requisition Item
	Purchase Requisition Date
	Document Type
	Purchasing Document Category
	Planned Delivery Time
	Latest Possible Goods Receipt
	Quantity
	Unit of Measure

Table	Description
Material Stocks	Client
	Material Number
	Plant
	Storage Location
	Stock in quality inspection
	Stock in transfer
	Stock in posting
	Stock of material provided to vendor
	Blocked Stock
	Returns Stock
Goods Receipts and Issues	Client
	Purchase Document Number
	Line Item in Purchase Document
	Sequential Number of Account Assignment
	Movement Type (Goods Receipt, Goods Issue)
	Fiscal Year
	Document Number
	Accounting Document Line
	Material Number
	Plant
	Reference Document Number
	Document Date in Document
	Posting Date in the Document
	Date of the Posting in the Document
	Time of the Posting in the Document
	Quantity
Sales Document Flows	Client
	Sales Document
	Sales Document Item
	Sales Document
	Sales Document Item
	Document Category of Subsequent Document
	Document Category of Preceding Document
	Document Date

Table	Description
Purchase Order Items	Purchase Order Number
	Purchase Order Item Number
	Material Number
	Plant
	Quantity
	Change Date
	Net Price
Purchase Order Documents	Purchase Document Number
	Record Creation Date
	Account Number of Vendor
	Purchase Order Date
	Purchasing Document Category
	Purchasing Document Type
	Blocking Indicator
Material Documents	Client
	Material Document Number
	Material Document Year
	Line Item
	Material Number
	Plant
	Storage Location
	Vendor's Account Number
	Customer Number
	Movement Type (Goods Receipt, Issue)
	Receiving Plant
	Quantity
	Posting Date in the Document
Order Suggestions	Order Number
	Order Position
	Article Number
	Order Quantity
	Date
	Order Date
	Delivery Date
	Plant

Fig. 3. List of tables considered from the relational data model.

Reorder Point (ROP) [29] triggers replenishment:

$$\mathrm{ROP}_m = d_m \times l_m + \mathrm{SS}_m,$$

where d_m is average demand.

Maximum Stock Level (Max) [29] limits excess stock:

$$\mathrm{Max}_m = \mathrm{EOQ}_m + \mathrm{SS}_m.$$

Overstock (OS) occurs when inventory surpasses the max level at a certain point in time (t). However, mathematical definitions of overstock vary in the literature [6,19,22,29]:

$$\mathrm{Overstock}_{m,t} = \text{Inventory Level}_{m,t} - \mathrm{Max}_m.$$

By applying these metrics, companies can maintain adequate service levels while controlling costs and avoiding excessive stock.

4 Approach (Data Extraction and Enrichment)

In this section, we start from the ERP database, applying the view (relational data model) described in Sect. 4.1 and extract a (single) object-centric event log (OCEL A) as described in Sect. 4.2. We then obtain in Sect. 4.3 a (single) "transformed" object-centric event log (OCEL B), in which the activities have been semantically enriched with inventory management exact mathematical optimization models (i.e. Min-Max method).

4.1 Relational Data Model

The relational data model presented in Fig. 3 can be extracted from various ERP systems, including SAP ERP and Oracle EBS. This model includes several key tables such as *Sales Order Documents*, *Sales Order Items*, *Materials*, *Purchase Requisitions*, *Material Stocks*, *Goods Receipts and Issues*, *Sales Document Flows*, *Purchase Order Items*, *Purchase Order Documents*, *Material Documents*, and *Order Suggestions*. Each table captures specific aspects, facilitating the management of O2C and P2P processes.

4.2 Object-Centric Data Model and Event Log (C1)

The proposed OCDM (Tables 1 and 2) includes various object types and event types, also known as activities, essential for tracking and analyzing business processes in an ERP system.

Object types mapped in Table 1 include *materials* and *plants*, along with specific identifiers for *purchase* and *sales order items*, *customers*, and *suppliers*. These object types link to related event tables, ensuring comprehensive data integration across materials, sales, purchases, and inventory activities.

Event types are mapped in Table 2 to specific source tables and timestamp fields, detailing key actions such as *Create Purchase Requisitions*, *Sales Order Items*, and *Purchase Order Items*, as well as recording *Goods Receipts* and *Issues*. For instance, the creation of a purchase requisition is logged in the *Purchase Requisitions* table with the associated date, while goods receipts and issues are captured in the *Goods Receipts* and *Issues* table with their posting dates.

More in detail, we assume that, for a considered table, its rows can be mapped to events having a constant activity (for example "Goods Receipt" or "Create Purchase Order Item") and a timestamp given by the values of a single column of the table. We also assume that a table's row provides the connection of the corresponding event to all the relevant object types. Therefore, if the proposed extraction procedure was restricted to a single table, we would still be able to extract an object-centric event log containing a single activity.

Starting from the object-centric data model, we use the OCEL 2.0 Object-Centric Event Log standard [16] for the storage of the event data. The format is well-supported by object-centric process mining tools like pm4py[1] and OC-PM[2].

[1] https://pm4py.fit.fraunhofer.de.
[2] https://www.ocpm.info/.

Table 1. Object types and related tables.

Object Type	Description	Related Tables	Essential Attributes
MAT	Material Number	Materials, Sales Order Items, Goods Receipts and Issues, Purchase Requisitions, Purchase Order Items, Material Documents	Stock Level, Material type, Material Group, Transport Group, Weight, Volume
PLA	Plant	Sales Order Items, Material Stocks, Purchase Requisitions, Goods Receipts and Issues, Purchase Order Items, Material Documents	
PO_ITEM	Purchase Document and Item Number	Purchase Order Items, Purchase Orders, Goods Receipts and Issues	Category, Type, Quantity, Price
SO_ITEM	Sales Order and Item Number	Sales Order Items, Sales Orders, Sales Document Flows	Document Type, Order Type, Quantity, Price
CUSTOMER	Customer Number	Sales Orders, Material Documents	
SUPPLIER	Supplier Number	Purchase Orders, Material Documents	

Table 2. Activities with related object types, source tables, and timestamp field.

Activity	Object Types	Source Event Table	Timestamp Field
Create Purchase Requisition	PO_ITEM, MAT, PLA, MAT_PLA	Purchase Requisitions	Purchase Requisition Date
Create Purchase Suggestion Item	PO_ITEM, MAT, PLA, MAT_PLA	Order Suggestions	Date
Create Purchase Order Item	PO_ITEM, MAT, PLA, MAT_PLA, SUPPLIER	Purchase Order Items	Purchase Order Date
Goods Receipt	PO_ITEM, MAT, PLA, MAT_PLA, SUPPLIER	Goods Receipts and Issues	Date of the Posting
Create Sales Order Item	SO_ITEM, MAT, PLA, MAT_PLA, CUSTOMER	Sales Order Items	Sales Order Date
Goods Issue	PO_ITEM, MAT, PLA, MAT_PLA, CUSTOMER	Goods Receipts and Issues	Date of the Posting

Each event, at the item level, is associated with at most one object per object type. The event log includes events/objects along with their attributes and event-to-object relationships essential for object-centric process mining. An important note is that each event of the proposed OCEL corresponds to an activity at the item level, so it is related to at most an object per object type. In principle, this allows for storing the object-centric event log in columnar databases for more performant querying and processing.

4.3 Computation of the Relevant Metrics (C2)

After storing the relevant information in an OCEL 2.0 event log, we can compute the relevant metrics of the *Min-Max* method. To calculate the *Economic Order Quantity* (EOQ_m), we compute D_m measuring the customer demand in each time unit (e.g. one month) by summing the amounts issued in events with the activity *Goods Issue*. S is the fixed cost per order. H is the holding cost per unit per year. In our case study, we used fixed values provided by the pet retail company. For S we used 100€ and for H 0.1€. To calculate the *Safety Stock* (SS_m), we compute the z-score z corresponding to the desired service level. For example, a service level of 99% indicates a 99% probability that demand will be fulfilled without stock-outs during the lead time. For the standard deviation of demand σ_m, we measure the customer demand in each time unit, and computed the average and standard deviation of this demand. The lead time, l_m, is defined as the average time passed between the event with *Create Purchase Order Item* and the corresponding *Goods Receipt.*

To calculate the *Reorder Point* (ROP_m), we compute d_m, l_m, and SS_m as described. For the *Maximum Stock level* (Max_m), we compute the EOQ_m and SS_m as reported. To calculate $Overstock_{m,t}$, we use the Inventory $\text{Level}_{m,t}$ given in the OCDM and compute the Max_m as described.

As an example calculation, we compute the proposed metrics for a single material of the pet retailer, a *toy for dogs* from the 'non-food' category, as follows:

- *Economic Order Quantity (EOQ)*:

$$\text{EOQ}_m = \sqrt{\frac{2D_m S}{H}}$$

  with $D_m = 281$ [packaging units], $S = 100$ €, $H = 0.1$ €.
  Thus,

$$EOQ_m = \sqrt{\frac{2 \times 281 \times 100}{0.1}} = 750 \text{ [packaging units]}.$$

- *Safety Stock (SS)*:

$$\text{SS}_m = z \times \sigma_m \times \sqrt{l_m}$$

  For a 99% service level, $z = 2.326$. Given $\sigma_m = 97$ [packaging units] and $l_m = 1.0$ [month]:

$$SS_m = 2.326 \times 97 \times \sqrt{1.0} = 226 \text{ [packaging units]}.$$

- *Reorder Point (ROP)*:

$$\text{ROP}_m = d_m \times l_m + \text{SS}_m$$

$$ROP_m = 281 \times 1.0 + 226 = 507 \text{ [packaging units]}.$$

- *Maximum Stock Level (Max)*:

$$\text{Max}_m = \text{EOQ}_m + \text{SS}_m = 750 + 226 = 976 \text{ [packaging units]}.$$

- *Overstock (OS)*:
  If Inventory $\text{Level}_{m,t} = 3017$, then

$$\text{Overstock}_{m,t} = \text{Inventory Level}_{m,t} - \text{Max}_m = 3017 - 976 = 2041 \text{ [packaging units]}.$$

To optimize the O2C and P2P processes, we transformed the activities of the object-centric event log by examining *Safety Stock (SS)*, *Overstock (OS)*, *Reorder*

Table 3. Performed transformation of the original activities. For simplicity, situations in which the stock fall to 0 are identified by 0 instead of *Understock*.

Activity	Stock Before	Stock After	ROP	EOQ	Transformed Activity	Assessment
Goods Receipt	< SS	< SS			Goods Receipt (Understock to Understock)	Negative
	< SS	≥ SS, < OS			Goods Receipt (Understock to Normal)	Positive
	< SS	≥ OS			Goods Receipt (Understock to Overstock)	Negative
	≥ SS, < OS	≥ SS, < OS			Goods Receipt (Normal to Normal)	Positive
	≥ SS, < OS	≥ OS			Goods Receipt (Normal to Overstock)	Negative
	≥ OS	≥ OS			Goods Receipt (Overstock to Overstock)	Negative
Goods Issue	< SS	< SS			Goods Issue (Understock to Understock)	Negative
	≥ SS, < OS	< SS			Goods Issue (Normal to Understock)	Negative
	≥ SS, < OS	≥ SS, < OS			Goods Issue (Normal to Normal)	Positive
	≥ OS	≥ SS, < OS			Goods Issue (Overstock to Normal)	Positive
	≥ OS	≥ OS			Goods Issue (Overstock to Overstock)	Positive
Create Purchase Order Item	< SS		≤	≤	Create Purchase Order Item (Understock) (<= EOQ)	Uncertain
	≥ SS, < OS		≤	≤	Create Purchase Order Item (Normal, ≤ ROP) (<= EOQ)	Uncertain
	≥ SS, < OS		>	≤	Create Purchase Order Item (Normal, > ROP) (<= EOQ)	Uncertain
	≥ OS		>	≤	Create Purchase Order Item (Overstock) (<= EOQ)	Negative
	< SS		≤	>	Create Purchase Order Item (Understock) (> EOQ)	Uncertain
	≥ SS, < OS		≤	>	Create Purchase Order Item (Normal, ≤ ROP) (> EOQ)	Uncertain
	≥ SS, < OS		>	>	Create Purchase Order Item (Normal, > ROP) (> EOQ)	Uncertain
	≥ OS		>	>	Create Purchase Order Item (Overstock) (> EOQ)	Negative
Create Purchase Suggestion Item	< SS		≤	≤	Create Purchase Suggestion Item (Understock) (<= EOQ)	Uncertain
	≥ SS, < OS		≤	≤	Create Purchase Suggestion Item (Normal, ≤ ROP) (<= EOQ)	Uncertain
	≥ SS, < OS		>	≤	Create Purchase Suggestion Item (Normal, > ROP) (<= EOQ)	Uncertain
	≥ OS		>	≤	Create Purchase Suggestion Item (Overstock) (<= EOQ)	Negative
	< SS		≤	>	Create Purchase Suggestion Item (Understock) (> EOQ)	Uncertain
	≥ SS, < OS		≤	>	Create Purchase Suggestion Item (Normal, ≤ ROP) (> EOQ)	Uncertain
	≥ SS, < OS		>	>	Create Purchase Suggestion Item (Normal, > ROP) (> EOQ)	Uncertain
	≥ OS		>	>	Create Purchase Suggestion Item (Overstock) (> EOQ)	Negative
Create Sales Order Item	≤ SS	≤ SS			Create Sales Order Item (Understock to Understock)	Negative
	≥ SS, < OS	≤ SS			Create Sales Order Item (Normal to Understock)	Negative
	> SS	≤ SS			Create Sales Order Item (Overstock to Understock)	Negative
	≥ SS, < OS	≥ SS, < OS			Create Sales Order Item (Normal to Normal)	Positive
	≥ OS	≥ SS, < OS			Create Sales Order Item (Overstock to Normal)	Positive
	≥ OS	≥ OS			Create Sales Order Item (Overstock to Overstock)	Positive

Point (ROP), and *Economic Order Quantity (EOQ)*. The transformation represented in Table 3 is necessary to enable the analysis of patterns in the related systems, highlighting how different states and thresholds impact operations. By categorizing activities based on inventory levels before and after executing the events, we could systematically evaluate the process behavior leading to positive (normal stock, reduction of understock or overstock), uncertain (risk of understock or overstock), and negative (understock or overstock) situations, and identify process-related causes. For example, when processing a goods receipt, we assessed if the stock before the event was *below SS*, *between SS and Max, i.e. Normal Stock*, or above *Max, i.e. Overstock*. This classification helped identify understock, normal stock, and overstock scenarios. Similarly, goods issue events were analyzed based on their impact on stock levels, revealing patterns in stock depletion.

5 Simulated OCDM and Implementation

Due to confidentiality, the original data and implementation cannot be disclosed. Instead, we provide a public repository[3] containing scripts to simulate and store inventory data following the OCDM described in Sect. 4. These scripts generate synthetic data mimicking key inventory management entities and demonstrate how to extract an OCEL (A) from the relational model, linking events, objects, and timestamps as in Table 2.

The repository also includes a post-processing script for OCEL A that enriches events with the metrics described in Sect. 4.3, producing OCEL B and reflecting the transformations in Table 3.

This simulated environment enables researchers and practitioners to experiment with the proposed techniques, deepening their understanding of the methodology and its applicability across various scenarios.

Table 4. Case study: simplified extract of the original object-centric event log (OCEL A) alongside the transformed activities (OCEL B). Both event logs are available at https://zenodo.org/records/13347782. We assume that for the material $M001$, $SS = 91$ and $OS = 150$.

Original Activity	Transformed Activity	Material	Supplier	Customer	PO Item	Quantity	Stock Before	Stock After
Goods Issue	Goods Issue (Normal to Normal)	M001		C001		2	100	98
Goods Issue	Goods Issue (Normal to Normal)	M001		C002		3	98	95
Create Purchase Suggestion Item	Create Purchase Suggestion Item (Normal, > ROP) (> EOQ)	M001	S001			100	95	95
Create Purchase Order Item	Create Purchase Order Item (Normal, > ROP) (> EOQ)	M001	S001		PO1	100	95	95
Goods Issue	Goods Issue (Normal to Understock)	M001		C003		5	95	90
Goods Receipt	Goods Receipt (Understock to Normal)	M001	S001		PO1	50	90	140
Goods Receipt	Goods Receipt (Normal to Overstock)	M001	S001		PO1	50	140	190
Goods Issue	Goods Issue (Overstock to Overstock)	M001		C001		10	190	180

[3] https://github.com/fit-alessandro-berti/inventory-management-simulation.

Table 5. Case study: classes of materials based on their stock level.

Material Stock	Percentage
Always normal	70.0 %
Understock (no Overstock)	14.7 %
Overstock (no Understock)	11.6 %
Understock and Overstock	3.7 %

Table 6. Case study: main activities in the proposed OCDM. Colors indicate effect on stock: positive, uncertain, negative.

Transformed Activity	Occurrences
Goods Issue (Normal to Normal)	381.307
Goods Issue (Overstock to Overstock)	68.059
Create Sales Order Item (Normal to Normal)	30.526
Create Purchase Suggestion Item (Normal, >ROP) (>EOQ)	2.512
Goods Receipt (Normal to Normal)	2.370
Create Purchase Order Item (Normal, >ROP) (>EOQ)	2.216
Create Purchase Suggestion Item (Overstock, <= EOQ)	703
Goods Receipt (Overstock to Overstock)	658

6 Case Study (C3)

We performed a case study with a leading European pet retailer. The pet retailer has a strong online and offline presence, achieving sales exceeding 4 billion euros in 2023. The company is confronted with inventory management challenges (see Table 5) in the seven-digit range and sees the potential of object-centric process mining to address these issues effectively.

The retailer's main goals included addressing key inventory management challenges: *maintaining optimal stock levels*, *reducing stock-outs and overstock*, and *improving the replenishment process*. Ensuring accurate *demand forecasting* was also critical to align stock levels with market demand, reduce costs, and increase supply chain responsiveness. The pet retailer uses SAP ERP for sales (O2C) and purchase orders (P2P), supplemented by Logomate for forecasting and replenishment (see Fig. 1). SAP tracks processes from procurement to sale. Logomate enhances management through data analytics and predictive modeling using historical and real-time data.

6.1 Main Results

Activities Analysis: We transformed the activities as defined in Table 3. Using the *Min-Max* method, we assessed the behavior in the processes and differentiated between positive (normal stock, reduction of understock or overstock), uncertain (risk of understock or overstock), and negative (understock or overstock) situations. As shown in Table 4, the original activities (OCEL A) are listed alongside their transformed counterparts (OCEL B). The transformations highlight when purchase orders and other actions deviate from the optimal reorder point or economic order quantity, leading to imbalances. For instance, we observe that orders are sometimes placed too early or too late, disrupting the desired stock levels. In some cases, the demand forecasting system suggests placing orders beyond the computed Reorder Point and at quantities exceeding the Economic Order Quantity, pushing the material into understock (as the stock is not replenished in time) and eventually overstock (as we receive more materials than needed). We further note instances of goods receipt events when the material was already in an overstock situation, exacerbating excessive inventory levels. This behavior can result in unnecessary holding costs and inefficiencies. The overall impact for all considered materials is summarized in Table 6 (the number of occurrences is approximated for confidentiality reasons).

Other Analyses: We considered applying traditional object-centric process mining analyses, such as the analysis of the control flow using object-centric DFGs [8] and the analysis of interactions between objects [9]. The control-flow analysis was not suitable, as the intertwining of the O2C and P2P processes leads to a quadratic number of flows between different activities. However, by studying interactions between objects, we observed that materials connected to several customers often contain understock, while materials connected to several suppliers are less prone to understock.

6.2 Process Improvements

Based on the findings, we suggested revising demand forecasting models to incorporate the number of customers and suppliers per material as standalone factors. The company should also enhance calculations of *Safety Stock*, *Reorder Points*, and *Economic Order Quantities* in Logomate and SAP to better align inventory with demand, reducing under- and overstock. Additionally, avoiding orders for materials already in overstock and adjusting supplier agreements on minimum order quantities and truck capacity can mitigate oversupply and holding costs.

7 Conclusion and Future Work

This paper proposed an integrated approach to optimize inventory management in retail by combining object-centric process mining with established exact mathematical optimization models.

RQ1 was addressed by introducing an Object-Centric Data Model (OCDM) that holistically captures the interplay of objects and events within O2C and P2P processes.

RQ2 was answered by enriching the OCDM with established inventory management exact mathematical optimization models, enabling data-driven analyses to pinpoint process-related drivers of inventory inefficiencies.

RQ3 was investigated through a case study at a European pet retailer, where the approach identified root causes of under- and overstock and informed actionable improvements, including refining demand forecasting and supplier management practices.

Future work includes integrating advanced data analytics and exploring applications beyond retail, further demonstrating the potential of integrating object-centric process mining and mathematical models for inventory optimization.

References

1. van der Aalst, W.M.P.: Process Mining - Data Science in Action, 2nd edn. Springer (2016)
2. van der Aalst, W.: Object-centric process mining: unraveling the fabric of real processes. Mathematics **11**(12), 1–11 (2023)
3. Alnahas, J.: Application of process mining in logistic processes of manufacturing organizations: a systematic review. Sustainability **15**(15), 11783 (2023)
4. Arnold, D., Isermann, H., Kuhn, A., Tempelmeier, H., Furmans, K.: Handbuch logistik. Springer (2008)
5. Asana, I., Radhitya, M., Widiartha, K., Santika, P., Wiguna, I.: Inventory control using ABC and min-max analysis on retail management information system. In: Journal of Physics: Conference Series, vol. 1469, p. 012097. IOP Publishing (2020)
6. Asdecker, B., Tscherner, M., Kurringer, N., Felch, V.: A dirty little secret? Conducting a systematic literature review regarding overstocks. In: Logistics Management Conference, pp. 229–247. Springer (2023)
7. Balkhi, B., Alshahrani, A., Khan, A.: Just-in-time approach in healthcare inventory management: does it really work? Saudi Pharm. J. **30**(12), 1830–1835 (2022)
8. Berti, A., van der Aalst, W.: OC-PM: analyzing object-centric event logs and process models. Int. J. Softw. Tools Technol. Transf. **25**(1), 1–17 (2023)
9. Berti, A., Herforth, J., Qafari, M.S., van der Aalst, W.M.: Graph-based feature extraction on object-centric event logs. Int. J. Data Sci. Anal. 1–17 (2023)
10. Brabänder, C.: Modell 4: Kontinuierliches Bestandsmanagement. Stochastisches Bestandsmanagement: Grundmodelle für Betriebswirte, pp. 75–92 (2020)
11. Dong, R., Su, F., Yang, S., Xu, L., Cheng, X., Chen, W.: Design and application on metadata management for information supply chain. In: ISCIT 2021, pp. 393–396. IEEE (2016)
12. Hasbullah, H., Santoso, Y.: Overstock improvement by combining forecasting, EOQ, and ROP. Jurnal PASTI **14**(3), 230–242 (2020)
13. Hsiao, J., Lin, C.: A buyer-vendor EOQ model with changeable lead-time in supply chain. Int. J. Adv. Manuf. Technol. **26**, 917–921 (2005)
14. Ivanov, D., Tsipoulanidis, A., Schönberger, J., et al.: Global supply chain and operations management. Springer (2021)
15. Jie, F.: Supply chain design for global competitiveness. In: Proceeding International Conference on Science and Engineering, vol. 1 (2017)
16. Koren, I., Adams, N., Berti, A.: OCEL 2.0 Resources - www.ocel-standard.org. CoRR (2024)
17. Kretzschmann, D., Park, G., Berti, A., van der Aalst, W.M.: Overstock problems in a purchase-to-pay process: an object-centric process mining case study. In: CAiSE 2024 Workshops, pp. 347–359. Springer (2024)
18. Kros, J.F., Falasca, M., Nadler, S.S.: Impact of just-in-time inventory systems on OEM suppliers. Ind. Manag. Data Syst. **106**(2), 224–241 (2006)
19. Kusuma, Y.A.: Supply arrangement of raw material and sugar stock to organize overstock risk in warehouse. In: Journal of Physics: Conference Series, vol. 1375, p. 012048. IOP Publishing (2019)
20. Ma, X., Zeyu, W., Ni, X., Ping, G.: Artificial intelligence-based inventory management for retail supply chain optimization: a case study of customer retention and revenue growth. J. Knowl. Learn. Sci. Technol. **3**(4), 260–273 (2024). ISSN: 2959-6386

21. Mali, N., Bojewar, S.: A survey of ETL tools. Int. J. Comput. Tech. **2**(5), 20–27 (2015)
22. Mekel, C., Anantadjaya, S.P., Lahindah, L.: Stock out analysis: an empirical study on forecasting, re-order point and safety stock level at PT Combiphar, Indonesia. RIBER Rev. Integr. Bus. Econ. Res. **3**(1), 52–64 (2014)
23. Meyer, J.C., Sander, U., Wetzchewald, P.: Bestände senken, Lieferservice steigern: Ansatzpunkt Bestandsmanagement. FIR e. V. an der RWTH Aachen (2019)
24. Mitta, N.R.: Leveraging AI for smart inventory management in retail: developing machine learning models for predictive replenishment, stock optimization, and demand-supply balancing. Aust. J. Mach. Learn. Res. Appl. **4**(2), 113–146 (2024). https://sydneyacademics.com/index.php/ajmlra/article/view/204
25. Pasupuleti, V., Thuraka, B., Kodete, C.S., Malisetty, S.: Enhancing supply chain agility and sustainability through machine learning: optimization techniques for logistics and inventory management. Logistics **8**(3) (2024). https://www.mdpi.com/2305-6290/8/3/73
26. Rahansyah, V.Z., Kusrini, E.: How to reduce overstock inventory: a case study. Int. J. Innov. Sci. Res. Technol. (2023)
27. Ravinder, H., Misra, R.B.: ABC analysis for inventory management: bridging the gap between research and classroom. Am. J. Bus. Educ. (2014)
28. Riza, M., Purba, H.H.: Mukhlisin: the implementation of economic order quantity for reducing inventory cost. Res. Logist. Prod. **8**(3), 207–216 (2018)
29. Rizqi, Z., Khairunisa, A.: Integration of deterministic and probabilistic inventory methods to optimize the balance between overstock and stockout. In: IOP Conference Series: Materials Science and Engineering, vol. 722, p. 012060. IOP Publishing (2020)
30. Sahu, M.K.: Ai-based supply chain optimization in manufacturing: enhancing demand forecasting and inventory management. J. Sci. Technol. **1**(1), 424–464 (2020)
31. Scarf, H.E., Arrow, K., Karlin, S.: A min-max solution of an inventory problem. Rand Corporation Santa Monica (1957)
32. Schuh, G., Gutzlaff, A., Cremer, S., Schoppen, M.: Understanding process mining for data-driven optimization of order processing. Procedia Manufact. **45**, 417–422 (2020)
33. Suryaputri, Z., Gabriel, D.S., Nurcahyo, R.: Integration of ABC-XYZ Analysis in Inventory Management Optimization: A Case Study in the Health Industry (2022)
34. Tempelmeier, H.: Bestandsmanagement in supply chains. BoD–Books on Demand (2005)
35. Waters, D.: Inventory Control and Management, 2nd edn. Wiley (2003)

Finance and Privacy

Cryptocurrency Wallet Security in the Era of Big Data and Quantum Computing: Brute Force Attacks and Blockchain Analysis with Google BigQuery

Patryk Szreder and Piotr Stolarski(✉)

Poznań University of Economics and Business, Poznań, Poland
piotr.stolarski@ue.poznan.pl

Abstract. Cryptocurrency wallets are essential for managing cryptographic keys and interacting with blockchain networks. This paper explores the security of Ethereum wallets, focusing on cryptographic foundations such as elliptic curve cryptography (SECP256K1) and the Keccak-256 hash function, while also analyzing threats posed by emerging technologies like Big Data analytics and quantum computing. Case studies, including the Wintermute hack exploiting vulnerabilities in tools like Profanity, highlight weaknesses in random number generation and vanity address creation. Using Google BigQuery, the study evaluates large-scale blockchain analysis for identifying private keys and examines GPU-based parallel processing for brute-force attacks. While Ethereum's vast key-address space provides strong theoretical security, advancements in quantum and cloud computing introduce new challenges, underscoring the need for continuous monitoring and adaptation.

Keywords: Cryptocurrency wallet security · Brute force attacks · Big Data in blockchain analysis · GPU-based cryptographic attacks · Ethereum vulnerabilities

1 Introduction

Cryptocurrency wallets serve as an essential interface between users and blockchain networks, enabling secure access to digital assets. Initially, these wallets were relatively simple applications, providing basic functionality to generate and store cryptographic keys - comprising private keys for transaction signing and public keys for receiving funds. However, as the cryptocurrency landscape evolved, particularly with the introduction of Ethereum, the role and complexity of wallets expanded significantly.

Ethereum introduced smart contracts, which operate as self-executing agreements encoded directly onto the blockchain. These contracts added a layer of functionality, allowing users to interact with decentralized applications (DApps) and manage a diverse range of digital assets. Consequently, modern wallets have evolved beyond storing cryptocurrencies like Ether (ETH) to managing tokens compliant with standards such as

K. Węcel (Ed.): BIS 2025, LNBIP 554, pp. 197–210, 2025.
https://doi.org/10.1007/978-3-031-94193-1_15

ERC-20 and ERC-721, supporting decentralized finance (DeFi), non-fungible tokens (NFTs), and other blockchain-based assets.

At the heart of all cryptocurrency wallets lies cryptographic key management. This involves deterministic key generation, allowing users to access their wallets from any device or location without loss of functionality. The cornerstone of wallet security is the private key - a randomly generated string of appropriate length that serves as the sole means of accessing stored assets. Paired with this private key is a public key, derived using public-key cryptography (PKC). This asymmetric cryptographic system relies on "trapdoor functions," which are easy to compute in one direction but computationally infeasible to reverse without knowledge of the private key.

As the cryptocurrency ecosystem continues to grow, so too does the need to ensure the security and robustness of wallet technologies. However, vulnerabilities in key generation processes, implementation flaws in tools, and the increasing computational power available for brute force attacks pose serious threats. These risks underscore the importance of ongoing research into the cryptographic underpinnings and practical implementations of cryptocurrency wallets to safeguard user assets and maintain trust in blockchain ecosystems.

1.1 Motivation for the Research

With the growing popularity of cryptocurrencies such as Ethereum and their increasing use in financial and business applications, the security of cryptocurrency wallets is becoming a key aspect of protecting user funds and the integrity of the entire blockchain ecosystem. Developments in technologies such as Big Data, GPUs and Quantum Computing, bring both new opportunities and risks. On the one hand, they enable more sophisticated blockchain analysis and security verification, while on the other hand, they can be used to launch attacks on a massive scale.

The article is motivated by the need to explore the extent to which modern tools and technologies can affect the security of cryptocurrency wallets. Of particular interest is to determine whether the theoretical security of cryptographic keys can be undermined by advanced computing and analytical techniques.

1.2 Research Questions

In view of the trends indicated above in the development of technology and the economic and business knowledge of blockchain applications, this article formulates four key research questions that form the basis for further analysis and discussion:

RQ1: How can Big Data tools, such as Google BigQuery, be used to analyze the blockchain to identify potential vulnerabilities in the key and address space?

RQ2: What are the limitations of these tools in a practical and technological context?

RQ3: What lessons can be drawn from case studies of known attacks, such as the use of the Profanity tool?

RQ4: Can quantum computing be used to launch attacks or to create more advanced protection mechanisms?

Big Data offers new opportunities for blockchain analysis due to its ability to process massive amounts of data in near real-time. Tools such as Google BigQuery allow searching transactions, detecting behavior patterns and potential vulnerabilities in the key and address space. It is crucial to explore whether Big Data can be used to develop effective methods to identify "orphaned" addresses or those that are vulnerable to attacks. The article examines examples of how these tools can be used in practice and their limitations due to the nature of decentralized blockchain.

Despite the promising capabilities of Big Data, these tools have limitations. In the blockchain context, issues such as scalability, performance and privacy pose significant challenges. Analyzing the key and address space requires enormous computing power and appropriate algorithms. In addition, the nature of blockchain's distributed registry makes real-time data acquisition difficult. This article discusses these technological barriers as well as the cost and infrastructure constraints required to effectively use Big Data tools.

The backbone of the security of cryptocurrency wallets are cryptographic algorithms, such as SECP256K1 or Keccak-256. Weaknesses in the implementation of these mechanisms, especially in the generation of private keys, can lead to the compromise of wallets. An example is the Profanity tool, which generated easy-to-predict addresses, which enabled the attack on Wintermute and the loss of $160 million. The analysis of this case provides key lessons on the importance of high-quality random number generators and rigorous testing of cryptographic tools.

Quantum computing is becoming a tool that can be used for both offensive and defensive purposes. These machines harness quantum mechanical properties to perform certain calculations exponentially faster than classical computers, which has profound implications for both attackers and defenders. On the offensive side, quantum computers could potentially break many current encryption standards by solving the underlying mathematical problems more efficiently. However, quantum computing also enables new forms of cryptography that are theoretically immune to quantum attacks, and can be used to develop more secure random number generation and key distribution systems.

The article is structured to address research questions and key findings. Sect. 2 provide background on cryptographic mechanisms, wallet addresses, and blockchain analysis techniques. Section 3 covers the methodology, including the key space analysis tool, its architecture, and the use of BigQuery, along with testing and ethical considerations. Section 4 examines vulnerabilities and attacks, including case studies. Section 5 presents analysis results and discusses quantum computing's impact on security. Section 6 suggests security improvements and future research directions.

2 Background

2.1 Cryptography and Address Creation

Ethereum wallets rely on cryptographic principles to secure transactions and assets. Two key mechanisms ensure this security: the SECP256K1 elliptic curve and the Keccak-256 hashing algorithm. SECP256K1 generates cryptographic key pairs, making it infeasible to derive a private key from a public key. Keccak-256, a cryptographic hash function, ensures data integrity and transforms input into a fixed-length hash, preventing reconstruction of the original data.

Ethereum wallet addresses are derived from a 256-bit random private key, which is transformed via SECP256K1 into a public key. The public key is then hashed with Keccak-256, and the last 20 bytes form the unique Ethereum address. This combination of elliptic curve cryptography and hashing guarantees the uniqueness, security, and integrity of Ethereum wallets.

Standards for Cryptocurrency Address Creation. Cryptocurrency wallets and addresses adhere to established standards to ensure security, usability, and interoperability. These standards have evolved to address technical challenges while simplifying user interactions.

Ethereum addresses consist of 42 characters, beginning with 0x, followed by 40 alphanumeric characters. Although initially case-insensitive, the Ethereum Improvement Proposal (EIP)-55 introduced a checksum mechanism for error detection [2]. This method uses a Keccak-256 hash to encode addresses in a case-sensitive format, helping users identify input errors and improving security.

To simplify wallet usage, BIP-39 standardized mnemonic phrases - 12 to 24 words from a predefined list of 2,048 [3]. These phrases encapsulate private keys in an easy-to-store format, with the final word serving as a checksum for validation. Later, wordlists in multiple languages were added, enhancing accessibility globally while maintaining security.

BIP-32 enabled hierarchical deterministic wallets with a single seed phrase, while BIP-44 expanded support to multiple blockchains. Though implementations vary—Ethereum uses Keccak-256, Bitcoin relies on SHA-256 and RIPEMD-160—standards like EIP-55 and BIP-39 ensure security and interoperability.

2.2 Technologies Supporting Blockchain Analysis

Big Data in Blockchain Analysis. The vast amount of data on blockchains like Ethereum creates opportunities for advanced analysis in areas such as transaction monitoring and vulnerability detection. Tools like Google BigQuery overcome the challenges of traditional methods, such as operating Ethereum nodes, which require significant storage (over 18TB) and suffer from slow data retrieval due to node communication delays.

Google BigQuery is a serverless data warehouse that enables efficient analysis of blockchain data. Since 2018, it has provided daily updates of Ethereum datasets, including a “balances” table with over 280 million records (65GB), offering comprehensive information on active addresses and ETH balances. This allows for streamlined analysis without the need for direct blockchain interaction.

BigQuery enables:

- Transaction Analysis: monitoring fund flows, calculating fees, and studying network congestion.
- Vulnerability Detection: identifying insecure wallets or compromised tools like Profanity by analyzing balance data and transaction patterns.

By offering both direct querying and data export options, BigQuery simplifies access to blockchain insights, making it an essential tool for scalable and efficient blockchain research.

The Role of the GPU in Brute Force Attacks. Graphics Processing Units (GPUs) have emerged as powerful tools for tasks beyond traditional graphics rendering, including computationally intensive operations such as brute force address generation. Their architecture, featuring thousands of cores, enables highly parallelized data processing, which is ideal for generating and testing large numbers of key-address pairs simultaneously. Additionally, GPUs utilize high-speed memory technologies like GDDR or HBM, offering significantly faster performance compared to standard DDR memory used in CPUs, thus further enhancing computational efficiency.

Key to leveraging GPUs for such tasks are frameworks like NVIDIA CUDA (Compute Unified Device Architecture), OpenCL, and ROCm. CUDA, developed by NVIDIA, is particularly notable for enabling direct programming of GPUs to execute numerical calculations with remarkable parallelism. This allows simultaneous execution of multiple operations, significantly accelerating tasks such as sequential private key generation and address hashing.

GPUs' high core count, fast memory, and atomic operations make them ideal for brute-force attacks, particularly in sequential private key generation. Effective use requires optimized parallel programming to maximize throughput. Compared to CPUs, GPUs offer superior performance but demand specialized knowledge. Properly leveraged, they can significantly enhance address generation and validation, underscoring both their potential and risks in cryptocurrency security.

3 Methodology

The proposed keyspace analysis tool is designed to evaluate the security of Ethereum wallets by systematically generating and analyzing key-address pairs. It uses a list of known Ethereum addresses exported from Google BigQuery, which is preprocessed to remove sensitive information like balances and convert the addresses into their corresponding xxHash64 values. This preprocessing ensures efficient storage and faster lookup operations during analysis.

The tool employs a multi-threaded approach, where each thread independently generates private keys, derives public keys and addresses, and compares the hashed addresses with the preloaded hash set of known addresses. If a match is found, the tool outputs the private key and corresponding address for further analysis.

The architecture of the solution consists of three main components:

- Preprocessing Module
- Parallel Key Generation Module
- Match Detection and Reporting

The first module processes and obtains the list of known Ethereum addresses from BigQuery:

```
SELECT address
FROM `bigquery-public-data.crypto_ethereum.balances`
WHERE eth_balance > 0
ORDER BY eth_balance DESC;
```

It then converts each address into its xxHash64 value and stores the resulting hash set in a file, to be loaded by the main analysis tool.

At runtime, the program initializes a parallel region with a specified number of threads. Each thread sequentially generates private keys within a predefined range, derives the associated public key and address, and hashes the address.

The hashed address is then compared against the preloaded hash set. If a hash match is found, the tool records the corresponding private key and address in the output. The tool ensures execution efficiency by avoiding reprocessing of already evaluated keys and halts after reaching the specified maximum range.

Use of BigQuery to Check Key-Address Pairs. Google BigQuery serves as a crucial component in this methodology by providing access to blockchain data in a structured and queryable format. A query is executed to extract a list of Ethereum addresses from the blockchain dataset, omitting balance data to maintain privacy. These addresses are subsequently converted to hash values, ensuring data efficiency and security.

The hashed addresses form the reference dataset against which the tool verifies its generated addresses. BigQuery's scalability and ability to handle large datasets ensure that even extensive blockchain address spaces can be processed and analyzed effectively. This integration enables the identification of vulnerabilities by systematically matching generated keys with real-world blockchain addresses, providing valuable insights into the feasibility of brute-force attacks.

The Process of Testing the Tool Under Laboratory Conditions. The keyspace analysis tool was extensively tested under controlled conditions across various hardware and software configurations. Separate CPU and GPU versions were evaluated to compare performance and resource utilization. Metrics such as key generation rate, memory usage, and computational efficiency were collected to ensure reproducibility. The Results section presents a detailed comparison of CPU and GPU implementations.

Ethical and Practical Limitations of the Study. This study explores cryptocurrency wallet vulnerabilities while adhering to strict ethical guidelines. Research was conducted responsibly, without malicious intent, using real-world blockchain data, though no funds or sensitive information were accessed. Computational limits restricted analysis to a fraction of the Ethereum key-address space, despite GPU acceleration. While controlled testing ensured reliability, real-world factors like network latency and hardware variability were not fully accounted for. These findings emphasize the need for ethical research practices and awareness of risks posed by advanced computational technologies.

4 Vulnerability Cases and Known Attacks

4.1 Analysis of the Profanity Tool and Its Vulnerabilities

The Profanity tool was designed to generate Ethereum vanity addresses - customized addresses containing specific character patterns. To achieve this, it employed a performance-optimized algorithm capable of generating large volumes of addresses rapidly. However, this optimization came at the cost of security, as the tool relied on a flawed random number generation (RNG) process. Instead of generating cryptographically secure private keys, Profanity used a deterministic approach that significantly reduced the effective keyspace, making it susceptible to brute-force attacks.

A major real-world exploit of this vulnerability occurred in September 2022, when the algorithmic trading firm Wintermute suffered a $160 million loss. Attackers were able to reconstruct private keys for addresses generated with Profanity, ultimately draining assets from compromised wallets. The root cause of this breach was the predictable nature of the random values used in key generation, which allowed attackers to precompute potential keys efficiently [4].

4.2 Other Known Attacks and Lessons Learned

In addition to the vulnerabilities exposed by the Profanity tool, several other attacks have targeted poorly secured cryptocurrency wallets. One of the most well-documented methods is brute force attacks on wallets that use weak private key generation mechanisms. These attacks exploit insufficient entropy in key creation, making it feasible to guess private keys through systematic or parallelized computations, often leveraging GPU acceleration.

A notable example is the case of wallets generated from predictable seeds, such as those derived from low-entropy sources like timestamps or user-defined passphrases. Attackers have successfully scanned the key space for such predictable wallets, draining their funds once a match is found.

The Profanity attack highlights the risks of weak RNGs and vanity addresses in cryptocurrency security. Wallets must use secure cryptographic libraries and TRNGs, while continuous audits and best practices in key management are essential. Developers should prioritize security, and users must avoid unverified key generation tools.

5 Results and Discussion

As mentioned earlier, each address within the Ethereum network is represented by a 160-bit string, defining the total address space. Under idealized conditions, assuming a uniform distribution of addresses, the expected number of attempts required to find a match can be approximated by the ratio of the total address space to the number of existing addresses. Equation (1) shows the expected value.

$$\frac{2^{160}}{2.8 \times 10^{8}} = 5.21 \times 10^{39} \quad (1)$$

Figure 1 illustrates the research pipeline designed for the efficient generation and verification of key-address pairs.

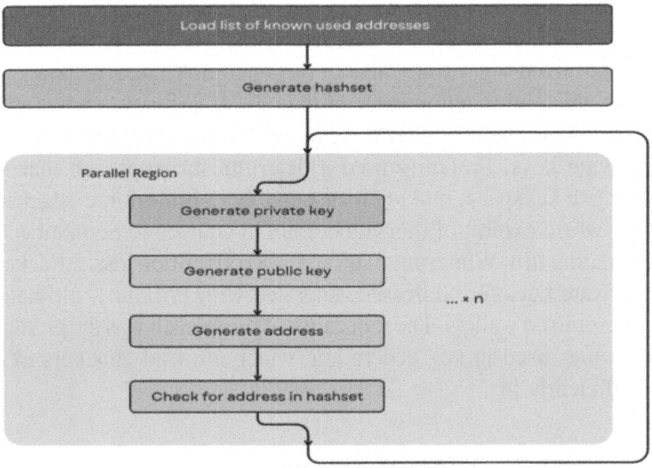

Fig. 1. The research pipeline for the efficient addresses brute-force analysis.

The pipeline is structured to implement the methodology outlined earlier, ensuring a systematic approach to security risks. It incorporates previously analyzed attack variants, allowing for a comprehensive assessment of potential threats and their impact on cryptographic key integrity within the Ethereum network.

The tests were conducted on two types of computational infrastructure: one relying solely on CPUs and another utilizing various GPUs. The results clearly indicate the superior efficiency of GPU-based computation in address generation.

The implementations were created in C and CUDA programming languages, which allow for code portability and relatively high speeds of execution. Additionally, a hash set code based on warpcore [5, 6] was implemented. The code was designed to be compatible with all hardware configurations that support a functional C compiler and dynamic memory allocation, provided that there is sufficient system memory available for the hash set.

5.1 GPU Tests

Table 1 presents the system configurations with various GPUs tested in this study.

The comparison of tested architectures is given in Fig. 2. Among the tested GPUs, the RTX 6000 Ada and RTX 4090 achieved the best results, generating 48.53 and 47.21

Table 1. GPU equipment and configurations used in the study

GPU	Operating System	Power draw (datasheet) [W]
RTX 2080 Super	Ubuntu 22.04 WSL	220
RTX A4000	Ubuntu 22.04	130
RTX A6000	Ubuntu 22.04 WSL	290
RTX 4090	Ubuntu 22.04 WSL	420
RTX 6000 Ada Generation	Ubuntu 22.04 WSL	300

million addresses per second, respectively. The RTX A6000, despite featuring 75% more CUDA cores than the A4000, only outperformed it by 65%, illustrating how other factors, such as memory bandwidth, clock speeds, and power limits, influence overall performance. Notably, the RTX 6000 Ada's potential was restricted by a lower power limit (300W vs. 420W for the RTX 4090), despite having 11% more cores. These findings underscore that while core count is a critical factor, energy constraints also play a crucial role in maximizing GPU efficiency. The best CPU tested, the i9-13900KF, while reaching 865 thousand addresses per second, was vastly outperformed by all the GPUs, even the slowest RTX 2080 Super.

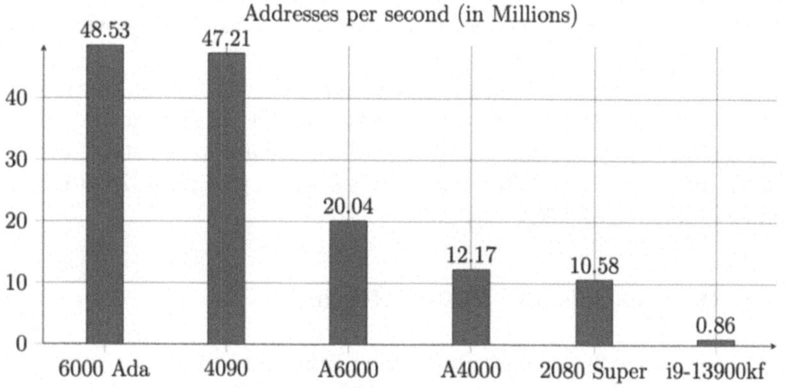

Fig. 2. Comparison of performance results obtained during brute force attack tests using GPU.

The results leave no illusions that accessing all the cryptocurrency wallets on the Ethereum network, is a major challenge. Assuming the use of the fastest of the tested devices, the RTX 6000 Ada Generation, checking all the possible wallet addresses that could be generated would take:

$$\frac{2^{160}}{48.53 \times 10^6} \approx 9.53 \times 10^{32}[yr] \quad (2)$$

The age of the universe has been estimated to be $13.8 \times 10^9 years$ [7] which means, that checking every possible Ethereum wallet address, using home-grade hardware, would take 23 orders of magnitude longer than the age of our universe.

5.2 FPGA and ASIC

ASIC (Application-Specific Integrated Circuit) and FPGA (Field-Programmable Gate Array) are specialized hardware devices designed for high-performance computing, with ASICs optimized for specific tasks and FPGAs offering reconfigurable logic for adaptable processing. They outperform CPUs and GPUs in power and efficiency by using specialized circuits tailored for cryptocurrency algorithms like SHA-256 (Bitcoin) and Ethash (Ethereum). Their capabilities could be leveraged to optimize key-address pair generation by efficiently implementing Keccak and secp256k1 algorithms.

When it comes to the Keccak algorithm, as it shares many similarities with SHA-3, it has been well-studied in the area of hardware optimization. As cryptocurrencies that utilize it as their hashing functions emerged, interest in making FPGA and ASIC solutions for efficient mining peaked. It has been shown that Keccak is well-suited for efficient hardware implementations. However, such an approach has also some drawbacks [8]. There are already FPGA solutions that allow the achieving of Keccak256 hashing with performance at the level of giga hashes per second [9].

Similar to Keccak, secp256k1, as it is fundamental in cryptocurrencies, has also been studied in terms of implementing it as FPGA/ASIC rather than computer code. It turns out, that elliptic curve cryptography can be implemented as an ASIC with little to no issues [10].

One of the world's largest FPGA manufacturers, AMD (formerly Xilinx) with their Vitis Libraries offers off-the-shelf solutions for multiple hashing and signing algorithms, including secp256k1 and keccak-256. Their documentation states, that signing a secp256k1 message, takes 0.2ms, and calculating a single keccak-256 (SHA-3 with appropriate padding) hash takes only 3.2ns.

6 Security Proposals and Future Research

6.1 Quantum Algorithms for Brute-Forcing Ethereum Addresses

Quantum computing leverages principles such as superposition, entanglement, and interference to solve specific computational problems more efficiently than classical systems. For the task of brute-forcing Ethereum private keys, Grover's algorithm offers a quadratic speedup, reducing the complexity of the search problem from $O(2^n)$ to $O(2^{\frac{n}{2}})$, where $n = 256$ for Ethereum private keys. This section explores the feasibility of using quantum algorithms to generate and verify Ethereum addresses against a list of known addresses.

Grover's Algorithm and Key Search. Grover's algorithm [11] addresses unstructured search problems by amplifying the probability of correct solutions through quantum superposition and interference. The process involves:

1. Initialization: A uniform superposition is created over all 2^{256} possible private keys.

2. Oracle: A quantum oracle flips the amplitude's sign for a "marked" state, representing a match with a known Ethereum address.
3. Amplitude Amplification: Grover's diffusion operator amplifies the amplitude of the marked state:

$$\text{New amplitude} = 2 \cdot \text{Mean amplitude} - \text{Current amplitude} \quad (3)$$

4. Measurement: After $O\left(\sqrt{2^{256}}\right) = 2^{128}$ iterations, the state is measured, collapsing the superposition to the most probable solution.

Address Verification Using Quantum Search. Once a private key is guessed, deriving its corresponding Ethereum address involves elliptic curve multiplication (SECP256k1) and hashing (Keccak-256). Verifying the derived address against a list of known Ethereum addresses can also benefit from Grover's algorithm:

1. For a list of M known addresses, Grover's algorithm reduces the search complexity from $O(M)$ to $O(\sqrt{M})$.
2. For large address databases ($M \sim 10^9$), this offers a practical advantage in address matching, assuming efficient quantum implementations of cryptographic functions.

Limitations and Practical Feasibility. Despite the theoretical advantages, several challenges hinder practical implementation:

- Resource Requirements: Grover's algorithm for 256-bit keys would require millions of logical qubits, far beyond the capabilities of current quantum computers [12].
- Elliptic Curve and Hashing Overhead: Quantum circuits for SECP256k1 and Keccak-256 are computationally expensive and do not inherently benefit from Grover's speedup.
- Energy Costs: The computational energy for 2^{128} operations remains prohibitive, even with advancements in quantum hardware.

While Grover's algorithm demonstrates a theoretical reduction in the computational complexity of brute-forcing Ethereum private keys and address verification, the practical application remains theoretically infeasible with current and near-future quantum computing technologies.

6.2 Mitigations Against Quantum Risks in Ethereum

While currently quantum computing is not feasible for the task of brute-forcing key-address space, the eventual development of scalable quantum computers poses a significant risk to cryptographic systems underlying Ethereum and other blockchain platforms. Specifically, Grover's algorithm can reduce brute-force complexity for private keys and address-space attacks, while Shor's algorithm threatens the foundational elliptic curve cryptography - it has been shown that multiple cryptocurrencies utilizing ECDSA are vulnerable to attacks using Shor and Grover algorithms [13, 14]. This section outlines possible mitigations designed to protect Ethereum's security against such threats.

Transition to Post-quantum Cryptography. The most effective long-term mitigation against quantum attacks is to adopt post-quantum cryptography (PQC) - cryptographic schemes resistant to both Grover's and Shor's algorithms. The National Institute of Standards and Technology (NIST) is actively standardizing PQC algorithms, including lattice-based, hash-based, and code-based approaches [15]. For Ethereum, replacing SECP256k1 with a post-quantum algorithm would require modifications at both the protocol and client levels. Some recommended PQC approaches include:

- Lattice-based Cryptography (e.g., CRYSTALS-Dilithium, Kyber): These schemes rely on the hardness of lattice problems and are efficient for both key generation and verification.
- Hash-based Cryptography (e.g., XMSS, SPHINCS +): Hash-based digital signatures offer provable security and minimal reliance on untested assumptions, making them a viable replacement for ECC.
- Hybrid Cryptographic Schemes: To ensure compatibility with current systems, hybrid approaches combining ECC with post-quantum algorithms can be deployed during a transition phase.

Challenges in adopting PQC include increased key sizes, signature sizes, and computational overhead, which may require further optimization to maintain the performance of blockchain systems.

In addition to updating cryptographic primitives, alternative blockchain designs tailored to quantum security, like quantum-secure consensus mechanisms can be implemented. Protocols like proof-of-space-time (PoST) and proof-of-stake (PoS) can be adapted to incorporate post-quantum algorithms in block validation and consensus.

Quantum attacks such as those facilitated by Shor's algorithm require access to the public key. For Ethereum, public keys are exposed when funds are spent from an address. To mitigate this risk, Ethereum users can employ a one-time address model, where each transaction uses a newly generated address - this ensures private keys are not vulnerable to quantum attacks until after their associated address has been used.

6.3 Hash Collisions

Ethereum, due to algorithm of address generation, is subject to collisions. The address space for Ethereum is 2^{160} whereas the private key space is $2,^{256}$ which means that 2^{96} private keys will produce collision. In other words, they will result in the same address that has been produced before. To gain access to all the wallets, a malicious actor theoretically has to check "only" 2^{160} private keys, but not necessarily so.

6.4 Literature Review

Recent studies highlight the growing security challenges in cryptocurrency wallets, particularly in the context of brute-force attacks, blockchain vulnerabilities, and emerging quantum threats. A systematic review categorizes cryptocurrency wallet vulnerabilities across multiple layers, revealing a gap between theoretical countermeasures and their practical adoption [20]. Research on password-based wallet security demonstrates how weak password policies can expose wallets to brute-force attacks, emphasizing the need for robust key management mechanisms and secure cryptographic implementations [16]. Further work explores the detection of brute-force attacks in blockchain networks and proposes modifications to cryptocurrency protocols to mitigate such threats [17]. In addition to brute-force attacks, studies on Bitcoin payment systems at retail locations examine double-spending risks and propose countermeasures to enhance transaction security [19]. Another critical concern is the potential impact of quantum computing, which could undermine existing cryptographic protections in blockchain networks, necessitating new security frameworks and transition strategies [18]. These findings underscore the necessity for continuous research and adaptation of security practices to address evolving threats in the cryptocurrency ecosystem.

6.5 Future Works

While current technology lacks the scale and performance to brute-force Ethereum keys, it might change with future advancements. Post-quantum cryptographic transitions, such as lattice-based or hybrid approaches, can be explored in order to improve blockchain security. Research might focus on optimizing quantum circuits for SECP256k1 and Keccak-256, reducing qubit requirements. Efficient FPGA/ASIC implementations could accelerate key-address generation, warranting further study. Finally, Ethereum's problem with hash collisions should be analyzed for potential mitigations in next-generation cryptographic designs.

Disclosure of Interests. The authors have no competing interests to declare that are relevant to the content of this article.

References

1. Bi, W., Jia, X., Zheng, M.: A secure multiple elliptic curves digital signature algorithm for Blockchain. arXiv preprint: arXiv:1808.02988 (2018)
2. Buterin, V., Van de Sande, A.: ERC-55: mixed-case checksum address encoding. Ethereum Improv. Proposals **55** (2016). https://eips.ethereum.org/EIPS/eip-55
3. Palatinus, M., Rusnak, P., Voisine, A., Bowe, S.: Mnemonic code for generating deterministic keys. GitHub (2013). https://github.com/bitcoin/bips/blob/master/bip-0039.mediawiki
4. Private key safety. GitHub Issue Comment (2022). https://github.com/johguse/profanity/issues/61#issuecomment-1017935267
5. Jünger, D., et al.: WarpCore: a library for fast hash tables on GPUs. In: 27th IEEE International Conference on High Performance Computing, Data, and Analytics (HiPC 2020), pp. 11–20. IEEE (2020). https://doi.org/10.1109/HiPC50609.2020.00015
6. Jünger, D., Hundt, C., Schmidt, B.: WarpDrive: massively parallel hashing on multi-GPU nodes. In: 2018 IEEE International Parallel and Distributed Processing Symposium (IPDPS 2018), pp. 441–450. IEEE (2018). https://doi.org/10.1109/IPDPS.2018.00054
7. Aghanim, N., et al.: Planck Collaboration. Planck 2018 results - VI. Cosmological parameters. A&A 641, A6 (2020). https://doi.org/10.1051/0004-6361/201833910
8. Honda, T., Guntur, H., Satoh, A.: FPGA implementation of new standard hash function Keccak. In: 2014 IEEE 3rd Global Conference on Consumer Electronics (GCCE 2014), pp. 275–279 (2015). https://doi.org/10.1109/GCCE.2014.7031105

9. Pereira, F.D., Ordonez, E.D.M., Sakai, I.D., Souza, A.M.: Exploiting parallelism on keccak: FPGA and GPU comparison. In: Conference Proceedings (2013). https://api.semanticscholar.org/CorpusID:18708592
10. Pirotte, N., Vliegen, J., Batina, L., Mentens, N.: Design of a fully balanced ASIC coprocessor implementing complete addition formulas on Weierstrass elliptic curves. In: 2018 21st Euromicro Conference on Digital System Design (DSD), pp. 545–552. IEEE (2018). https://doi.org/10.1109/DSD.2018.00095
11. Grover, L.K.: A fast quantum mechanical algorithm for database search. In: Proceedings of the 28th Annual ACM Symposium on Theory of Computing (STOC '96), pp. 212–219. ACM (1996). https://doi.org/10.1145/237814.237866
12. Gidney, C., Ekerå, M.: How to factor 2048-bit RSA integers in 8 hours using 20 million noisy qubits. Quantum **5**, 433 (2021). arXiv:1905.09749 [quant-ph], https://doi.org/10.48550/arXiv.1905.09749
13. Kearney, J.J., Perez-Delgado, C.A.: Vulnerability of Blockchain technologies to quantum attacks. Array **10**, 100065 (2021). https://doi.org/10.1016/j.array.2021.100065
14. Holmes, S., Chen, L.: Assessment of quantum threat to bitcoin and derived cryptocurrencies. Cryptology ePrint Archive, Paper 2021/967 (2021). https://eprint.iacr.org/2021/967
15. National Institute of Standards and Technology (NIST): Post-Quantum Cryptography Standardization Project (2022). https://csrc.nist.gov/projects/post-quantum-cryptography
16. Byun, H., Kim, J., Jeong, Y., Seok, B., Gong, S., Lee, C.: A security analysis of cryptocurrency wallets against password brute-force attacks. Electronics **13**(13), 2433 (2024). https://doi.org/10.3390/electronics13132433
17. Kiktenko, E.O., Kudinov, M.A., Fedorov, A.K.: Detecting brute-force attacks on cryptocurrency wallets. In: International Conference on Business Information Systems, pp. 232–242. Springer International Publishing, Cham (2019)
18. Schärer, K., Comuzzi, M.: The quantum threat to Blockchain: summary and timeline analysis. Quantum Mach. Intell. **5**, 19 (2023). https://doi.org/10.1007/s42484-023-00105-4
19. Masteika, S., Rebždys, E., Driaunys, K., Šapkauskienė, A., Mačerinskienė, A., Krampas, E.: Bitcoin double-spending risk and countermeasures at physical retail locations. Int. J. Inf. Manage. **79**, 102727 (2024). https://doi.org/10.1016/j.ijinfomgt.2023.102727
20. Houy, S., Schmid, P., Bartel, A.: Security aspects of cryptocurrency wallets—a systematic literature review. ACM Comput. Surv. **56**(1), 1–31 (2023)

Compliance On-Chain Framework for Decentralized Finance to Meet AML/CTF Requirements While Preserving Privacy, Security and Decentralization

Gintarė Košubienė(✉) and Saulius Masteika

Vilnius University, Muitines 8, 44280 Kaunas, Lithuania
{gintare.kosubiene,saulius.masteika}@knf.vu.lt

Abstract. This paper presents the Compliance On-Chain Framework (CCF), a hybrid compliance solution designed to fulfill Anti-Money Laundering (AML) and Counter-Terrorism Financing (CTF) requirements in decentralized finance (DeFi) while upholding privacy, security, and decentralization principles. The framework leverages decentralized identity ERC-734/ERC-735 standards, to integrate off-chain trusted data providers (TDPs) and on-chain mechanisms for issuing verifiable credentials. A key innovation of the CCF is its emphasis on continuous compliance through automated validity checks and real-time updates, addressing critical gaps in current compliance approaches. The framework's user-centric design empowers users to demonstrate regulatory adherence without compromising decentralization. Initial testing on an EVM-compatible testnet validates the framework's feasibility, scalability, and potential for broader real-world adoption, paving the way for innovative compliance solutions in the DeFi ecosystem.

Keywords: AML/CTF Compliance · Decentralized Finance · ERC-734/ERC-735

1 Introduction

Decentralized Finance (DeFi) utilizes distributed ledger technology (DLT) and smart contracts to enable peer-to-peer financial transactions without intermediaries, offering users enhanced autonomy and global accessibility [1]. With a compound annual growth rate (CAGR) of 46%, DeFi is projected to surpass $450 billion by 2030 [2], addressing inefficiencies in traditional finance such as high costs and exclusion of underbanked populations [3].

However, this rapid growth has also heightened regulatory concerns, including money laundering and terrorism financing. Regulatory bodies are increasingly scrutinizing DeFi platforms to combat financial crime, ensure market integrity, and protect investors through measures like Know Your Customer (KYC) and Anti-Money Laundering (AML) requirements. The International Organization of Securities Commissions

K. Węcel (Ed.): BIS 2025, LNBIP 554, pp. 211–224, 2025.
https://doi.org/10.1007/978-3-031-94193-1_16

(IOSCO) emphasizes global regulatory consistency, publishing 2023 guidelines Recommendation 3 – Achieve Common Standards of Regulatory Outcomes to address market integrity and investor protection [1]. Similarly, the Financial Action Task Force (FATF) high-lights compliance challenges in DeFi, particularly for systems involving centralized elements, urging adherence to AML/CTF standards and reporting obligations [4]. In the EU, the 2023 Markets in Crypto-Assets (MiCA) regulation offers a comprehensive framework for crypto-asset service providers, promoting transparency and compliance while supporting innovation [5].

Despite these efforts, uncertainties persist due to jurisdictional complexities, risks from smart contracts, and the rapid pace of DeFi innovation [3]. The following section will examine the AML/CTF requirements, and the challenges associated with meeting them in Decentralized Finance.

1.1 AML/CTF Requirements and Challenges Meeting Them in DeFi

The following AML/CTF requirements (R1-R9) were derived from The Association of Certified Anti-Money Laundering Specialists (ACAMS) guidance [8] and refined through collaboration with a legal firm specializing in crypto compliance. To ensure clarity and alignment with industry standards, we group these requirements into five categories recommended by ACAMS. Each category captures a distinct function within a comprehensive compliance framework:

- **Risk-management practices** encompass strategic-level controls that shape an organization's approach to AML/CTF, including AML policy creation (R1), high-risk customer handling or enhanced due diligence (EDD) (R2) and termination of non-compliant relationship (R3).
- **Internal controls** involve technical and operational mechanisms used to detect and prevent illicit activity in real time, such as transaction monitoring (R4) and sanction screening (R5).
- **Customer due diligence (CDD)** refers to both initial Know Your Customer (KYC) (R6) and ongoing identity verification and risk assessment processes (R7), which are critical in mitigating exposure to money laundering.
- **Reporting and record-keeping systems** are essential for transparency and accountability, it includes reporting suspicious transactions to authorities (SARs) (R8) and secure storage of customer and transaction records for audits and investigations (R9).
- **Monitoring and managing compliance** serve as the overarching function ensuring that policies, controls, and due diligence efforts are consistently enforced and updated.

While these components are well-addressed in traditional finance, DeFi faces significant difficulties in balancing AML/CTF requirements with main issues (I1-I3) illustrated in Fig. 1: decentralization and trusted data availability, data security, and privacy.

I1. **Decentralization, Trusted Data Availability**. DeFi's decentralized architecture lacks intermediaries, making it difficult to manage off-chain data [9]. Embedding compliance-related information in smart contracts requires mechanisms that ensure data integrity while preserving decentralization. It directly affects requirements relying on external data sources, such as transaction monitoring (R4), sanction screening (R5), ongoing identity verification (R7) and EDD (R2).

I2. **Security Vulnerabilities**. The immutable and transparent nature of blockchains exposes sensitive on-chain data to potential breaches [12, 13]. It compromises record keeping (R9), KYC (R6) and ongoing CDD (R7).

I3. **Privacy Concerns**. Users value anonymity and often resist sharing personal data. Compliance must ensure disclosures occur only under clear protocols for legal investigations, balancing user trust with regulatory obligations [16]. Privacy concerns create tension with identity-related requirements like EDD (R2), initial KYC (R6), and ongoing CDD (R7), where users may resist disclosing personal information.

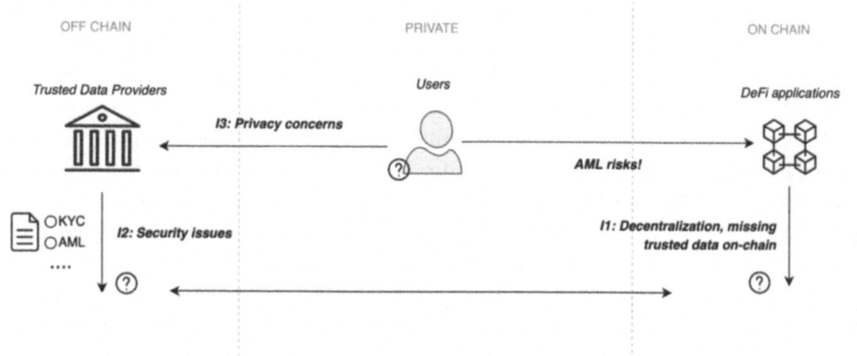

Fig. 1. Key issues in DeFi compliance: data security, privacy concerns and decentralization.

These interdependencies highlight the need for privacy-preserving, secure, and decentralized compliance solutions in DeFi. In the next section, we will review existing compliance technologies and frameworks that aim to address the outlined AML/CTF challenges in decentralized finance, highlighting their key contributions, limitations, and areas requiring further innovation.

1.2 Related Works

Recent advancements in AML/CTF compliance technologies have emphasized innovations in Know Your Customer (KYC) processes and Self-Sovereign Identity (SSI) frameworks [2, 7, 10, 14]. These approaches aim to enhance security, efficiency, and user privacy in identity verification while reducing redundancies.

For example, Glucksman (2024) presents a comparative analysis of existing AML technologies in DeFi, including solutions like SelfKey, uPort, and Sovrn [6]. His research also discusses privacy-preserving technologies such as Zero-Knowledge Proofs (ZKPs) and their integration into smart contracts. However, Glucksman concludes that most existing frameworks focus primarily on KYC, lacking comprehensive coverage of broader AML processes tailored to decentralized platforms.

Sun N. (2024) proposed a universal schema for privacy-preserving KYC compliance across public blockchains [14]. Their Ethereum-based solution employs Merkle trees to establish a unified identity structure supporting interoperability. While this represents a step forward, the framework's narrow focus on KYC identification leaves significant gaps in continuous AML compliance (Sect. 1.1., R2, R4, R5, R7).

Burleson, Korver, and Boneh (2024) explore the application of ZKPs for bridging privacy with regulatory requirements [15]. Their approach incorporates deposit and withdrawal screening mechanisms alongside blocklists and allowlists, facilitating real-time compliance (Sect. 1.1., R4) while preserving user privacy. A key innovation in their model is selective de-anonymization, which grants regulators-controlled access to transaction details without exposing broader user data.

Vitalik Buterin's concept of Privacy Pools (2024) demonstrates the potential of ZK-SNARKs in balancing privacy and compliance [11]. This protocol allows users to prove their funds are not linked to illicit activity by associating their transactions with "good" sets while excluding them from "bad" sets. However, practical challenges include the reliability of association set providers (ASPs) and the risk of centralization, which could undermine the protocol's alignment with decentralized principles.

Despite these innovations, existing frameworks often focus on isolated AML/CTF aspects, lacking comprehensive solutions for DeFi. KYC processes and privacy mechanisms often dominate, with limited attention to constant transaction monitoring or continuous compliance updates. Furthermore, the decentralized and pseudonymous nature of DeFi creates barriers to integrating traditional compliance measures effectively.

To address these gaps, this paper proposes the Compliance On-Chain Framework (CCF), a hybrid solution that combines decentralized identity standards with automated compliance mechanisms. By incorporating both on-chain and off-chain components, the CCF aims to deliver a comprehensive, scalable, and privacy-preserving solution for AML/CTF compliance in DeFi.

1.3 Hypothesis and Scope of This Paper

Decentralized Finance (DeFi) leverages blockchain technology to create autonomous, transparent ecosystems. Despite its rapid growth and potential, DeFi faces regulatory challenges in meeting AML/CTF requirements (Sect. 1.1). Its non-custodial, decentralized structure complicates compliance, raising issues of data integrity, privacy, and security (Sect. 1.1). Regulatory frameworks such as those proposed by FATF, MiCA, and IOSCO emphasize the need for mechanisms that balance compliance with DeFi's core principles. However, technical and operational challenges for comprehensive ongoing compliance processes remain unresolved (Sect. 1.2), necessitating innovative solutions. These concerns about balancing regulatory compliance with decentralization principles prompt the following hypothesis.

Hypothesis. A blockchain-based compliance framework combining both on-chain and off-chain mechanisms can meet AML/CTF requirements in DeFi while preserving privacy, data security, and decentralization ethos.

The methodology for evaluating the proposed Compliance On-Chain Framework (CCF) is structured to analyze its ability to meet AML/CTF requirements in DeFi while addressing key challenges of privacy, security, and decentralization (Sect. 1.1).

The proposed CCF is designed to address these challenges by providing a comprehensive solution that ensures AML/CTF compliance in DeFi ecosystems. The methodology begins with a comparative review of existing tools to identify gaps and evaluate their advantages in addressing AML/CTF requirements (Sect. 2). Key metrics include coverage of AML/CTF requirements (R1-R9) and the resolution of key issues (I1-I3) to DeFi.

To address the identified gaps, the CCF is developed (Sect. 3) and tested via simulations on the Arbitrum Sepolia Testnet (Sect. 4), focusing on core AML/CTF activities: initial KYC, sanctions screening management, ongoing CDD, and identity integrity are evaluated for robustness, scalability, and reliability.

2 Overview of Existing Compliance Tools

The steady growth of blockchain-based AML solutions is driven not only by increasing regulatory scrutiny and the broader adoption of cryptocurrencies but also by the critical need to safeguard individual cryptoassets [4]. These solutions vary in design, functionality, and coverage of certain AML/CTF activities, illustrating a fragmented compliance landscape. Table 1 categorizes and highlights key approaches.

Table 1. Categories of existing compliance tools and their approaches.

Category	Solutions	Key Features	Differentiation	AML/CTF Requirements	Issues for DeFi
Blockchain data analysis & intelligence	Chainalysis, Scorechain, Elliptic, Crystal Blockchain	Integration with financial institutions, real-time transaction monitoring, manual investigation tool	Prioritize regulatory alignment, may trade off user's privacy and decentralization	For financial institutions mainly: R1, R3–R5, R9	I1, I3
Self-Sovereign identity frameworks	Sovrin, uPort, KYCT, Fractal ID (idos), OxKYC.id, Outdid, zk.me, HashKeyDID	Verifiable credentials, privacy-preserving mechanisms	Excels in user's privacy	For CeDeFi mainly: R6	I2
Hybrid solutions	Quadrata, Nexera, Anima, Violet, micapass	DID issuance, KYC/AML person and crypto AML compatibility	Balance privacy and compliance requirements	For DeFi and CeDeFi partially: R1-R8	Underdeveloped, not fully ready yet

Blockchain data analysis solutions prioritize regulatory alignment and transaction monitoring but often compromise privacy (I1) and decentralization (I3) (see Table 1). These tools rely on centralized mechanisms that conflict with the decentralized ethos of blockchain systems, leading to concerns about trusted data availability and user privacy.

Conversely, SSI frameworks employ verifiable credentials and privacy-preserving mechanisms to enhance user control over personal data. These solutions excel in protecting user privacy but often fail to integrate comprehensive AML/CTF processes, limiting their effectiveness for broader compliance needs (see Table 1).

Hybrid solutions combine on-chain and off-chain mechanisms to balance privacy and compliance. These models represent a promising middle ground, integrating decentralized identity solutions with AML/CTF processes to meet regulatory requirements while respecting user autonomy (see Table 1). On another hand, hybrid solutions remain underdeveloped, requiring further innovation to address the operational complexities of DeFi, such as real-time compliance monitoring and seamless scalability.

The review of existing compliance tools reveals a significant reliance on centralized mechanisms that prioritize regulatory alignment at the expense of privacy and decentralization. Hybrid models appear to hold the most promise for addressing these gaps, yet their evolution requires further research and refinement. Achieving comprehensive AML/CTF compliance in DeFi demands not only technological innovation but also thoughtful integration of these emerging solutions into a unified framework.

The Compliance On-Chain Framework (CCF) leverages ERC-734/ERC-735 standards, combining SSI frameworks and hybrid solutions. It ensures continuous compliance through automated checks and scalability, aligning regulatory needs with decentralized principles. The design and implementation are detailed in the following sections.

3 Proposed Compliance On-Chain Framework

In this section we provide a high-level technical architecture, discuss smart contracts building blocks and general process flow of our proposed Compliance On-Chain Framework (CCF) tailored specifically for decentralized finance participants.

3.1 High Level User Centric Architecture and Process Flow

The architecture of proposed Compliance On-Chain Framework (CCF) (see Fig. 2) ensures verified credentials exchanges among off-chain trusted data providers (TDPs), private user-controlled Self-Sovereign Identity (SSI), and decentralized on-chain applications (dApps).

In the **public off-chain area**, TDPs such as Ondato or Chainalysis act as issuers of verified credentials (VCs) required for compliance. Users begin by creating a self-sovereign digital identity linked to their crypto wallet (step 1). This identity serves as the foundation for all compliance-related activities. Users select their preferred TDPs (e.g., there could be many providers for KYC) based on their trust or jurisdictional requirements (step 2). The chosen TDPs verify user identity and compliance data, such as KYC or AML checks, and issue credentials via CCF (step 3). The framework ensures that proofs of verification (rather than raw data) are stored on-chain as verifiable credentials until the user decides to save them into his digital identity (step 4). This approach enables transparency while safeguarding sensitive information.

The **private ecosystem** leverages SSI, enabling users to own and control their data (see Fig. 2). The SSI is created only once and serves as the central repository for credentials, ensuring that users retain full ownership of their identity (step 1). Through selective data sharing (step 4), users can manage proofs of compliance without exposing raw or sensitive data. For example, users may share evidence of KYC compliance or AML risk scoring without revealing their full personal information. The SSI acts as a bridge between off-chain TDPs and on-chain dApps, facilitating seamless interaction while preserving privacy and decentralization principles.

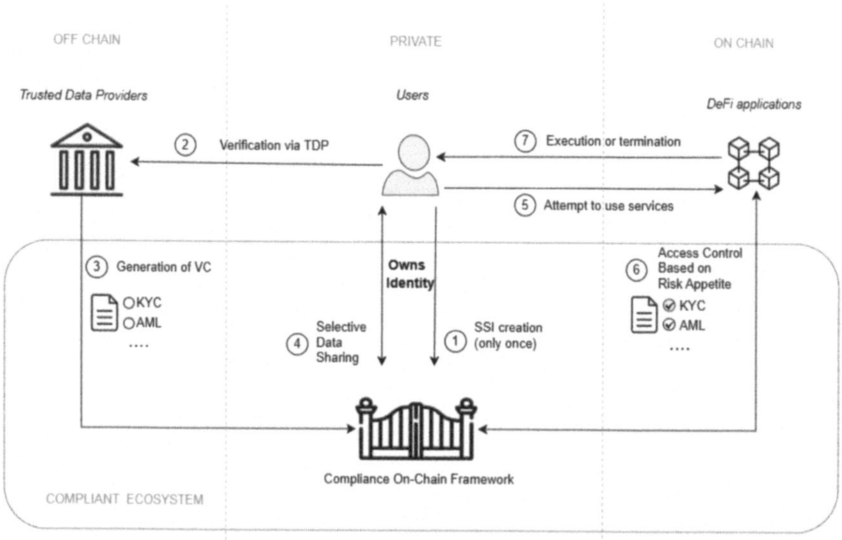

Fig. 2. High level architecture of Compliance On-Chain Framework.

In the **on-chain ecosystem** (see Fig. 2), dApps evaluate the user's credentials to determine access permissions (step 5, 7). This process enables real-time compliance while ensuring that sensitive data remains off-chain. dApps apply risk-based access control mechanisms (step 6) to allow or deny user interaction based on the shared credentials, ensuring alignment with regulatory requirements. For instance, small deposits may require sanctions screening, while withdrawals above a threshold might necessitate full CDD. These rules are customizable and vary depending on the type of DeFi application, jurisdiction, DAO governance, or other. In exceptional cases, such as regulatory investigations, the framework allows controlled user data disclosure to authorized regulators with consent from the user, CCF, and credential issuer, ensuring privacy compliance and trust.

The core architectural components of the Compliance On-Chain Framework (CCF) include a backend server, a database, and a web3 library for seamless frontend integration. These components and their primary functionalities are summarized in the sequence diagram (see Fig. 5) without delving into details. As the framework's core functionality relies heavily on on-chain mechanisms, the following section provides an exploration of the CCF's smart contract architecture.

3.2 CCF Smart Contracts Based on Identity Standards ERC-734/ERC-735

The Compliance On-Chain Framework (CCF) is built on ERC-734/ERC-735 standards. Its modular design ensures adaptability to evolving regulations, aligning with ERC-734's goal of providing a robust, widely compatible solution [18]. Simplified representations of CCF core entities are shown in the schema for clarity (see Fig. 3.).

As shown in Fig. 3, the `CCF`[1] smart contract acts as a factory for deploying and managing identities and associated wallets. It maintains mappings to store the relationships between deployed identities and wallets, also lists of trusted service providers (issuers) and supported claim topics.

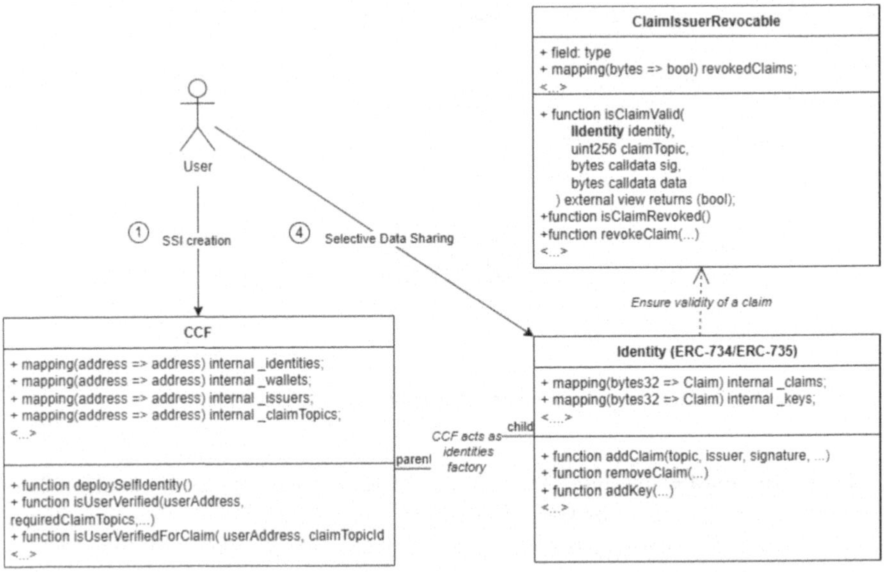

Fig. 3. Simplified CCF architecture based on identity standards.

The `Identity`[2] smart contract, built on ERC-734/ERC-735, serves as the core for managing decentralized identities and includes mappings for claims. ERC-734 standard interface allows users to maintain and govern their own identity data while providing functionality to authenticate other users and sign transactions on their behalf [18]. Identity management works via proxy smart contracts and multikey control enabling individuals or groups to manage cryptographic keys for essential operations such as authentication, transactions, and access permissions [18].

`ClaimIssuerRevocable`[3] smart contract ensures the validity of claims associated with an `Identity`. Claims are pieces of information issued by an external entity about the identity holder [18]. These claims, verified using cryptographic signatures, include details such as identity attributes or other information relevant to the holder's identity.

ERC-735 allows claims to be added or updated over time, enabling digital identities to adapt dynamically to the evolving needs of both identity holders and verifiers. For instance, if a claim issued by the CCF becomes invalid due to the expiration of an identity

1 https://arbiscan.io/address/0xE56dCcf1f2E98A4C13b760A790bcE0bfE7FAE093%23code.

2 https://arbiscan.io/address/0x75B94C3393D48cfF750687B8e532C1Fea28b2013%23code.

3 https://arbiscan.io/address/0x56Aa5eFbbfCEAC2d07184A433c0E6052BA57E5FB%23code.

document or inclusion on a sanctions list, the claim can be revoked. This flexibility significantly enhances the practicality and reliability of digital identities.

The integration of these standards (see Fig. 3) provides a scalable and secure foundation for identity and compliance management within decentralized ecosystems, aligning with the principles of decentralization, privacy, and security.

3.3 Claim Lifecycle in CCF

The claim lifecycle in CCF ensures that compliance requirements are met not just at the time of verification but continuously throughout the lifecycle of the claim. As illustrated in Fig. 4, this lifecycle includes two phases: creation and ongoing validity checks.

Phase 1: Claim Creation. In the initial phase, a TDP generates a unique identifier and link, enabling the user to begin the verification process. The user then uploads required documents and completes necessary actions, such as liveness checks, guided by the TDP. The TDP reviews the submitted information, typically within seconds, to confirm its validity. Upon successful verification, the proof is finalized and remains off-chain. Users may choose to save this proof as part of their decentralized identity, at which point it becomes an immutable digital credential. Termination statuses, such as expired links, voluntary termination, or failed verification due to invalid documentation, ensure that the process is secure and robust.

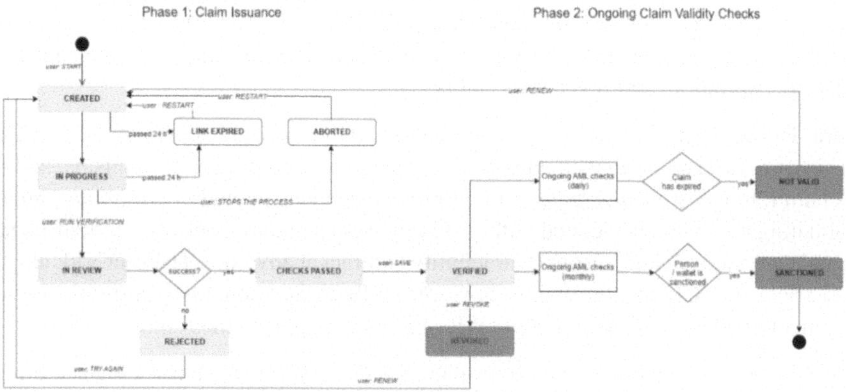

Fig. 4. Claim lifecycle in Compliance On-Chain Framework.

Phase 2: Ongoing Validity Checks. This phase focuses on continuous monitoring to maintain compliance over time, representing the most innovative aspect of the CCF lifecycle. Once a claim is saved on-chain, the framework performs regular off-chain checks to validate its ongoing accuracy. These include daily screenings against updated sanctions and Politically Exposed Persons (PEP) lists, as well as monthly reviews to monitor expiring documents and detect changes in AML-related statuses. If issues such as document expiration or sanctions flags arise, the framework automatically revokes the claim, ensuring it remains trustworthy.

By addressing the critical need for continuous compliance, phase 2 differentiates the framework from existing solutions and underscores its potential to provide secure, efficient compliance across decentralized applications.

3.4 Claim Issuance and Management

This section provides an overview of KYC claim issuance and management within the CCF framework. Other claim types (e.g. sanctions screening, AML risk score) follow similar sequential principles and are therefore not detailed further.

Initialization. The system begins with the deployment of `Identity` and `ClaimIssuerRevocable` smart contracts onto the blockchain. The `ClaimIssuerRevocable` contract is initialized with management keys, claim issuer public keys, and revocation keys owned by CCF. These keys are used to verify claims securely. The `CCF` smart contract is deployed with an initial configuration specifying the `Identity` contract address, fee collector address, and supported claims.

KYC Claim Issuance. The KYC claim issuance flow (see Fig. 5.) starts with the verification of the user's identity and active claim statuses through backend calls. If the KYC status is absent, revoked, or expired, users are redirected to an external KYC provider for verification. Once the process is complete, the KYC provider notifies the CCF backend, which issues a digitally signed claim using its private key. This claim is presented to the user on the CCF dApp and can be saved on the blockchain by invoking the `Identity.addClaim()` function on their deployed `Identity` contract. The `ClaimIssuerRevocable` contract (see Fig. 3) verifies the claim's validity by checking its signature, expiration, and revocation status.

Claim Revocation. To ensure comprehensive claim management, the `ClaimIssuerRevocable` contract enables claim revocation through two mechanisms: adding the claim ID to a list `revokedClaims` or removing the signing key from the contract, invalidating all claims associated with that key. Users maintain control over their identities and can remove claims from their identity contract at any time. The user's identity is not restricted to CCF-specific claims; however, callback functionality is currently limited to claims issued by CCF, with potential for broader support in the future.

KYC claim data includes key information such as approval status, issuance timestamp, and expiration timestamp. This structured approach ensures a secure, efficient, and decentralized method for managing KYC compliance, addressing AML/CTF requirements while respecting user autonomy and privacy.

Claim Status Verification. The verification of a user's claim status within the CCF framework can be performed using two primary methods on-chain. The first method involves smart contract verification, where other dApps or smart contracts invoke the `isUserVerifiedForClaim(<user's address>, <claim_topic_id>)` function on the *CCF* smart contract. This process retrieves the user's `Identity` contract address, the `ClaimIssuerRevocable` address, and validates the claim by checking for revocation, expiration, and signature validity. The second method lets developers or applications manually verify claims by inspecting

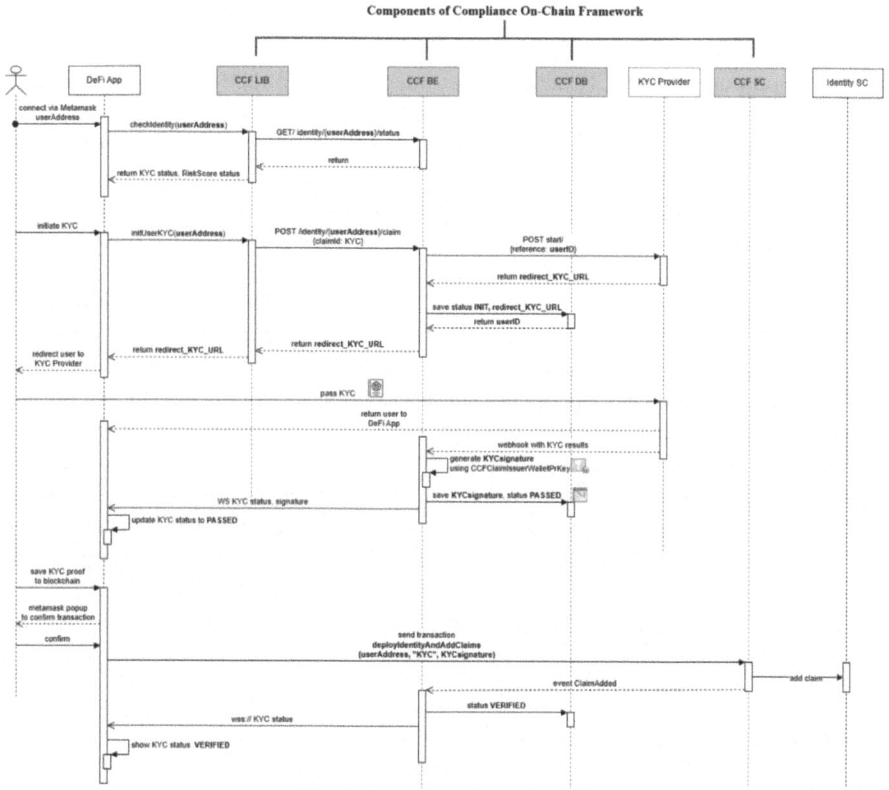

Fig. 5. Sequence diagram of KYC proof issuance.

the user's identity and CCF issuer contracts, ensuring seamless integration and robust verification.

Storage and Signing Keys. The management of claims and signing keys within the CCF is designed to prioritize security and user autonomy. Each user's `Identity` contract stores all associated claims, ensuring decentralized management and ownership. Claims can be revoked by marking them as invalid in the `ClaimIssuerRevocable` contract, which guarantees that subsequent validation checks for revoked claims will fail. To enhance security, multiple issuers can be deployed, and signing keys can be rotated periodically, minimizing potential damage in the event of a key compromise.

4 Scenario Based Simulation

The proposed Compliance On-Chain Framework (CCF) was deployed on the Ethereum Virtual Machine (EVM)-based Arbitrum Sepolia Testnet to validate its functionality through three primary scenarios: (1) deployment of identities, (2) issuance of KYC and sanctions related claims, and (3) revocation of claims.

The implementation of the CCF followed the steps outlined in Sect. 3.4 (Fig. 5), detailing the process for successful interaction with the framework for KYC compliance. For testing purposes, the CCF backend was integrated with the Ondato API for KYC verification. Additionally, a testing dApp was developed to test the integration of the CCF library, and the Chainalysis Free Sanctions API [17] was integrated to facilitate sanctions-related claim issuance and verification.

The deployment and testing phase yielded several significant results. Over 350 identities were successfully deployed, each accompanied by a KYC claim verified through the Ondato API. Most of these identities also included claims certifying the wallet as sanctions-free, leveraging data from the Chainalysis API. Continuous compliance features, such as monitoring for expired documents or sanctioned wallets, were also simulated to evaluate the framework's ability to detect and revoke invalid claims. These tests confirmed that the CCF effectively handles continuous compliance by monitoring for document validity and sanctions updates.

While the primary features of the CCF were successfully validated, several challenges were identified during testing. Delays in fetching information from the blockchain via Ethereum RPC (ERPC) providers affected the overall user experience. Additionally, continuous on-chain validation of claims proved to be resource intensive. Optimizations are necessary to reduce costs, such as selectively targeting identities for revalidation. For instance, if no new wallets are added to sanctions lists, issued claims may not require immediate checks.

5 Future Considerations and Discussions

Enhancing the Compliance On-Chain Framework (CCF) requires addressing both technical and regulatory challenges. A key concern is the jurisdictional variability of AML/CTF laws. While implementations differ, ACAMS notes that major jurisdictions align with FATF recommendations, aiming for consistent regulatory outcomes across similar activities. Differences lie primarily in application and supervision. CCF partially addresses this through configurable risk thresholds for certain DeFi functions on-chain, but further support for extraterritorial obligations (e.g., sanctions, PEP screening) is needed. This can be achieved via modular rule sets or verifiable credentials tailored to local contexts, while maintaining a global compliance baseline.

Technically, improvements may include dynamic validation triggered by material changes (e.g., sanction updates), RPC optimization via caching, and minimizing on-chain operations to reduce gas costs. In addition, risk-based monitoring can further improve efficiency by focusing on high-risk entities.

Future work includes: broader testing across diverse DeFi applications and longitudinal studies to evaluate long-term compliance performance. These enhancements aim to establish CCF as a more scalable and jurisdiction-aware compliance solution for decentralized finance applications.

6 Conclusion

Decentralized finance (DeFi) faces significant challenges in meeting AML/CTF requirements while maintaining privacy, security, and decentralization. Traditional compliance solutions lack compatibility with DeFi's decentralized nature, creating gaps in continuous monitoring and comprehensive compliance frameworks.

This paper proposed the Compliance On-Chain Framework (CCF), a hybrid solution combining off-chain trusted data providers with on-chain mechanisms based on ERC-734/ERC-735 standards. The framework supports automated credential issuance, revocation, and validation, enabling dynamic and scalable compliance processes. Initial testing on the Arbitrum Sopolia Testnet demonstrated the feasibility of identity deployment and claim evaluation.

While the CCF addresses key compliance gaps in DeFi, challenges such as blockchain data latency, gas costs, and jurisdictional variability remain. To mitigate these, we proposed optimizations in validation logic, smart contract efficiency, and modular compliance design. With further testing and refinement, CCF can serve as a foundation for interoperable, scalable, and jurisdiction-aware AML/CTF compliance in decentralized financial systems.

Disclosure of Interests. The authors have no competing interests to declare that are relevant to the content of this article.

References

1. The Board of the International Organization of Securities Commissions: Final Report with Policy Recommendations for Decentralized Finance (DeFi). https://www.iosco.org/library/pubdocs/pdf/IOSCOPD754.pdf, Recommendation 3 – Achieve Common Standards of Regulatory Outcomes, p. 29. Accessed 30 Dec 2024
2. Grand View Research: Decentralized Finance Market Size and Share Report (2030). https://www.grandviewresearch.com/industry-analysis/decentralized-finance-market-report. Accessed 30 Dec 2024
3. Ajayi, A., Emmanuel, I., Soyele, A., Enyejo J.: Enhancing digital identity and financial security in decentralized finance (Defi) through zero-knowledge Proofs (ZKPs) and Blockchain solutions for regulatory compliance and privacy. Iconic Res. Eng. J. **8**(4), 373–394 (2024)
4. Financial Action Task Force: Virtual Assets: Targeted Update on Implementation of the FATF Standards on Virtual Assets and Virtual Asset Service Providers, p. 31, no. 39 (2023). https://www.fatf-gafi.org/content/fatf-gafi/en/publications/Fatfrecommendations/targeted-update-virtual-assets-vasps-2023.html. Accessed 30 Dec 2024
5. European Union: Regulation (EU) 2023/1114 of the European Parliament and of the Council of 31 May 2023 on markets in crypto-assets. In: Official Journal of the European Union. European Union Law, ELI (2024). http://data.europa.eu/eli/reg/2023/1114/oj
6. Sak, M.: KYC/AML Technologies in Decentralized Finance (DeFi). The Leonard N. Stern School of Business Glucksman Institute (2024). https://www.stern.nyu.edu/sites/default/files/2024-07/Glucksman_Sak_2024.pdf. Accessed 30 Dec 2024
7. Schlatt, V., Sedlmeir, J., Feulner, S., Urbach, N.: Designing a framework for digital KYC processes built on Blockchain-based self-sovereign identity. Inf. Manage. **59**(7) (2022). https://doi.org/10.1016/j.im.2021.103553. ISSN 0378–7206

8. The Association of Certified Anti-Money Laundering Specialists: Risk Management Programs for Cryptoasset and Blockchain. Study Guide. Version 1.03. (2022)
9. Truong, N., Lee, G.M., Sun, K., Guitton, F., Guo, Y.: A blockchain-based trust system for decentralised applications: when trustless needs trust. Fut. Gener. Comput. Syst. **124**, 68–79 (2021). https://doi.org/10.1016/j.future.2021.05.025
10. Pauwels, P.: zkKYC A solution concept for KYC without knowing your customer, leveraging self-sovereign identity and zero-knowledge proofs. In: Paper 2021/907 Preprint. MINOR revision (2021). https://ia.cr/2021/907
11. Buterin, V., Illum, J., Nadler, M., Fabian, S., Ameen, S.: Blockchain privacy and regulatory compliance: towards a practical equilibrium (2023). https://doi.org/10.2139/ssrn.4563364
12. Fractal ID: Data Breach Post Mortem (2024). https://web.fractal.id/fractal-id-data-breach-post-mortem. Accessed 30 Dec 2024
13. Perlman, L., Gurung, N.: Focus note: the use of eKYC for customer identity and verification and AML (2019). https://doi.org/10.2139/ssrn.3370665
14. Sun, N., Zhang, Y., Liu, Y. A privacy-preserving KYC-compliant identity scheme for accounts on all public Blockchains. Sustainability **14**, 14584 (2022). https://doi.org/ https://doi.org/10.3390/su142114584
15. Burleson, J., Korver, M., Boneh, D.: Privacy-protecting regulatory solutions using zero-knowledge proofs (2022). https://api.a16zcrypto.com/wp-content/uploads/2022/11/ZKPs-and-Regulatory-Compliant-Privacy.pd
16. Trask, A.: Beyond privacy trade-offs with structured transparency. In: University of Oxford, OpenMined, arXiv: 2012.08347v1 [cs.DS] (2022)
17. Chainalysis Compliance. Free Cryptocurrency sanctions screening tool. https://www.chainalysis.com/free-cryptocurrency-sanctions-screening-tools/. Accessed 30 Dec 2024
18. ERC-725 Alliance. https://erc725alliance.org. Accessed 30 Dec 2024

Integrating Genetic Algorithms and Ensemble Learning for Improved and Transparent Credit Scoring

Helmi Ayari[1(✉)] and Ramzi Guetari[1,2]

[1] SERCOM Laboratory, Polytechnic School of Tunisia, University of Carthage, La Marsa, Tunisia
helmi.ayari@ept.rnu.tn, ramzi.guetari@ept.ucar.tn
[2] Novobit Group GmbH, theodor heuss straße, 38122 Braunschweig, Germany
https://www.ept.tn/, https://www.novobit.eu/

Abstract. Credit scoring is essential in the financial sector, where accurate predictions help mitigate financial risks for lending institutions and ensure fair access to credit for applicants. The traditional credit scoring models often struggle with balancing predictive accuracy and interpretability, two crucial factors in real-world applications. In this paper, we present a novel model that integrates genetic algorithms for feature selection with ensemble learning techniques to enhance the performance. The model addresses data imbalance using ADASYN oversampling and employs stacking strategies to benefit from the strengths of multiple base classifiers. Interpretability is provided through SHAP, which evaluates the contributions of base learners within the ensemble to ensure transparency. The proposed model was evaluated on three credit scoring datasets, achieving 86.00% accuracy on the german dataset, 92.03% on the australian dataset, and 92.39% on the polish dataset, outperforming benchmark classifiers. While deep learning models showed strong performance on the larger polish dataset, our approach delivers competitive results with superior interpretability and transparency. This study highlights the importance of combining feature selection, balancing techniques, ensemble methods, and interpretability for a trustworthy credit scoring model.

Keywords: Credit Scoring · Genetic Algorithms · Ensemble Learning · Interpretability · SHAP · ADASYN · Stacking

1 Introduction

Credit scoring is vital in the financial sector, helping assess the creditworthiness of individuals and organizations. Accurate predictions reduce financial risks for lenders while ensuring fair access to credit. However, building reliable credit scoring models is challenging, especially when balancing accuracy with interpretability.

K. Węcel (Ed.): BIS 2025, LNBIP 554, pp. 225–238, 2025.
https://doi.org/10.1007/978-3-031-94193-1_17

Highly accurate models tend to be complex and difficult to understand, yet interpretability is essential for fostering trust and transparency in the financial domain. A major challenge is the imbalance in credit scoring datasets, where instances of "good" and "bad" credit are often disproportionate, leading to biased predictions that favor the dominant class and compromise model reliability.

Feature selection also plays a key role in enhancing both accuracy and interpretability. Identifying the most relevant features improves predictive performance and reduces model complexity, making the decision-making process more transparent.

This paper proposes a hybrid approach that combines the ADASYN technique for balancing data, genetic algorithms for feature selection, ensemble learning methods, and SHAP for model interpretability. SHAP is used to gain insights into the contribution of each base learner in the stacking ensemble, providing a better understanding of how classifiers contribute to the final prediction. The proposed method aims to improve both model accuracy and transparency in credit scoring.

The paper is structured as follows: Section 2 reviews related work. Section 3 outlines the methodology. Section 4 describes the experimental framework. Section 5 presents the results. Section 6 presents the limitations, and section 7 concludes the paper.

2 Related Works

Recent advancements in credit scoring have incorporated machine learning methods to address the limitations of traditional models. While logistic regression and Decision Trees (DTs) are commonly used, ensemble techniques like stacking have shown promise in improving accuracy and robustness [1].

Stacking, which combines multiple base models into a meta-model, has been effectively used in credit scoring. For instance, Rofik et al. [2] applied stacking with Random Forest (RF), Support Vector Machine (SVM), Extra-Tree, and Extreme Gradient Boosting (XGBoost), achieving 83.21% accuracy and 91.78% recall. Similarly, Muslim et al. [1] used KNN, SVM, and RF as base learners, achieving nearly 100% accuracy on a P2P lending dataset. In [3], a multi-stage ensemble model combining a hybrid genetic algorithm with instance hardness threshold and majority voting addressed class imbalance in credit prediction, achieving 93.79% recall on the polish 2 credit dataset. Raimundo et al. [4] demonstrated that stacking ensembles outperform single classifiers in credit scoring, particularly for predicting defaults by achieving 83.99% F1-score on the lending club dataset. Anil Kumar et al. [5] proposed a heterogeneous ensemble classifier with six base classifiers and three meta-classifiers using stacking, achieving 91.56% accuracy on the australian dataset and 84.35% on the german dataset. These studies demonstrate the ability of stacking to improve predictive performance by combining diverse models.

Feature selection is crucial in credit scoring, with genetic algorithms (GAs) widely used. Lenka et al. [6] showed that GA-based CatBoost outperformed other models, achieving 88.40% accuracy on the australian dataset. Zhang et al. [7] proposed a hybrid GA for feature and classifier selection, reaching 93.52% accuracy on the GMSC dataset. Jadhav et al. [8] introduced IGDFS, reducing complexity and improving accuracy to 90.75% on the australian dataset. Kozeny et al. [9] found a variable bitmask GA function to be most effective (75.85% accuracy, 53.85% specificity). Ilter et al. [10] combined GAs with an information complexity criterion, achieving 88% accuracy on the australian dataset and 80% on the german dataset.

Effectively handling imbalanced datasets is essential in credit scoring, with methods like SMOTE and cost-sensitive learning being widely used. Abedin et al. [11] proposed WSMOTE-ensemble, achieving 98.97% minority class accuracy on a small business credit dataset. Eko et al. [12] found SMOTE beneficial for detecting high-risk mortgage customers, reaching 65.96% specificity. A comparative study [13] highlighted ADASYN as a superior balancing technique, enhancing ensemble classifier performance, with XGBoost achieving 80.28% accuracy.

Ensemble methods improve accuracy but often lack interpretability, making post-hoc explanation techniques like Shapley Additive ExPlanations (SHAP) [14] crucial for transparency. Kong et al. [15] applied SHAP to XGBoost for credit scoring, achieving 87% accuracy and 95% AUC. Hjelkrem et al. [16] showed a deep learning model outperformed BERT for credit scoring with 90.90% AUC. Bussmann et al. [17] used LIME and SHAP, achieving 93% AUROC on the lending club dataset. Ariza-Garzón et al. [18] found logistic regression outperformed other models on the lending club dataset, using SHAP to capture complex feature relationships. These studies highlight the significance of interpretability in credit scoring.

3 Proposed Approach

This study proposes a novel framework for credit scoring, which emphasizes both accuracy and interpretability. As illustrated in Fig. 1, the methodology progresses through different stages: preprocessing the dataset, addressing the class imbalance, selecting relevant features using GAs, building a stacking ensemble for improved performance, and employing SHAP values for interpretability. Each stage is outlined below to emphasize its function within the framework.

3.1 Data Preprocessing

The preprocessing begins with handling missing values in the german, australian, and polish credit datasets, imputing with the median or mode. Numerical features are normalized, and categorical variables are encoded using one-hot encoding for compatibility with machine learning algorithms. To address the class imbalance, we apply ADASYN [19], which generates synthetic samples for the minority class, focusing on harder-to-classify instances. This helps balance the class distribution, reduces bias towards the majority class, and improves model fairness and performance.

3.2 Feature Selection Using Genetic Algorithms

After addressing the class imbalance, we proceed with feature selection to reduce dataset complexity and enhance model performance. To achieve this, we employ GAs, which are a global search method inspired by natural evolution [3]. The algorithm simulates natural selection, in which a population of candidate feature subsets evolves through selection, crossover, and mutation.

1. The fitness function evaluates the contribution of feature subsets to model performance, aiming to maximize metrics like accuracy, precision, recall, or F1-score. Each individual is represented by a binary vector indicating whether a feature is included or excluded. The model is trained and tested on the selected features, and higher-performing solutions are more likely to reproduce in subsequent generations [9].
2. Crossover combines two parent solutions by swapping feature subsets at a random point in the binary vector, allowing exploration of different combinations.
3. Mutation introduces randomness by flipping bits to include or exclude features, preventing premature convergence and promoting diverse feature subsets.

The output of the GA-based feature selection process is a refined set of features, which is then used in the next stage to improve predictive performance.

3.3 Stacking Ensemble

The next stage uses a stacking ensemble to combine multiple base models into a stronger meta-model [1]. Stacking first trains base learners on the dataset and then uses their predictions as features for a second-layer meta-learner, which generates the final output. Stacking was chosen for its ability to leverage the strengths of multiple classifiers, improving model performance by combining diverse base learners. This contrasts with single-model classifiers that may struggle with complex patterns. To ensure diversity, we use classifiers such as logistic regression [20], SVM [21], DT [22], XGBoost [23], and KNN [24] as base learners. Logistic regression serves as the meta-learner, effectively combining base model outputs and improving accuracy.

3.4 Model Interpretability Using SHAP

The final stage focuses on model interpretability, crucial for building trust in machine learning models, especially in sensitive areas like credit scoring. We use SHAP, not for analyzing individual feature contributions but to assess the influence of each base learner in the stacking ensemble model. By calculating SHAP values, we identify which base learner most influences the final prediction. This analysis reveals the importance of base learners like logistic regression, SVM, XGBoost, RF, and DT in determining credit classification. Understanding each learner's contribution helps optimize the stacking algorithm and enhances trust in the model's outcomes.

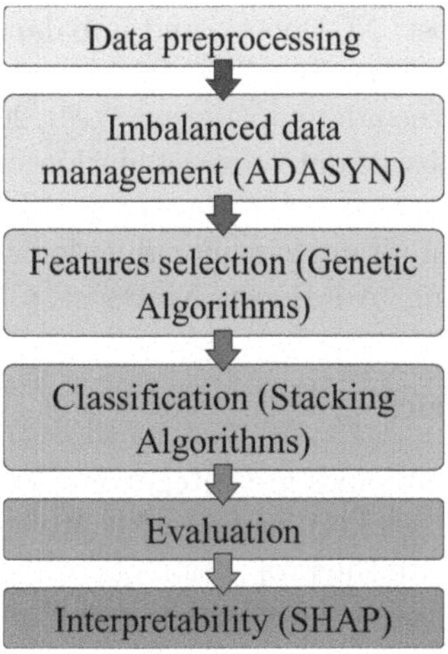

Fig. 1. Pipeline of the proposed model.

4 Research Methodology

4.1 Credit Dataset

This paper evaluates the proposed credit scoring model using the german, australian, and polish credit datasets from the UCI Machine Learning Repository (Table 1). These datasets were chosen due to their relevance to real-world credit scoring tasks and class imbalance. The german dataset contains 1,000 records (700 reputable, 300 non-reputable) with 20 features. The australian dataset has 690 records (307 creditworthy, 383 non-creditworthy) with 14 features. The polish dataset includes 43,405 records (41,314 bad credit, 2,091 good credit) with 65 features. For user security, all datasets have been anonymized by encrypting values and replacing feature names with meaningless symbols.

Table 1. Dataset information

Name of Data	Total Instances	No. of Attributes	Bad (Bankrupt)/Good (Non-Bankrupt)
German	1000	20	300/700
Australian	690	14	307/383
Polish	43405	64	41314/2091

4.2 Experimental Setup

The experiments were conducted on a desktop system with an Intel Core i7-7700HQ processor, 16GB of RAM, and Windows 10 Professional, using Python 3.12.8 and PyCharm. The primary libraries for model implementation included NumPy, Pandas, Scikit-learn (for logistic regression, SVM, DTs, RF, KNN, gradient boosting, AdaBoost, XGBoost), and imbalanced-learn (for ADASYN). TensorFlow with Keras was used for CNN, RNN, and MLP models, which were trained for 50 epochs with the adam optimizer. A 80/20 train-test split was used, and models were evaluated without cross-validation or hyperparameter tuning. Only SVM had adjusted parameters (probability=True). The setup aimed to compare the performance of all models, including logistic regression, SVM, DTs, RF, KNN, gradient boosting, AdaBoost, XGBoost, CNN, RNN, and MLP.

4.3 Performance Metrics

In a classification model, outcomes are categorized as True Positive (TP), False Positive (FP), True Negative (TN), or False Negative (FN). Accuracy is a commonly used metric to measure the proportion of correctly classified instances out of all instances. However, it can be misleading for imbalanced datasets, where one class dominates [25]. To address this, additional metrics such as precision, recall, F1-Score, specificity, and AUC are often used. Precision measures the proportion of true positives among predicted positives, while recall (or sensitivity) evaluates the ratio of correctly identified positives to all actual positives. The F1-Score balances precision and recall, especially with uneven class distribution. Specificity measures the proportion of correctly identified negatives, and the ROC curve shows the trade-off between sensitivity and specificity at different thresholds, with AUC quantifying the model's ability to distinguish between classes.

5 Experimental Results and Analysis

This section evaluates the performance of the proposed model using the german, australian and polish datasets. Additionally, SHAP analysis is conducted to explore the interpretability of the ensemble model, offering insights into the contributions of individual base learners.

5.1 Model Results and Analysis for the German Dataset

As shown in Table 2, the proposed model demonstrates superior performance on the german dataset, achieving an accuracy of 86.00%, the highest among all models. It also excels in precision with 82.98%, indicating its strong ability to correctly identify positive instances. The recall is 66.10%, which reflects a balanced capacity for detecting minority class instances. The F1-score of 73.58% highlights the model's balanced performance between precision and recall, while the specificity of 94.33% emphasizes its ability to correctly identify the majority class. Furthermore, the AUC of 86.78% showcases the model's ability to effectively distinguish between classes.

Compared to traditional machine learning models such as RF and XGBoost, which achieve accuracies of 82.50% and 84.50%, respectively, the proposed model consistently outperforms them across key metrics. Deep learning models like

CNN, RNN, and MLP exhibit lower accuracy (71.00%, 68.00%, and 76.50% respectively), demonstrating the superior balance and robustness of the proposed approach.

The proposed model demonstrates superior performance on the german dataset compared to traditional machine learning and deep learning models. It achieves the highest accuracy, alongside balanced metrics across precision, recall, and specificity, highlighting the model's robustness in credit scoring.

5.2 Model Results and Analysis for the Australian Credit Dataset

On the australian credit dataset (Table 3), the proposed model achieves the highest overall performance across all metrics, with an accuracy of 92.03%, surpassing other models, including XGBoost (89.86%) and RF (89.86%). In terms of precision, the proposed model leads with 93.48%, recall of 84.31%, while the F1 score of 88.66% further reflects its balanced performance between precision and recall. Additionally, the proposed model achieves the highest specificity at 96.55%, meaning it excels in correctly identifying the majority class without false positives. The AUC of 93.08% underscores its excellent ability to differentiate between the classes, solidifying the proposed model's effectiveness on this dataset. In comparison, the deep learning models, including RNN, CNN, and MLP, perform well but fall short of the proposed model's overall performance. The MLP achieves an accuracy of 87.68%, while both CNN and RNN reach 86.23%.

These results confirm the robustness of our model, as it outperforms benchmark classifiers across both german and australian datasets with distinct characteristics.

Table 2. Performance comparison on german dataset

Model	Acc. (%)	Prec. (%)	Rec. (%)	F1 (%)	Spec. (%)	AUC (%)
RNN	68.00	42.42	23.73	30.43	86.52	63.07
CNN	71.00	50.67	64.41	56.72	73.76	76.10
MLP	76.50	59.38	64.41	61.79	81.56	79.83
KNN	67.50	47.12	83.05	60.12	60.99	77.59
Log. Regression	69.50	48.86	72.88	58.50	68.09	78.41
DT	72.00	52.38	55.93	54.10	78.72	67.33
AdaBoost	76.50	59.09	66.10	62.40	80.85	82.39
SVM	77.50	60.94	66.10	63.41	82.27	80.53
Gradient Boosting	79.00	66.04	59.32	62.50	87.23	83.11
RF	82.50	75.00	61.02	67.29	91.49	84.72
XGBoost	84.50	79.17	64.41	71.03	92.91	84.29
Proposed Model	**86.00**	**82.98**	**66.10**	**73.58**	**94.33**	**86.78**

Table 3. Performance comparison on australian dataset

Model	Acc. (%)	Prec. (%)	Rec. (%)	F1 (%)	Spec. (%)	AUC (%)
RNN	86.23	80.77	82.35	81.55	88.51	89.79
CNN	86.23	80.77	82.35	81.55	88.51	90.78
MLP	87.68	86.96	78.43	82.47	93.10	90.58
KNN	89.86	87.76	84.31	86.00	93.10	91.75
Log. Regression	85.51	78.18	84.31	81.13	86.21	90.38
DT	89.13	86.00	84.31	85.15	91.95	88.45
AdaBoost	85.51	79.25	82.35	80.77	87.36	89.31
SVM	86.23	76.67	90.20	82.88	83.91	92.68
Gradient Boosting	84.78	81.25	76.47	78.79	89.66	90.07
RF	89.86	91.11	80.39	85.42	95.40	92.89
XGBoost	89.86	93.02	78.43	85.11	96.55	90.74
Proposed Model	**92.03**	**93.48**	**84.31**	**88.66**	**96.55**	**93.08**

5.3 Model Results and Analysis for the Polish Dataset

Table 4. Performance comparison on polish dataset

Model	Acc. (%)	Prec. (%)	Rec. (%)	F1 (%)	Spec. (%)	AUC (%)
RNN	95.51	86.96	13.61	23.53	99.89	81.64
CNN	96.20	88.28	29.02	43.69	99.79	91.59
MLP	96.06	76.19	32.65	45.71	99.45	90.01
KNN	75.75	11.37	55.56	18.88	76.83	70.27
Log. Regression	73.24	10.86	59.18	18.35	73.99	72.66
DT	83.96	13.34	39.23	19.91	86.36	62.79
AdaBoost	75.10	11.98	61.45	20.04	75.83	75.19
SVM	73.24	10.50	56.69	17.71	74.13	71.81
Gradient Boosting	78.56	13.85	61.68	22.62	79.47	79.17
RF	91.76	26.86	36.05	30.78	94.75	80.68
XGBoost	90.23	26.09	50.34	34.37	92.37	82.19
Proposed Model	**92.39**	**26.35**	**49.66**	**34.43**	**92.57**	**82.55**

On the polish dataset (Table 4), deep learning models like CNN, MLP, and RNN achieved higher accuracy, with CNN performing the best at 96.20%. The larger size of the dataset (43,405 instances) contributes to the superior performance of deep learning models, which generally perform better on larger datasets. However, these models exhibit imbalanced performance, particularly in recall. The proposed model, with an accuracy of 92.39%, outperforms traditional machine learning models and is competitive with deep learning models like CNN (96.20%) and MLP (96.06%). Its strength lies in its balanced performance, with a specificity of 92.57% and an AUC of 82.55%. However, its precision is relatively low at 26.35%, indicating a higher number of false positives. This trend is also observed in traditional machine learning models compared to deep learning, which showed better precision. Despite this, the proposed model offers solid overall performance and provides a well-balanced solution for both majority and minority class predictions, making it a strong choice compared to other models.

5.4 SHAP Analysis and Model Interpretability

To better understand the interpretability of the proposed stacking ensemble model, SHAP analysis was conducted to assess the contributions of the base learners (XGBoost, KNN, DT, logistic regression, and SVM) to the meta-learner's predictions on german, australian, and polish dataset.

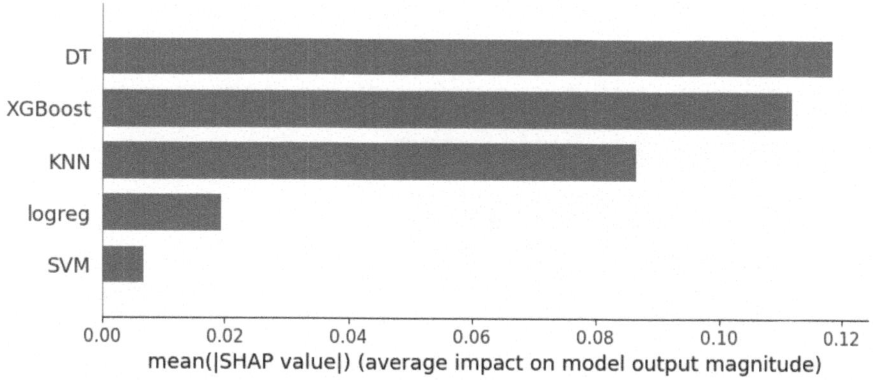

Fig. 2. Base learner importance bar chart for german dataset.

Base Learner Importance in the German Dataset: Figure 2 shows the global importance of input variables, determined by the mean absolute SHAP value for each base learner. Higher SHAP values indicate a greater influence on the model's predictions. The DT has the highest mean SHAP value, confirming its key role in the ensemble model for this dataset. XGBoost and KNN follow with similar contributions, while Logistic regression has a moderate impact and SVM shows the smallest value, highlighting its limited influence on the model's decisions.

The SHAP summary plot in Fig. 3 shows each base learner's contribution to the model's predictions. Positive SHAP values indicate a positive influence, while negative values suggest a negative impact. The DT is the most influential base learner, with its dominant position on the plot and its wide range of SHAP values indicating its significant role in both positive and negative predictions. XGBoost and KNN follow with secondary contributions, while Logistic regression shows a moderate impact. SVM has the smallest influence, reflected in its narrow SHAP value range. These results highlight the ensemble's reliance on the strengths of individual learners, with XGBoost and DT being the most impactful.

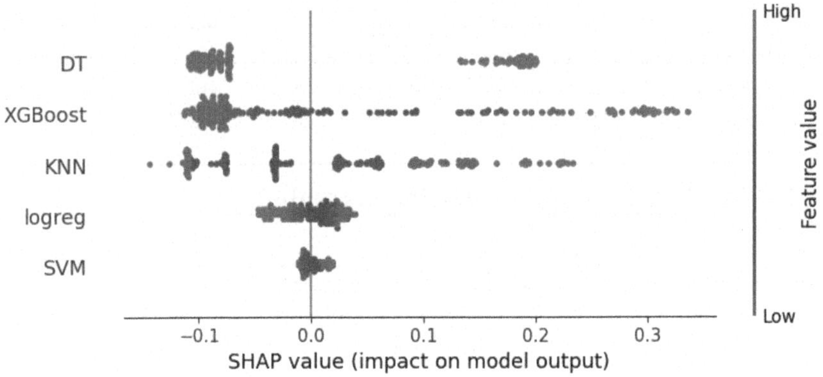

Fig. 3. SHAP summary plot of base learners for german dataset.

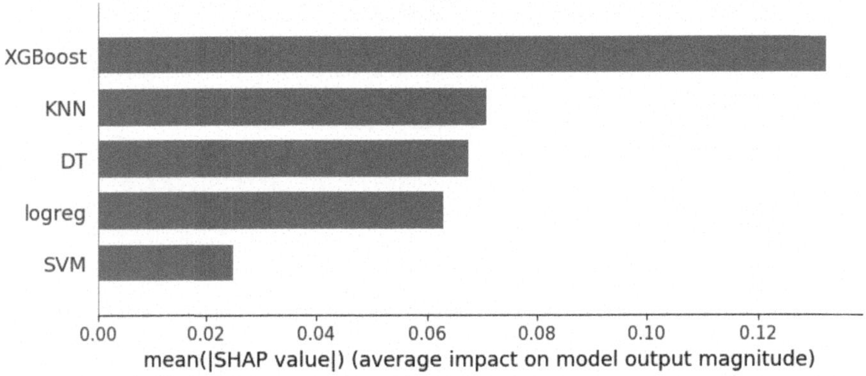

Fig. 4. Base learner importance bar chart for australian dataset.

Base Learner Importance in the Australian Dataset: Figure 4 highlights that XGBoost is the most dominant base learner, with the highest mean SHAP value. KNN and DT follow, contributing significantly to the ensemble but less than XGBoost. Logistic regression shows a moderate level of impact, while SVM has the smallest mean SHAP value similar to its role in the german dataset.

The summary plot of the australian dataset (Fig. 5) reveals that the XGBoost is the most influential base learner. KNN and DT follow indicating their supporting roles in complementing XGBoost. Logistic regression shows good impact, while SVM has the least impact, with a concentration of SHAP values near the center and limited spread, highlighting its specialized but restricted role.

Base Learner Importance in the Polish Dataset: Figure 6 illustrates that XGBoost is the most influential base learner, exhibiting the highest mean SHAP value, surpassing 0.10. DT follows, though with a significantly lower contribution, just below 0.04. KNN ranks slightly lower than DT, while logistic regression shows a more limited impact, falling below 0.02 but still outperforming SVM,

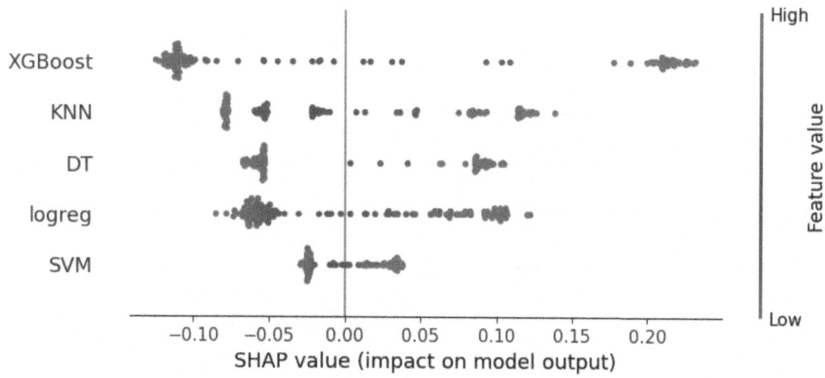

Fig. 5. SHAP summary plot of base learner for australian dataset.

which has the lowest mean SHAP value, indicating its minimal influence on the model's predictions.

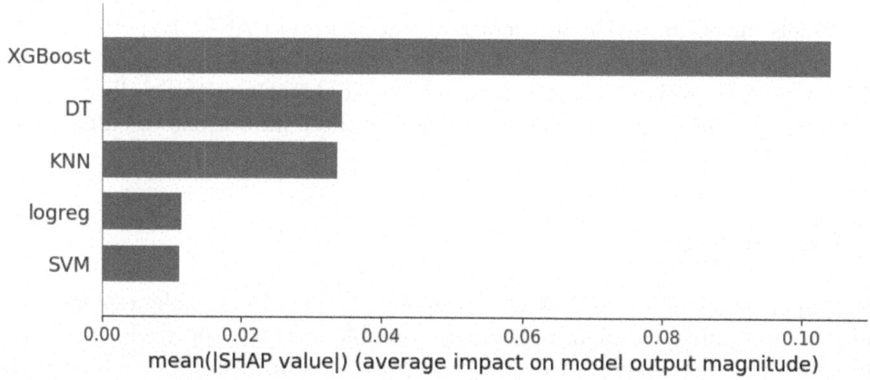

Fig. 6. Base learner importance bar chart for polish dataset.

The summary plot of the polish dataset (Fig. 7) illustrates that XGBoost is the most dominant base learner. DT and KNN follow, acting as complementary models that reinforce overall performance. Logistic regression and SVM demonstrate the least influence, with SHAP values clustered around the center and minimal dispersion, reflecting their narrow yet distinct contributions.

This analysis validates the effectiveness of our approach, as it clearly highlights the key learners that drive the model's performance. By identifying these contributions, we provide transparency into the inner workings of the model, which is an important step towards building more interpretable machine learning models in credit scoring.

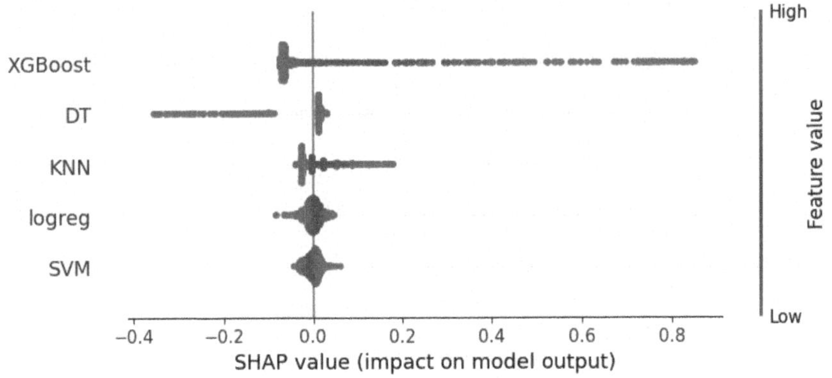

Fig. 7. SHAP summary plot of base learner for polish dataset.

6 Limitations

While the model performed well on smaller datasets like the german and australian datasets, it struggled with the larger polish datasets, where deep learning models outperformed it in accuracy, precision, and AUC. Hyperparameter tuning was also not applied, which could improve performance. Additionally, SHAP was only used on base learners, not the final stacking model, limiting the full interpretability. Future work could address these limitations to enhance the model.

7 Conclusion

This paper presents an innovative credit scoring model that balances accuracy and interpretability through a stacking ensemble learning approach. ADASYN addresses class imbalance, while GAs improve feature selection. These techniques, when combined, significantly enhance predictive performance.

Experimental results on widely used credit scoring datasets show that the proposed model outperforms benchmark classifiers in terms of accuracy and other evaluation metrics. It excels on small datasets and remains competitive on larger ones, where deep learning models generally perform better. SHAP analysis provides insights into the contributions of individual base learners, with XGBoost and DT being the most influential, improving both accuracy and transparency. This balance between performance and transparency is essential for fostering trust in financial applications.

The implications of this research are significant for both scientific and practical domains. Scientifically, the study demonstrates how ensemble learning techniques can improve predictive power and explainability, while also emphasizing the role of SHAP in enhancing model transparency. Practically, the model provides a viable and interpretable solution for credit risk assessment, offering robustness and supporting regulatory compliance. However, further validation on larger and more diverse datasets is needed for real-world adoption.

Future research could focus on incorporating unsupervised learning techniques or expanding the model to address challenges such as evolving credit behaviors. Overall, this work contributes to the development of more transparent and fair credit scoring solutions, improving financial decision-making practices.

References

1. Muslim, M.A., Nikmah, T.L., Pertiwi, D., Dasril, Y.: New model combination meta-learner to improve accuracy prediction P2P lending with stacking ensemble learning. Intell. Syst. Appl. **18**, 200204 (2023). https://doi.org/10.1016/j.iswa.2023.200204
2. Rofik, R., Aulia, R., Musaadah, K., Ardyani, S.S.F., Hakim, A.A.: The optimization of credit scoring model using stacking ensemble learning and oversampling techniques. J. Inf. Syst. Explor. Res. 2(1), (2024). https://doi.org/10.52465/joiser.v2i1.203
3. Jin, Y., Zhang, W., Wu, X., Liu, Y., Hu, Z.: A novel multi-stage ensemble model with a hybrid genetic algorithm for credit scoring on imbalanced data. IEEE Access **9**, 143593–143607 (2021). https://doi.org/10.1109/ACCESS.2021.3120086
4. Raimundo, B., and Bravo, J. M.: Credit risk scoring: a stacking generalization approach. In: World Conference on Information Systems and Technologies, pp. 382–396. Springer, Cham (2023). https://doi.org/10.1007/978-3-031-45642-8-38
5. Anil Kumar, C.J., Raghavendra, B.K., Raghavendra, S.: A credit scoring heterogeneous ensemble model using stacking and voting. Indian J. Sci. Technol. **15**(7), 300–308 (2022). https://doi.org/10.17485/IJST/v15i7.1715
6. Lenka, S.R., Bisoy, S.K., Priyadarshini, R., Sain, M.: Empirical analysis of ensemble learning for imbalanced credit scoring datasets: a systematic review. Wireless Commun. Mobile Comput. **2022**(1), 6584352 (2022). https://doi.org/10.1155/2022/6584352
7. Zhang, W., He, H., Zhang, S.: A novel multi-stage hybrid model with enhanced multi-population niche genetic algorithm: an application in credit scoring. Expert Syst. Appl. **121**, 221–232 (2019). https://doi.org/10.1016/j.eswa.2018.12.020
8. Jadhav, S., He, H., Jenkins, K.: Information gain directed genetic algorithm wrapper feature selection for credit rating. Appl. Soft Comput. **69**, 541–553 (2018). https://doi.org/10.1016/j.asoc.2018.04.033
9. Kozeny, V.: Genetic algorithms for credit scoring: alternative fitness function performance comparison. Expert Syst. with Appl. **42**(6), 2998–3004 (2015). https://doi.org/10.1016/j.eswa.2014.11.028
10. Ilter, D., Deniz, E., Kocadagli, O.: Hybridized artificial neural network classifiers with a novel feature selection procedure based genetic algorithms and information complexity in credit scoring. Appl. Stochastic Models Bus. Ind. **37**(2), 203–228 (2021). https://doi.org/10.1002/asmb.2614
11. Abedin, M.Z., Guotai, C., Hajek, P., Zhang, T.: Combining weighted SMOTE with ensemble learning for the class-imbalanced prediction of small business credit risk. Complex Intell. Syst., 1–21 (2021). https://doi.org/10.1007/s40747-021-00614-4
12. Eko, W.H., Heti, M., Teguh, S.I.: Improving credit scoring model of mortgage financing with SMOTE methods in sharia banking. Russ. J. Agric. Socio-Econ. Sci. **92**(8), 56–67 (2019). https://doi.org/10.18551/rjoas.2019-08.07

13. Zhao, Z., Aumeboonsuke, V.: Imbalanced credit risk prediction in ensemble learning classifiers: a comparative analysis of SMOTE, ADASYN, SMOTETomek, and cluster centroids. J. Arts Manage. **7**(3), 959–984 (2023)
14. Lundberg, S.: A unified approach to interpreting model predictions. arXiv preprint: arXiv:1705.07874 (2017)
15. Kong, Y., Wang, Y., Sun, S., Wang, J.: XGB and SHAP credit scoring model based on Bayesian optimization. J. Comput. Electron. Inf. Manag. **10**(1), 46–53 (2023). https://doi.org/10.54097/jceim.v10i1.5368
16. Hjelkrem, L.O., Lange, P.: Explaining deep learning models for credit scoring with SHAP: s case study using open banking data. J. Risk Financial Manag. **16**(4), 221 (2023). https://doi.org/10.3390/jrfm16040221
17. Bussmann, N., Giudici, P., Marinelli, D., Papenbrock, J.: Explainable machine learning in credit risk management. Comput. Econ. **57**(1), 203–216 (2020). https://doi.org/10.1007/s10614-020-10042-0
18. Ariza-Garzón, M.J., Arroyo, J., Caparrini, A., Segovia-Vargas, M.J.: Explainability of a machine learning granting scoring model in peer-to-peer lending. IEEE Access **8**, 64873–64890 (2020). https://doi.org/10.1109/ACCESS.2020.2984412
19. He, H., Bai, Y., Garcia, E.A., Li, S.: ADASYN: adaptive synthetic sampling approach for imbalanced learning. In: Proceedings of the 2008 IEEE International Joint Conference Neural Networks, pp. 1322–1328 (2008). https://doi.org/10.1109/IJCNN.2008.4633969
20. LaValley, M.P.: Logistic regression. Circulation **117**(18), 2395–2399 (2008)
21. Noble, W. S.: What is a support vector machine? Nat. Biotechnol. **24**(12), 1565–1567 (2006). https://doi.org/10.1038/nbt1206-1565
22. Safavian, S.R., Landgrebe, D.: A survey of decision tree classifier methodology. IEEE Trans. Syst., Man, Cybernet. **21**(3), 660–674 (1991). https://doi.org/10.1109/21.97458
23. Chen, T., Guestrin, C.: XGBoost: a scalable tree boosting system. In: Proceedings of the 22nd ACM SIGKDD International Conference on Knowledge Discovery and Data Mining, pp. 785–794 (2016). https://doi.org/10.1145/2939672.293978
24. Guo, G., Wang, H., Bell, D., Bi, Y., and Greer, K.: KNN model-based approach in classification. In: On The Move to Meaningful Internet Systems 2003: CoopIS, DOA, and ODBASE: OTM Confederated International Conferences, Catania, Sicily, Italy, vol. 2888, pp. 986–996. Springer (2003). https://doi.org/10.1007/978-3-540-39964-3-62
25. Guetari, R., Ayari, H., Sakly, H.: Computer-aided diagnosis systems: a comparative study of classical machine learning versus deep learning-based approaches. Knowl. Inf. Syst. **65**(10), 3881–3921 (2023). https://doi.org/10.1007/s10115-023-01894-7

Data Platforms and Deployment

Architectures and Implementations of Data Lakehouses: Case Studies from Industrial Practice

Jan Schneider[1(✉)], Christoph Gröger[2], Arnold Lutsch[2], Holger Schwarz[1], and Bernhard Mitschang[1]

[1] University of Stuttgart, Universitätsstraße 38, 70569 Stuttgart, Germany
{Jan.Schneider,Holger.Schwarz,Bernhard.Mitschang}@ipvs.uni-stuttgart.de
[2] Robert Bosch GmbH, Borsigstraße 4, 70469 Stuttgart, Germany
{Christoph.Groger,Arnold.Lutsch}@de.bosch.com

Abstract. Data analytics and AI constitute key drivers for business innovation and are hence indispensable for ensuring the competitiveness of enterprises in the digital age. Yet, they can only be leveraged when sufficient data management is carried out, which requires the design and development of suitable data platforms. In recent years, so-called data lakehouses found their way into practice, which promise to combine the benefits of data warehouses and data lakes. However, only little is known about the data and technology architectures that are utilized for these data platforms in practice, as well as the experiences that enterprises have made with them. To address this gap, we conducted four case studies on large-scale, real-world data lakehouse implementations from the industrial sector that are used for various kinds of analytics and AI applications. This paper presents our within-case and cross-case results, which provide insights on common architectural decisions, practical experiences and observed challenges. They outline directions for future research and can support enterprises in the design of suitable data platforms.

Keywords: Data Lakehouse · Data Platforms · AI Applications · Case Studies · Industry Experience

1 Introduction

Data analytics and Artificial Intelligence (AI) constitute key ingredients for the digital transformation of enterprises, as they allow to derive insights from data and thus enable data-driven business decisions. Recent advances in the field of AI are particularly promising for industrial enterprises, as they generate large amounts of heterogeneous data across the industrial value chain that provide huge potentials for business process optimization and business model innovation. Examples include predictive maintenance [6], AI-based optical inspection on the factory shop floor and data-driven engineering [28]. In order to be able to leverage these analytics- and AI-driven business opportunities, data management

K. Węcel (Ed.): BIS 2025, LNBIP 554, pp. 241–255, 2025.
https://doi.org/10.1007/978-3-031-94193-1_18

is crucial. In industrial practice, data management accounts for up to 80% of the implementation efforts for AI use cases, with data platforms constituting the core IT components for it [10]. *Data warehouses* [12] represent the most conventional type of data platform and are primarily utilized for reporting and Online Analytical Processing (OLAP) [3]. However, their rigid data models and limitations in the support of advanced analytics [5] led to the increased utilization of *data lakes* [11], which store all incoming data in its raw format on scalable and directly accessible storage systems. In general, data warehouses and data lakes possess rather diverging properties and support different kinds of analytics [20].

Since around 2020, so-called *data lakehouses* [2] have found their way into practice. They pursue to combine benefits of data warehouses and data lakes and are typically implemented with open-source frameworks such as Delta Lake, Apache Hudi and Apache Iceberg [1,14,21]. As illustrated in Fig. 1, these frameworks control the read and write access of processing engines for data that resides on highly-scalable storage systems. By managing technical metadata such as log files, they provide additional features for data processing, e.g. relational semantics, ACID guarantees and an increased performance for batch and stream processing [1,14]. Since these frameworks allow to store data similarly to data lakes, but still provide typical capabilities of data warehouses, the resulting data platforms can cover analytical workloads from both worlds.

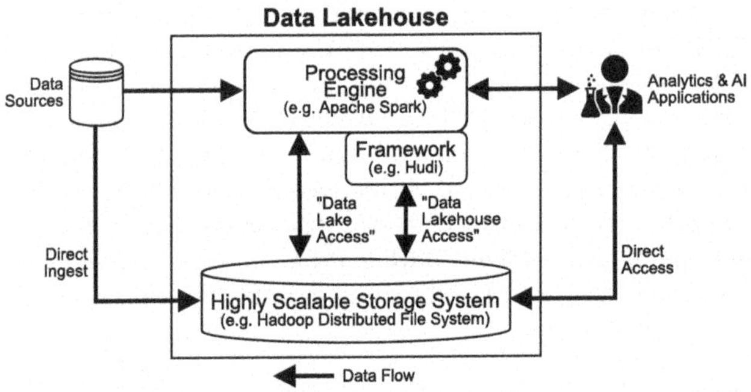

Fig. 1. High-level view on the components of a typical data lakehouse.

However, little is known about data and technology architectures of data lakehouse implementations in industrial practice, e.g. in terms of data modeling and technologies. Also, there is a lack of insights on practical experiences and indications for future research. This leads to the following research questions:

RQ1: What architectures, i.e. data and technology architectures, are used for data lakehouse implementations in practice?
RQ2: What practical experiences have enterprises made during the development and operation of data lakehouse implementations?
RQ3: What directions for future research result from the practical experiences?

To address these questions, we studied four real-world cases at a globally operating manufacturing group, in which data lakehouse implementations have been developed, tested and operated for between one and three years in different productive settings with huge data volumes and different analytical use cases, including AI applications. For each case, we examined its context, systematically analyzed the data and technology architecture of the corresponding implementation and collected practical experiences from the responsible analytics teams. This paper presents our within-case and cross-case results of the case studies. It builds on our previous preliminary work [22], in which we briefly described one exemplary implementation of a data lakehouse in practice. The paper at hands takes this as a starting point to make the following contributions: It a) *reviews related work* from literature (cf. Sect. 2), b) *introduces the method* that we applied, as well as the four cases and their contexts (cf. Sect. 3), c) systematically *analyzes and compares the data and technology architectures* of the investigated implementations (cf. Sect. 4.1), d) derives a *generalized data lakehouse architecture* from the four cases and highlights *interesting findings* (cf. Sect. 4.2), e) discusses the *practical experiences* gathered from the analytics teams (cf. Sect. 5) and f) points to *future research directions* (cf. Sect. 6).

2 Related Work

In literature, several implementations of data lakehouses are presented, spanning different domains like healthcare (cf. [24,29]), cyber security (cf. [1,27]) and telecommunication (cf. [26]). However, these works are not suitable for answering the research questions RQ1-RQ3 due to several reasons:

Lack of Cross-Case Comparisons: To the best of our knowledge, there are no works available yet that compare multiple independent data lakehouse implementations. Instead, the existing works focus only on implementations from individual settings, which limits the generalizability of the insights (cf. [26,27]).

Heterogeneous Architecture Assessments: The architecture assessments in literature differ strongly in their purpose, scope, granularity and form of representation: Some of them are detailed and cover far-reaching properties like data processes and data formats (cf. [4,24]), while others stay high-level and only outline superficial aspects, such as the involved technologies (cf. [15,27]). This heterogeneity impedes a structured analysis and comparison of the architectures due to the lack of an universally applicable architecture framework.

Unclear Level of Maturity: Several works do not explicitly state whether the presented data lakehouses constitute conceptual designs, prototypes or data platforms in productive use (cf. [18,27]), While for some it can be assumed that they are used in production (cf. [4,30]), it remains unclear for how long these data platforms have been operated and hence how stable the insights are.

Missing Practical Experiences: Prolonged experiences and challenges that occurred during the development and operation of a data lakehouse implementation are barely discussed in literature. Consequently, no cross-case comparisons are conducted either, preventing the derivation of future research directions.

To address these shortcomings in our work, we a) designed the four case studies with the goal of performing cross-case comparisons in mind (cf. Sect. 3.1), b) selected and applied an architecture framework as guidance for analyzing and comparing the architectures of the data lakehouse implementations in a systematic manner (cf. Sect. 4), c) only investigated data platforms that have been in productive use for between one and three years and d) explicitly gathered practical experiences and challenges in the course of interviews (cf. Sect. 5).

3 Method and Case Overview

This section first explains the method and scope that we applied in the course of our case studies. Afterwards, the investigated cases are introduced.

3.1 Method and Scope of Study

We decided to conduct case studies [7], as this method is suitable for answering questions of "what" and "how" [19]. Furthermore, case studies are recognized as suitable method for complex topics where context needs to be taken into consideration [7]. These properties apply to the domain of our research questions, as architectures of data platforms are subject to a wide range of organizational and technical requirements, while the practical experiences are likely to depend on contextual factors such as data volumes and the types of analytical use cases.

In the scope of this research, we studied four real-world cases at a globally operating manufacturing group with a world-wide network of factories and suppliers. In each case, one analytics team was in charge of designing, developing and operating a data platform that corresponds to a data lakehouse as outlined in Sect. 1 and has been in productive use for between one and three years. The four analytics teams belong to completely different business units with different employees, processes and IT system landscapes and thus developed their solutions independently of each other for their requirements, without centrally defined standards. Consequently, our approach can be considered a multiple case study [19] that assesses several cases in different settings and hence may lead to more generalizable results in comparison to a single case study.

For each case, we conducted interviews (45–60 minutes) with members of the responsible analytics team, who described themselves as either data engineers, data architects or solution architects and already had prior experience in the development of data warehouses and data lakes. In total, six interviews were conducted, which followed a semi-structured approach with questions prepared in advance[1]. The interviewers also responded dynamically to answers of the participants with suitable follow-up questions. To enable a systematic comparison

[1] For details details on the interview structure and prepared questions, please see: https://gist.github.com/schneijan/41cb39804a7eaa5dac84d14432390b90.

of the data and technology architectures across the cases (cf. RQ1), we utilized the *Data Lake Architecture Framework* by Giebler et al. [8] as reference for the interview questions and guidance for the within- and cross-case analyses. The framework is defined on a generalized level, facilitating its application to various kinds of data-lake-like data platforms, including data lakehouses.

Table 1. Overview of business and IT aspects of the four investigated cases.

		C1: Production Analytics	**C2: MES Analytics**	**C3: Product Lifecycle Analytics**	**C4: Telemetry Analytics**
Business Asp.	**Product Lifecycle Phases**	Production Execution	Production Execution	All	Operations and Service
Business Asp.	**Users**	Data Analysts, Data Scientists	Data Analysts, Data Scientists	Data Analysts, Data Scientists	Data Analysts, Data Scientists, Service Engin.
IT Aspects	**Source Systems**	MES, Image Repository	MES	IoT Devices, CRM, ERP, Social Media	IoT Devices
IT Aspects	**Data Types**	Structured, Unstructured	Structured	Structured	Structured
IT Aspects	**Data Volume**	Terabytes	Petabytes	Terabytes	Terabytes
IT Aspects	**Analytical Workloads**	Reporting, OLAP, DM/ML, NRT Reporting	Reporting, OLAP, DM/ML, Str. Analytics	Reporting, DM/ML	Reporting, OLAP, DM/ML, Exploration
IT Aspects	**Types of Analytics**	Descriptive, Diagnostic, Predictive			

Engin.: Engineers; MES: Manufacturing Execution System; CRM: Customer Relationship Management System; ERP: Enterprise Resource Planning System; DM/ML: Data Mining/Machine Learning; NRT: Near-realtime; Str.: Streaming

3.2 Overview of the Cases

In the following, we provide an overview of the investigated cases at the manufacturing group. For reasons of confidentiality, we generalize certain aspects. Table 1 contains business and IT aspects of the four cases C1-C4. The first business aspect refers to the phases of the industrial product lifecycle [25], for which the data lakehouse is utilized, while the second one lists the roles of the involved users. The IT aspects cover details about the source systems, the types and volumes of the managed data, as well as the envisaged analytical workloads.

Production Analytics (C1): In this case, different electronic components for vehicles are produced along a production line. The involved machines and sensors generate data about the quality of produced components and the condition of the machines. The goal of the analytics team was the construction of a data platform that is capable of collecting, managing and preparing this manufacturing data for various analyses. This also includes near-time reporting, since some analysis results are supposed to be displayed on dashboards along the shop floor. Furthermore, machine learning for the detection of faulty components needs to be carried out. In our previous work [22], we described this case in detail.

MES Analytics (C2): This case deals with the mass production of engine parts. On the shop floor, data from machines and sensors is generated and collected in a Manufacturing Execution System (MES) and then ingested into a data platform. Its goal is to enable the processing of the generated data for flexible ad-hoc analytics, including the generation of Key Performance Indicators (KPIs) and visualizations, data mining and machine learning activities for optimizing the manufacturing processes, as well as streaming analytics.

Product Lifecycle Analytics (C3): In the business unit of this case, different types of appliances for consumers are developed, produced and distributed. Across all phases of the industrial product lifecycle, primarily structured data about the different business processes is collected, including data about products, suppliers and resellers. This data is stored in operational IT systems, such as Customer Relationship Management (CRM) and Enterprise Resource Planning (ERP) systems. Additionally, field data from IoT-enabled appliances is collected and KPIs about social media activities are gathered. All this data needs to be ingested into a data platform to be processed and analyzed. The overall goal is to reduce the development time for new analytics applications (reporting, data mining and machine learning) by creating a single source of truth that provides quick and easy access to all the available data.

Telemetry Analytics (C4): The data platform implementation of this case supports the analyses of data from network-enabled IoT devices that are used by customers. The goal is to enable self-service analytics for reporting, OLAP, data mining and machine learning workloads on structured data. Parts of the data also represent technical logs of the IoT devices, which should be made available to service engineers who need it for identifying problems at the customer site.

4 Architectures of Data Lakehouses in Practice

This section addresses RQ1 by analyzing the data and technology architectures of the four data lakehouse implementations within and across cases. For this purpose, the Data Lake Architecture Framework by Giebler et al. [8] is utilized. This framework breaks down the architecture of a data platform into several aspects, of which the aspects *Data Organization* (A1), *Data Modeling* (A2) and *Data Flow* (A3) belong to the data architecture, while the aspects *Data Storage* (A4) and *Infrastructure* (A5) describe the technology architecture.

In the following, Sect. 4.1 assesses and compares the four data lakehouse implementations regarding these aspects. Subsequently, Sect. 4.2 derives a generalized data lakehouse architecture from the investigated implementations that combines their characteristics and points to interesting findings.

4.1 Analysis of the Data and Technology Architectures

Table 2 characterizes the investigated data lakehouse implementations in terms of the aspects A1-A5 from the architecture framework. The following subsections describe the individual aspects in detail and compare them across the cases.

Table 2. Data & technology architectures of the data lakehouse implementations.

	Aspects				**C1: Production Analytics**	**C2: MES Analytics**	**C3: Product Lifecycle Analytics**	**C4: Telemetry Analytics**
Data Architecture	**A1**	**Data Organization (Data Zones)**			Raw, Harmonization, Delivery, ML	Bronze, Silver, Gold	Raw, Curated, Aggregated, Mach. Learning	Raw, Metadata, Harmonization, Delivery, Tech
	A2	**Data Modeling**			Use-case-drv., Denormalized	Use-case-drv., Normalized	Use-case-drv., Multi-dim.	Use-case-drv., Normalized
	A3		**Data Flow**	Ingestion	Stream, Batch Ingest	Stream	Stream, Batch Ingest	Stream
				Processing	Stream, Batch	Stream, Batch	Batch	Stream
				Consumption	Batch Queries, Stream Sink, DM/ML Tools	Batch Queries, Stream Sink, DM/ML Tools	Batch Queries, DM/ML Tools	Batch Queries, Manual Access
Tech. Arch.	**A4**	**Data Storage**			Object Store			
	A5		**Infrastr.**	Deployment	Azure Cloud	Azure Cloud	Azure Cloud	AWS Cloud
				Storage	ADLS2	ADLS2	ADLS2	S3
				Processing	Databricks	Databricks	Databricks	Amazon EMR
				Framework	Delta Lake	Delta Lake	Delta Lake	Apache Hudi

Use-case-drv.: Use-case-driven; DM/ML: Data Mining/Machine Learning

Data Organization (A1): All data lakehouse implementations leverage the concept of data zones [23] for organizing data of different granularity, quality and application-specificity. However, the numbers and names of the zones differ between the cases. The implementations of C1, C3 and C4 all possess a "Raw" zone. In addition, most of them have a zone that manages consolidated datasets ("Harmonization" in C1 and C4, "Curated" in C3). For reporting and OLAP, C1 and C4 provide a "Delivery" zone, while their "Machine Learning" zones are dedicated to data mining and machine learning activities. Despite its name, the "Metadata" zone of C4 is not related to metadata management, but rather corresponds to another "Delivery" zone containing master data about IoT devices. The zones "Aggregated" of C3, which contains pre-processed use-case-specific data, and "Tech" of C4, which holds log files for the service engineers, have no equivalent zones in the other cases. C2 constitutes an exception with regards to this architectural aspect, as its implementation uses the more generic medallion model [16], in which raw data is stored in the *Bronze* zone, processed and cleansed data in the *Silver* zone and use-case-specific data in the *Gold* zone.

Data Modeling (A2): In all four implementations, a schema-on-read approach with use-case-driven data modeling is carried out, which provides a high degree of flexibility. According to the interview participants, they have no central guidelines that would enforce the usage of certain modeling techniques, such as Data Vault [17]. However, in the cases C2 and C4, the data is already normalized, as it is extracted from the source systems in this shape. In addition, multi-dimensional data models [13] are found in C3. Interestingly, the data of C1 is intentionally de-normalized, which introduces additional redundancy, but increases the query performance, as expensive join operations can be avoided. According to the analytics team, this approach is economically reasonable, as computing power in the cloud is often more expensive than costs for additional storage space.

Data Flow (A3): For data ingestion, an event hub like Apache Kafka is utilized in all four cases, which temporarily stores and buffers stream data from the source systems before it is moved to the data lakehouse. In parallel, C1 and C2 also employ batch processing for data ingestion, such as for image data that needs to be prepared for machine learning. The raw data that is stored on the data platforms is subsequently cleansed and prepared for the different types of analytics applications by using either batch or stream processing. Depending on the types of analytical workloads, the stored data is consumed in different ways: For reporting and OLAP, query engines are utilized to query the available data with SQL statements on behalf of reporting and visualization tools. Data that must be analyzed in near-realtime (cf. C1 and C2) is consumed by appropriate processing engines, such as Apache Spark or Apache Flink, and then made available in dedicated sinks for dashboards or similar applications. For data mining and machine learning, various data science tools like the MLlib module of Apache Spark can be used to work on the stored data. In C4, the service engineers can also access and explore log data directly via the underlying object store, for example in order to search for potential causes of device failures.

Data Storage and Infrastructure (A4, A5): All four investigated implementations are entirely deployed on public clouds. The implementations of C1-C3 leverage similar technology stacks, including Microsoft Azure as cloud provider, Azure Data Lake Storage Gen2 as storage system and the Databricks runtime as processing engine in combination with the Delta Lake framework. Only C4 uses a different stack, which runs on Amazon Web Services and comprises the Amazon Simple Storage Service (S3) as storage system, together with Apache Spark on Amazon EMR and the framework Apache Hudi. Consequently, all four implementations employ highly-scalable object stores for persisting the data.

4.2 Generalized Data Lakehouse Architecture and Findings

To provide a more concise answer to RQ1, we consolidated the architectures of the investigated data lakehouse implementations into a generalized architecture that is illustrated in Fig. 2. It combines the architectural characteristics of the four implementations and was derived by iteratively abstracting and merging their architectures. Hence, the generalized architecture can be seen as a template, of which each of the four data lakehouse implementations is an instance.

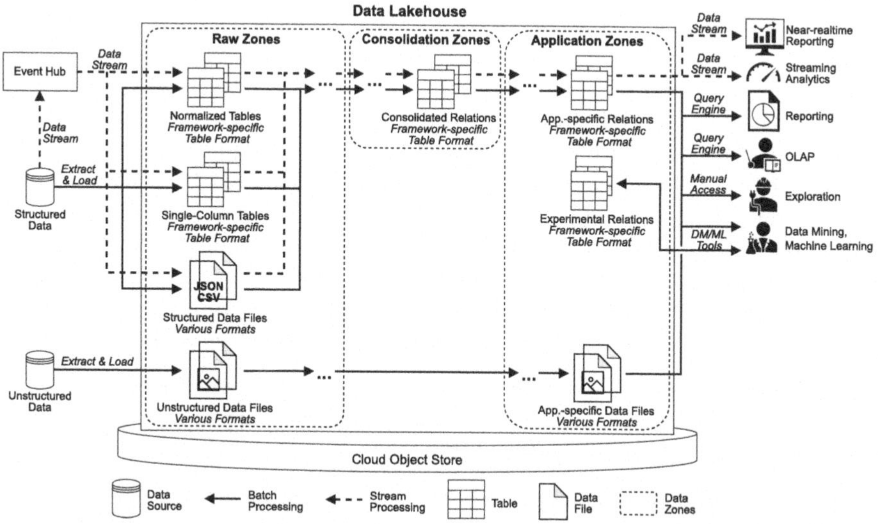

Fig. 2. Generalized data lakehouse architecture as derived from the four cases.

Two types of source data can be distinguished in the generalized architecture: *Structured Data* corresponds to data where the structure is already known in advance, such as measurement data from machines that complies to a pre-defined schema. In contrast, the structure of *Unstructured Data* is not known in advance, which includes image data (cf. C1). Semi-structured data does not occur, as all data in the four cases can be classified as either structured or unstructured data. While the unstructured data is only extracted as batches (cf. C1), the structured data can either be ingested as batches or as unbounded data streams. In the latter case, an event hub like Apache Kafka serves as a buffer.

The data lakehouse itself is based on a *cloud object store* and can be divided into three parts, where each part consists of one or multiple data zones. The first part embodies the *Raw Zones*, which hold raw and only slightly cleansed data. In the architectures of the four examined implementations, we observed three different approaches of how the structured raw data is stored: With *Normalized Tables*, the records of the new data are normalized during the ingestion and added to a table with columns that mostly match the structure of the data. In contrast, *Single-Column Tables* provide only one column, e.g. of type "text", in which the entire data record is stored. Finally, in the *Structured Data Files* approach, the new data is stored as files of its raw format, e.g. as CSV or JSON files. If the data is managed in tables, this is done in the framework-specific table format that is utilized, such as Delta or Hudi tables.

In contrast to the structured data, the unstructured data is always stored in its raw format. Depending on the storage approach that is chosen for the raw data, different processing steps follow, which are either carried out via batch or stream processing. In any case, the structured data eventually arrives in the

Consolidation Zones part of the data platform, which includes zones for the management of harmonized and integrated data that provide a standardized, holistic view (cf. [9]). In these zones, the data is solely stored as tables of the framework-specific format. In the subsequent processing steps, the data is prepared for specific analytics applications. The resulting application-specific data then resides in *Application Zones*, where it can be accessed from users or processes for various analytical purposes. In addition, data scientists can derive new experimental data and store it as new tables in these zones. In the examined data lakehouse implementations, unstructured data is solely processed as batches and only rudimentary prepared before it is made available in an application zone.

In summary, several architectural observations can be made to answer RQ1:

Data Zones for Data Lakehouses: The concept of data zones [11], which has originally been developed for data lakes, is applied in all four cases, which indicates that it has also proven useful in the context of data lakehouses.

Several Storage Approaches for Structured Raw Data: We identified three different approaches for the storage of structured raw data, which is either managed as framework-specific tables or plain data files.

Combination of Batch and Stream Processing: In industrial practice, both batch and stream processing seem to play an important role for the preparation of data. While data lakehouses support both modes in a flexible manner, still additional components for stream processing are required, such as an event hub.

Non-uniform Data Architectures: The four investigated data lakehouse implementations are all deployed in the cloud and leverage similar technology architectures, consisting of an object store, batch and stream processing engines, an event hub and a framework. However, in terms of the data architectures, they diverge noticeably. Based on our observations during the case studies, it can be concluded that apparently no best practices have yet emerged that would recommend modeling patterns for various data types and zones.

5 Practical Experiences

This section addresses research question RQ2 by presenting our cross-case analysis results regarding practical experiences with the data lakehouse approach.

In general, all interview participants had a positive opinion on the data lakehouse concept. Furthermore, they were finally satisfied with their implementation in terms of functional demands, performance, data freshness, interoperability and costs. They also indicated that they would use the same approach again for their present setting, as they saw benefits in comparison to their previously used implementations. Nevertheless, they also described several technical challenges that arose during the development and operation of the data lakehouse implementations, which are summarized in Table 3 and described below.

Table 3. Technical challenges that were encountered during the development and operation of the investigated data lakehouse implementations.

	Encountered Challenges	C1: Production Analytics	C2: MES Analytics	C3: Product Lifecycle Analytics	C4: Telemetry Analytics
CH1	Configuration & Optimization	✗	✗	✗	✗
CH2	Processing Latency	✗	✗	✗	
CH3	Data Management	✗	✗		✗
CH4	Technology Selection		✗		✗

Configuration & Optimization (CH1): In the examined implementations, the frameworks Delta Lake and Apache Hudi constitute key components (cf. Sect. 4). Both frameworks are highly configurable, providing hundreds of configuration parameters that allow to control different aspects, such as the reading and writing behavior, partitioning strategies and background processes. For example, Fig. 3 visualizes a series of measurements that we collected from the implementation of case C2 over a period of 30 days. This metric describes the total volume of storage that has been occupied on the underlying cloud object storage at the end of each day. Accordingly, the storage addressed around 2.69 petabytes of data storage on average, with a negative trend that is caused by public holidays on which the production was reduced. A pattern can be discovered, according to which the data volume continuously increases over the course of each week and then abruptly decreases each Saturday. While the increases can be attributed to new data arriving at the data platform, the decreases can be explained by optimization procedures that are carried out in the background. In particular, they likely reflect the impact of the `VACUUM` command[2] from the Delta Lake framework, which deletes data files that are no longer referenced and have exceeded a retention time, shrinking the data volume. A suitable configuration of this command is crucial to optimize the data platform in terms of performance and costs and has an influence on its operational behavior.

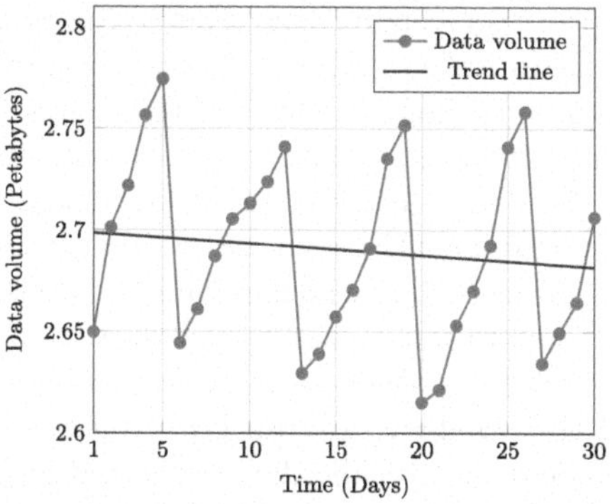

Fig. 3. Total volume of occupied storage space on the cloud object store of case C2 at the end of each day over a period of 30 days.

[2] https://docs.databricks.com/en/delta/vacuum.html.

However, interview participants from all cases reported that finding a suitable configuration constituted a major challenge, as the impact, side-effects and interactions of the parameters were sometimes hard to predict. As a result, the developers were forced to test various configurations through empirical testing in a trial-and-error manner, which was time-consuming and required difficult trade-offs to be made, e.g. with respect to the read and write performance of tables. In addition, they stated that the optimization strongly differed from the tuning of traditional databases and required more technical knowledge.

Processing Latency (CH2): In three cases, latency issues were reported, which affected either SQL queries for reporting and OLAP workloads or stream processing jobs. The participants from C1 and C2 both stated that due to high latencies, data processing was only possible to a limited extent in near-realtime and suspected that the micro-batching approach of Apache Spark may constitute a bottleneck. However, as there are no strict timing requirements, these limitations could be tolerated. In C1, C2 and C3, latency problems were also noticeable when querying the data; however, these problems could be overcome with tunings at the frameworks and adjustments at the utilized compute cluster.

Data Platform Management (CH3): During the interviews for the cases C1, C2 and C4, challenges related to the management of the data lakehouse implementations and the handling of the stored data were mentioned. This pertains various activities at an operational level, such as the discovery and exploration of relevant datasets in different zones, the development, maintenance and documentation of data pipelines, the description of data semantics, the monitoring of data quality, the enforcement of access control, as well as activities related to the collaboration between data producers and consumers. From the view of the interview participants, the data lakehouse approach and the associated technologies constitute primarily a technical solution, but do not offer much support at the operational level. Although external data catalogs can be used for this purpose, they are considered to not sufficiently integrate with the data platform.

Technology Selection (CH4): During the development, it was not clear to the analytics teams which processing engine, e.g. Apache Spark or Apache Flink, and which framework, e.g. Delta Lake or Apache Iceberg, were most suitable for the envisaged use cases and according to which criteria these technologies should be selected. As a result, in both C2 and C4, two frameworks were tried out in parallel, and finally the decision was made in favor of Delta Lake for C2 and Apache Hudi for C4. To cope with these uncertainties, a technology-independent approach was employed in case C4, in which incoming data is stored as structured data files (cf. Sect. 4.2) before it is transformed to a framework-specific table. This allows them to switch to another framework at a later point.

It can be observed that challenges arose both during development and the operation of the implementations. While they generated high satisfaction among the analytics teams, the missing support for data platform management, the selection of suitable technologies and the configuration remain as open challenges.

6 Conclusion and Future Research Directions

Based on our cross-case study of four real-world implementations, we conclude that data lakehouses have proven as data platforms in large-scale industrial settings with petabytes of data, broad user bases and various analytical workloads. Hence, they can be considered state of the art in industrial practice. We found that the architectures of real-world implementations commonly apply data zones, leverage different approaches for the management of raw data, interleave batch and stream processing, use an event hub as buffer for streaming data, comprise similar technology stacks and are typically deployed to public clouds.

Based on our architectural findings and the collected practical experiences and challenges, we derive directions for future research (RQ3): To address the uncertainties regarding the data architecture (cf. Sect. 4.1), *design patterns for data architectures* are required. In addition, the selection of appropriate technologies (cf. challenge CH4), e.g. in terms of processing engines like Apache Spark and frameworks like Apache Hudi, raises the need for *decision guidance*, i.e. criteria that allow to evaluate technologies for different settings. On a similar note, a systematic *method for the configuration and optimization of data lakehouse implementations* (cf. challenges CH1 and CH2) in terms of the utilized frameworks is necessary to ease their adaption in practice. Finally, we currently see the greatest need for research in the area of *data platform management* (cf. challenge CH3). Here, many different activities for various user groups need to be supported and tightly integrated with the data platform to improve its usability and to close the gap between technological solutions and operational processes.

Acknowledgments. We sincerely thank Thomas Kohler for his valuable inputs and profound support during the conduction of the case studies.

References

1. Armbrust, M., Das, T., Sun, L., et al.: Delta lake: high-performance ACID table storage over cloud object stores. Proc. VLDB Endowment **13**(12), 3411–3424 (2020)
2. Armbrust, M., et al.: Lakehouse: a new generation of open platforms that unify data warehousing and advanced analytics. In: 11th Conference on Innovative Data Systems Research (CIDR), Online Proceedings (2021)
3. Baars, H., Kemper, H.G.: Business Intelligence & Analytics. Springer Fachmedien Wiesbaden, Wiesbaden (2021)
4. Begoli, E., Goethert, I., Knight, K.: A lakehouse architecture for the management and analysis of heterogeneous data for biomedical research and mega-biobanks. In: 2021 IEEE International Conference on Big Data, pp. 4643–4651. IEEE (2021)
5. Bose, R.: Advanced analytics: opportunities and challenges. Ind. Manage. Data Syst. **109**(2), 155–172 (2009)
6. Dogan, A., Birant, D.: Machine learning and data mining in manufacturing. Expert Syst. Appl. **166**, 114060 (2021)

7. Dul, J., Hak, T.: Case Study Methodology in Business Research. Routledge, London and New York (2008)
8. Giebler, C., Gröger, C., Hoos, E., Eichler, R., Schwarz, H., Mitschang, B.: The data lake architecture framework: a foundation for building a comprehensive data lake architecture. In: Conference for Database Systems for Business, Technology and Web (BTW), vol. 70469 (2021)
9. Giebler, C., et al.: A zone reference model for enterprise-grade data lake management. In: 2020 IEEE 24th International Enterprise Distributed Object Computing Conference (EDOC), pp. 57–66. IEEE (2020)
10. Gröger, C.: There is no AI without data. Commun. ACM **64**(11), 98–108 (2021)
11. Hai, R., Koutras, C., Quix, C., et al.: Data lakes: a survey of functions and systems. IEEE Trans. Knowl. Data Eng. **35**(12), 12571–12590 (2023)
12. Inmon, W.H.: Building the Data Warehouse. Wiley technology publishing Timely, practical, reliable, Wiley, Indianapolis, Ind., 4th edition edn. (2005)
13. Iqbal, M., et al.: A review of star schema and snowflakes schema. In: Intelligent Technologies and Applications, Communications in Computer and Information Science, vol. 1198, pp. 129–140. Springer Singapore (2020)
14. Jain, P., et al.: Analyzing and comparing lakehouse storage systems. In: Proceedings of the 13th Annual Conference on Innovative Data Systems Research (2023)
15. Kumar, D., Li, S.: Separating storage and compute with the databricks lakehouse platform. In: 2022 IEEE 9th International Conference on Data Science and Advanced Analytics (DSAA), pp. 1–2. IEEE (2022)
16. L'Esteve, R.C.: The Cloud Leader's Handbook. Apress, Berkeley, CA (2023)
17. Linstedt, D., Olschimke, M.: Building a Scalable Data Warehouse with Data Vault 2.0. Morgan Kaufmann (2015)
18. Liu, G., et al.: IoT Lakehouse: a new data management paradigm for AIoT. In: Big Data - BigData 2023, Lecture Notes in Computer Science, vol. 14203, pp. 34–47. Springer Nature Switzerland (2023)
19. Meredith, J.: Building operations management theory through case and field research. J. Oper. Manag. **16**(4), 441–454 (1998)
20. Nambiar, A., Mundra, D.: An overview of data warehouse and data lake in modern enterprise data management. Big Data Cogn. Comput. **6**(4), 132 (2022)
21. Schneider, J., et al.: The lakehouse: state of the art on concepts and technologies. SN Comput. Sci. **5**(5) (2024)
22. Schneider, J., et al.: First experiences on the application of lakehouses in industrial practice. In: Proceedings of the 35th GI-Workshop Grundlagen von Datenbanken, vol. 3710, pp. 3–8 (2024)
23. Sharma, B.: Architecting Data Lakes: Data Management Architectures for Advanced Business Use Cases. O'Reilly Media, Sebastopol, CA, second edition edn. (2018)
24. Siddiqi, M.H., Idris, M., Alruwaili, M.: FAIR health informatics: a health informatics framework for verifiable and explainable data analysis. Healthcare (Basel, Switzerland) **11**(12) (2023)
25. Stark, J.: Product Lifecycle Management (Vol. 1): 21st Century Paradigm for Product Realisation. Decision Engineering, Springer International, 4th edn. (2020)
26. Tang, X., et al.: Separation is for better reunion: data lake storage at Huawei. In: 2024 IEEE 40th International Conference on Data Engineering (ICDE), pp. 5142–5155. IEEE (2024)
27. Tovarnak, D., Racek, M., Velan, P.: Cloud native data platform for network telemetry and analytics. In: 2021 17th International Conference on Network and Service Management (CNSM), pp. 394–396. IEEE (2021)

28. Wang, J., Xu, C., Zhang, J., Zhong, R.: Big data analytics for intelligent manufacturing systems: a review. J. Manuf. Syst. **62**, 738–752 (2022)
29. Xiao, Q., et al.: MHDML: construction of a medical lakehouse for multi-source heterogeneous data. In: Health Information Science, Lecture Notes in Computer Science, vol. 13705, pp. 127–135 (2022)
30. Zhang, Y., Peng, B., Du, Y., Su, J.: GeoLake: bringing geospatial support to lakehouses. IEEE Access **11**, 143037–143049 (2023)

Reducing Risks of Continuous Deployments for SaaS Excellence: A Hybrid Risk-Model Matrix Approach Implementation of Continuous Deployments

Dagmara Makula[1,2] and Torben Tambo[1(✉)]

[1] Aarhus University, 7400 Herning, Denmark
dagamakula@gmail.com, torbento@btech.au.dk
[2] Stibo Systems A/S, Axel Kiers Vej 11, 8270 Højbjerg, Denmark

Abstract. This study addresses risk management in the transition to continuous deployment at Stibo Systems, a company evolving as a Software-as-a-Service (SaaS) provider. The purpose is to identify risks associated with the newly adopted quarterly update policy, evaluate their impact, and propose mitigation strategies. Adopting a pragmatic case study approach, qualitative data from expert interviews and quantitative survey results inform the analysis. Using risk assessment frameworks such as Failure Mode and Effects Analysis (FMEA), the study classifies risks into internal and external dimensions, highlighting issues like organizational misalignment and customer adaptation challenges. This research's originality lies in applying established risk frameworks to a SaaS transition context. Key findings include identifying process inefficiencies and service disruptions during updates. The study provides actionable recommendations for enhancing organizational resilience and customer satisfaction, contributing to smoother SaaS adoption. These insights can guide similar transitions in the IT sector.

Keywords: Software-as-a-Service · Continuous Deployment · Risk Management · Risk Matrix Approach · Business Excellence · Cloud Computing

1 Introduction

The shift towards cloud-based solutions has profoundly transformed the software industry, with Software as a Service (SaaS) emerging as a dominant delivery model. SaaS offers organizations the ability to access software through subscription-based services hosted on the cloud, reducing infrastructure costs and improving scalability [15, 2]. This transformation not only simplifies IT management but also enables businesses to focus on their core operations while benefiting from continuous software updates and enhanced collaboration [14]. However, transitioning to SaaS involves challenges, including data security, vendor lock-in, and the need for organizational change management [6].

Continuous deployment (CD) is a critical practice in SaaS delivery, enabling frequent, automated updates that improve software quality and reduce time-to-market [11]. By

K. Węcel (Ed.): BIS 2025, LNBIP 554, pp. 256–268, 2025.
https://doi.org/10.1007/978-3-031-94193-1_19

deploying incremental changes directly to production, CD fosters agility, adaptability, and rapid feedback cycles [24]. However, adopting CD requires significant changes in organizational culture, technical infrastructure, and risk management practices [3]. The need to align these elements highlights the complexity of achieving a "true SaaS provider" status.

Stibo Systems, a global provider of Master Data Management software, has undergone a substantial transition from delivering on-premises solutions to offering a cloud-hosted SaaS product. In its latest strategic initiative, the company announced a "Quarterly Update" policy to enforce regular updates, signaling a commitment to continuous deployment. While this approach promises operational efficiencies and enhanced customer experience, it also introduces risks such as system disruptions, resource overload, and customer resistance. Effective risk management is essential to address these challenges and ensure the success of the policy [4, 13].

This research seeks to explore how Stibo Systems can reduce risks associated with implementing its continuous update policy. To achieve this, the study adopts a multidisciplinary approach, combining theories of cloud computing, SaaS, continuous deployment, and risk analysis.

Research Question: How can Stibo Systems effectively reduce the risks associated with implementing a continuous update policy?

Subquestions: (1) What are the key risks faced by Stibo Systems when adopting a continuous update policy? (2) How can these risks be prioritized and evaluated? (3) What strategies can mitigate the identified risks and support effective implementation?

This study aims to provide actionable insights to facilitate Stibo Systems' transition toward SaaS excellence while minimizing associated risks.

2 Theoretical Background

This theoretical review synthesizes existing literature to establish a foundation for analyzing the transition to Software as a Service (SaaS) and continuous deployment. It examines cloud computing, SaaS, continuous deployment, and risk management, emphasizing both their benefits and inherent challenges.

2.1 Cloud Computing

Cloud computing is a transformative IT model that delivers computing resources over networks on demand [2]. It supports resource sharing, dynamic scalability, and cost-effective operations, replacing significant upfront capital investments with operational expenses [15]. The National Institute of Standards and Technology highlights its core characteristics: self-service, device independence, multitenancy, and resource pooling [33].

The four types of cloud environments—public, private, community, and hybrid—cater to varying organizational needs [15]. Public clouds, offered by third-party providers, are cost-effective for SMEs, while private clouds provide more control for large enterprises. Hybrid models combine the two, offering flexibility for non-critical and business-critical data management [12].

Cloud computing's benefits include rapid scalability, cost reduction, and operational efficiency [2]. However, concerns over security, governance, and vendor lock-in remain significant barriers [22]. Effective change management, as emphasized by Dehmlow, requires organizational adjustments, from articulating the need for change to involving leadership in decision-making.

2.2 Software as a Service (SaaS)

SaaS, a prominent cloud computing model, enables users to access software and data via the internet without requiring local installations [18]. This model supports scalability, cost efficiency, and improved collaboration [14]. SaaS is particularly advantageous for its flexibility, central management of updates, and reduced operational costs.

Historically, SaaS adoption surged due to globalization, cost-cutting pressures, and the growing complexity of internal IT infrastructure [19]. SaaS's core value lies in treating IT as an expense rather than an asset, simplifying cost structures, and integrating services like customization and maintenance [21].

Despite its benefits, SaaS presents challenges such as data security, platform reliability, and performance issues. Customers often hesitate due to concerns about proprietary information and system failures [6]. Building trust through robust security and transparent communication is critical for addressing these challenges [14].

2.3 Continuous Deployment and Integration

Continuous deployment (CD) expands on continuous integration (CI) by automating the testing and deployment of incremental changes into production environments [11]. CI emphasizes daily code integration and testing, while CD ensures the software is always production-ready [9].

The key benefits of continuous deployment include:

- Shortened Time-to-Market: Smaller, frequent updates reduce release cycles, enhancing responsiveness to market demands [23].
- Improved Quality: Regular, focused testing of smaller code batches minimizes bugs [20].
- Enhanced Customer Satisfaction: Rapid issue resolution and feature delivery align with user expectations [28].

However, CD adoption is challenging, requiring agile practices, robust testing infrastructure, and cultural shifts toward transparency and collaboration [24]. Organizations must also address architectural and tooling limitations to enable efficient deployment pipelines [3].

2.4 Risk Analysis and Management

Risk management is a critical element in SaaS and CD transitions, encompassing risk identification, evaluation, and mitigation [4]. Effective risk management reduces project failures and aligns strategies with organizational objectives [13].

Risk Matrix Approach (RMA). RMA quantifies risk as the product of probability and impact, offering a logical method for prioritization [17]. It's graphical representation simplifies risk comprehension, though it struggles with complex or interdependent risks [27].

Failure Mode and Effects Analysis (FMEA). FMEA evaluates risks using the Risk Priority Number (RPN), calculated from severity, occurrence, and detection scores [26]. While widely used, FMEA's rigid scoring system may overlook nuanced interactions between risk factors [25].

Fault Tree Analysis (FTA). FTA employs a deductive approach to trace root causes of system failures. It visualizes logical event sequences, aiding in system optimization and reliability improvements [29]. Despite its benefits, FTA's manual effort and inability to model dynamic systems limit its application [30].

These methods underscore the need for prioritization, integrating risk analysis into broader strategies, and aligning mitigation efforts with organizational objectives [32].

2.5 SaaS, DevOps Culture and Continuous Practices

Adopting DevOps principles is essential for successful continuous deployment. DevOps fosters collaboration, trust, and shared responsibility, integrating people, processes, and technology [31]. Automation plays a key role, enabling frequent and reliable deployments while reducing the likelihood of defects [5].

Transitioning to SaaS requires balancing technical innovation with organizational adaptation. Businesses must address risks, establish trust with customers, and restructure internal processes [14]. Theories on cloud computing, SaaS, and continuous deployment provide a roadmap for managing these complexities, supporting resilience and competitiveness in dynamic markets.

This theoretical framework highlights the interplay between cloud computing, SaaS, and continuous deployment, emphasizing their transformative potential alongside associated challenges. By leveraging robust risk management frameworks and embracing DevOps culture, organizations can navigate the transition effectively, ensuring long-term success in the SaaS-driven era.

3 Methodological Considerations

The research employs a structured methodology inspired by the "Research Onion" framework [16], integrating qualitative and quantitative approaches to investigate the risks associated with Stibo Systems' transition to a continuous update policy. A pragmatic research philosophy underpins the study, emphasizing actionable outcomes over theoretical abstraction [7].

A deductive approach guides the research, beginning with established theories in SaaS, continuous deployment, and risk management, followed by empirical data collection to analyze and apply these concepts [11, 4].

The study adopts a case study design, focusing on Stibo Systems to capture its specific context and challenges [8].

Data collection combines qualitative semi-structured interviews and quantitative surveys. Interviews with 20 stakeholders, including executives and subject matter experts, provided in-depth insights into risks and strategies [1]. Quantitative surveys complemented these findings by prioritizing risks based on impact and likelihood, aligning with frameworks like the Risk Matrix Approach [17].

The Gioia methodology [10] was used to analyze qualitative data, categorizing risks into first-order concepts and higher-order dimensions. This iterative process ensured robust data triangulation, combining stakeholder perspectives with theoretical insights to generate actionable recommendations.

This mixed-methods approach ensures a comprehensive understanding of the problem and informs targeted risk mitigation strategies. Triangulation was further assured with the cross-organizational representation of respondents and the depth and breadth of the questions.

4 Empirical Work

4.1 Interview and Stakeholder Data Elicitation

Stakeholder interviews were the primary data collection. Initial interviews where weld with key stakeholders to understand Stibo Systems' SaaS journey and challenges. Subsequently risk interviews were held with Subject matter experts (SMEs) from different departments to identify potential risks associated with the quarterly update policy. Afterwards data was processed in the following steps:

- Risk Identification and Categorization: Interview findings were analyzed using qualitative methodologies to identify and group risks.
- Risk Evaluation: Quantitative surveys and internal documents were used to prioritize risks based on impact and likelihood.
- Risk Analysis and Mitigation: Strategies for managing key risks were proposed, considering both theoretical and practical insights.

Furthermore, the initial interviews provided a comprehensive overview of Stibo Systems' transition from on-premises solutions to SaaS. Five experts from leadership roles were interviewed, revealing the advantages of SaaS, including scalability, cost efficiency, and enhanced customer experience. However, the interviews also highlighted the challenges of becoming a "true SaaS provider," particularly the need for organizational changes and the adoption of new operational practices.

Key insights included: The criticality of aligning technical and organizational strategies. Also, the significance of continuous deployment in achieving SaaS excellence.

The potential for customer resistance and internal misalignment during major transitions.

The interview process used 15 fifteen SMEs (Fig. 1.) representing different departments participated in individual interviews to identify risks associated with the quarterly update policy. These interviews revealed a range of concerns, categorized as internal and external risks. These were further divided into classes using the Gioia principles. The SMEs emphasized the interconnected nature of these risks, highlighting the need for a comprehensive risk management strategy.

A1	Internal Risk, Process,	Updates negatively impact Stibo's ongoing projects with customers	D2	Internal Risks, Software,	Testing period is too short to ensure successful releases
A2	Internal Risk, Process,	The number of Jira tickets will substantially increase	D3	Internal Risks, Software,	Custom extensions don't get updated before each STEP release
A3	Internal Risk, Process,	Service levels and response time will decrease due to overload of resources and increased Jira tickets	D4	Internal Risks, Software,	Self-service portal is not working properly
B1	Internal Risk, Planning & Strategy,	Employees unsure what strategy to follow in case of exception handling	H1	External Risks, Process,	Customers are not ready for the upgrade by the deadline
B2	Internal Risk, Planning & Strategy,	Employees unsure what to prioritize in cases of overload due to no clear strategy	H2	External Risks, Process,	Many customers wait until the last moment (Upgrading at the same time)
B3	Internal Risk, Planning & Strategy,	Stibo cannot follow up and monitor the progress of the new policy, because data is inconsistent	H3	External Risks, Process,	Customers do not self-upgrade
B4	Internal Risk, Planning & Strategy,	No one monitors the cross-department progress of the policy	H4	External Risks, Process,	Customers are not upgrading on time in accordance to the policy, and Stibo must do it for them
B5	Internal Risk, Planning & Strategy,	No one monitors the post-upgrade feedback & lessons learned	H5	External Risks, Process,	Customers try to self-upgrade and the upgrade crashes
B6	Internal Risk, Planning & Strategy,	No overall project plan gets prepared and ownership (PM) to coordinate between departments	H6	External Risks, Process,	Customers are not aware that they must upgrade their self-developed custom extensions
B7	Internal Risk, Planning & Strategy,	No clear schedule gets prepared for customers and their upgrades	H7	External Risks, Process,	Customers' first forced upgrade is a big upgrade
C1	Internal Risk, Internal Communication	Stibo's departments work on upgrade aspects in silos and do not communicate with each other	H8	External Risks, Process,	Customers will not use preview sandboxes
C2	Internal Risk, Internal Communication	Employees get confused with upgrade policy due to incomplete or insufficient information	H9	External Risks, Process,	Upgrades fall in a wrong time
C3	Internal Risks, Internal Communication,	Departments aren't prepared for impacts of the upgrade due to lack of communication	H10	External Risks, Process,	Upgrades affect customers' ongoing internal projects
C4	Internal Risks, Internal Communication,	Overall resource priorities are not clear communicated	H11	External Risks, Process,	Partner systems updates affect ongoing activities
F1	Internal Risks, External Communication,	Information provided to customers proves to be insufficient	H12	External Risks, Process,	External integration services fail
F2	Internal Risks, External Communication,	Information regarding the upgrade does not reach the customer	G1	External Risks, Customer Relations,	Customers are dissatisfied with the new policy
F3	Internal Risks, External Communication,	Inconsistency in Stibo's communication to customers, confusing the customers	G2	External Risks, Customer Relations,	Customers raise many escalations
F4	Internal Risks, External Communication,	Inconsistency in Stibo's communication and actions	G3	External Risks, Customer Relations,	Customers leave Stibo
F5	Internal Risks, External Communication,	Stibo's documentation doesn't get updated with each release	I1	External Risks, Policy,	Regulated customers refuse to follow the policy (legal compliance)
F6	Internal Risks, External Communication,	Stibo's marketing materials are not up to date with every release	I2	External Risks, Policy,	On the deadline of the upgrade customers refuse to update, claiming they are not contractually bound to follow the policy
E1	Internal Risks, Resources,	Support Team becomes overloaded with tickets and escalations	I3	External Risks, Policy,	Customers do not want to move from on-prem to SaaS due to the upgrade policy for SaaS customers
E2	Internal Risks, Resources,	Global Technology Services department becomes overloaded	I4	External Risks, Policy,	Customers hesitant to follow the new policy due to lack of trust in reliability of the STEP upgrades
E3	Internal Risks, Resources,	Not enough resources to run more on-prem migration projects	I5	External Risks, Policy,	Customers ask for exceptions and want to upgrade less frequently
E4	Internal Risks, Resources,	PS Resources (Consultants and Architects) become overloaded	I6	External Risks, Policy,	Customers refuse to follow the policy and strongly push back
D1	Internal Risks, Software,	New releases have quality issues or are unstable			

Fig. 1. Risk groups and factors

4.2 Risk Identification and Categorization

The qualitative data from the interviews were analyzed using the Gioia methodology, which organizes findings into three levels: representative quotes, first-order concepts, and higher-order dimensions. This approach identified the 49 risks above, grouped into nine categories across internal and external dimensions. Internal Risks were dominant, accounting for 28 of the identified risks. They were categorized into areas such as process inefficiencies, planning challenges, and resource constraints. External Risks encompassed 21 risks, focusing on customer-related issues and market dynamics. A visual mapping of these risks was created to illustrate their relationships and potential impacts, providing a structured framework for subsequent analysis.

The identified risks were evaluated using quantitative surveys distributed to stakeholders. Risks were scored based on their **probability, impact, detectability, and recoverability**. This prioritization process provided a clear ranking of risks, enabling the development of targeted mitigation strategies.

Key findings included:

- Process-related risks, such as increased Jira tickets and resource overload, ranked highest due to their immediate operational impacts.
- Planning and strategy risks were significant, particularly the lack of coordinated monitoring and exception-handling strategies.
- Customer-related risks, while external, were deemed critical due to their potential to affect customer trust and satisfaction.

5 Data Analysis and Risk Recommendations

5.1 Data Analysis and Risk Evaluation

To determine the ultimate score for each characteristic of every risk, a mean average is applied. For enhanced comprehension and analysis of the data, the minimum and maximum values assigned to each risk have been incorporated. Additionally, a median is established for each risk, offering insights into how closely the average score aligns with the central value of the numerical set in the results.

For each of the three dimensions, three types of graphs will be presented, to visualize the data: bar chart, pie chart, and a box plot chart (Fig. 2.). The bar chart will enable a quick graphic representation of the score of each risk, allowing to promptly spot any outliers. The pie chart will gather the scores of the risk in three categories: low (score below 2), medium (score between 2 and 4), and high (score above 4), showing the proportions of the scores. Finally, a box plot chart will be presented. A box plot chart facilitates a deeper understanding of the details of the scores provided by the interviewees for each risk and thus enhances data interpretation. In the case of this survey, it is substantial, as fourteen people were evaluating forty-nine risks. A box plot chart graphically represents information such as: IQR, range, mean, median and outliers. The whiskers (lines extending from the boxes) represent the range. The boxes represent central tendency – IQR (Interquartile Range). IQR is a statistic measure used in data analysis, which shows the difference between the third quartile (Q3), which represents 75% of the data set, and the first quartile (Q1), which represents 25% of the data set. It is used for representing the central tendency in the dataset, and shows another angle to the simple range, because it is less affected by the potential outliers. Inside of the boxes, a straight line can be found, which represents the median. The mean is represented by the "X" symbol.

To enable risk evaluation, a survey was plotted, including all 49 risks. Applying the principles of the Risk Matrix Approach (Fig. 3.), the 15 interviewees were tasked with assessing risks on a scale from 1 to 5 on three dimensions: impact, probability, following the RMA theory, and detection, responding to the third dimension – FMEA. The numerical understanding of the scoring was presented as follows: 1 (very low), 2 (low), 3 (medium), 4 (high), 5 (very high). Despite FMEA usually using the score from 1–10, for the purposes of alignment with the RMA, and for the ease and visual simplicity and speed for the interviewees, the detection dimension was also scored on a scale from 1 to 5.

The Risk Priority Number (RPN) (Fig. 4.) is a value used in the FMEA risk analysis approach, which helps with risk prioritizing. The RPN is calculated by multiplying three

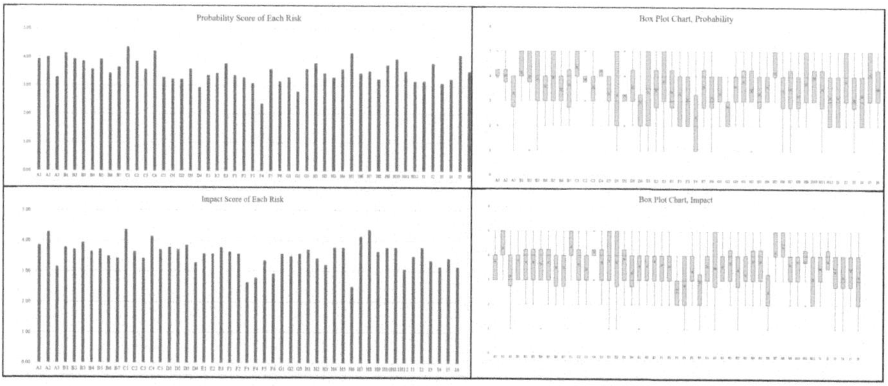

Fig. 2. Impact and probability for each factor with box plot

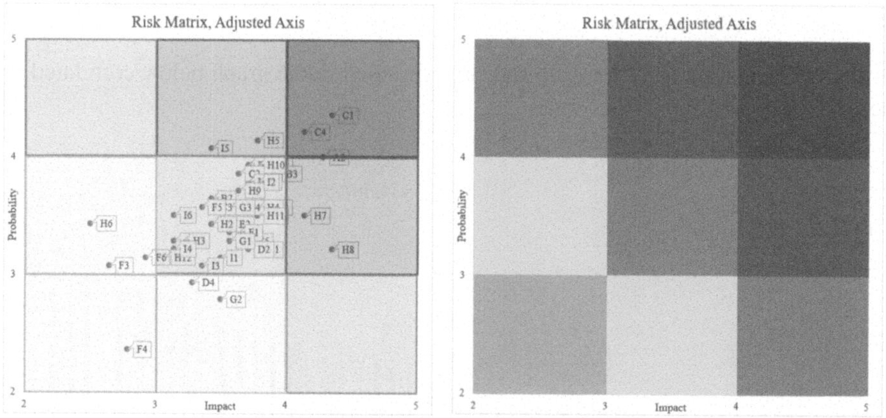

Fig. 3. Risk Matrix with classification

factors: Severity (Impact), Occurrence (Probability) (O), and Detection (D). The formula for calculating Risk Priority Number (RPN) is:

$$\mathbf{RPN} = \mathbf{S} * \mathbf{O} * \mathbf{D} \text{ or } \mathbf{RPN} = \mathbf{I} * \mathbf{P} * \mathbf{O}$$

Once RPN values are calculated for all identified risks, they can be used to prioritize risks. The higher the RPN, the higher the priority of addressing that particular risk. The values for Impact, Probability and Detection have been gathered in the risk evaluation survey.

It is important to highlight that, given the main choice for the risk analysis in this report is the Risk Matrix Approach (Fig. 5.), using a scale of 1–5 is preferred. This choice extends to the evaluation of detection, simplifying the process for interviewees by maintaining consistency with the 1–5 scale instead of the conventional 1–10 scale. Despite this adjustment, the outcomes remain proportionate to what they would be on a higher scale, offering valid insights into identifying risks with the highest priority. The

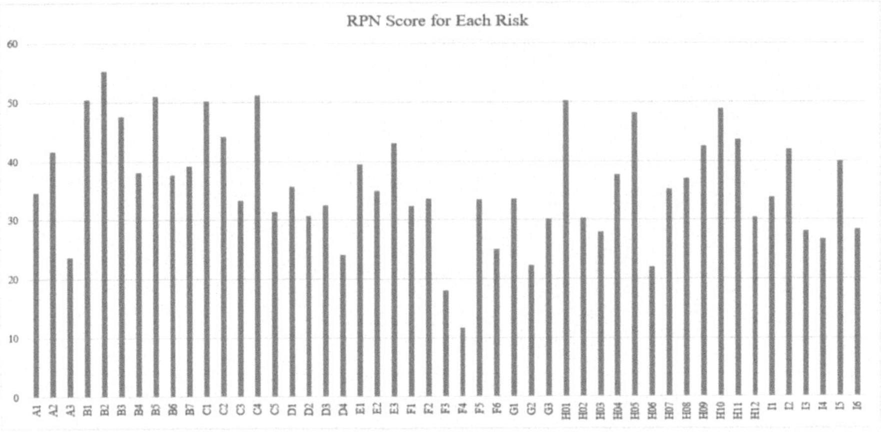

Fig. 4. Risk Priority Number

results of calculating RPN for each risk are presented in the graph below correlated to the risk matrix coding.

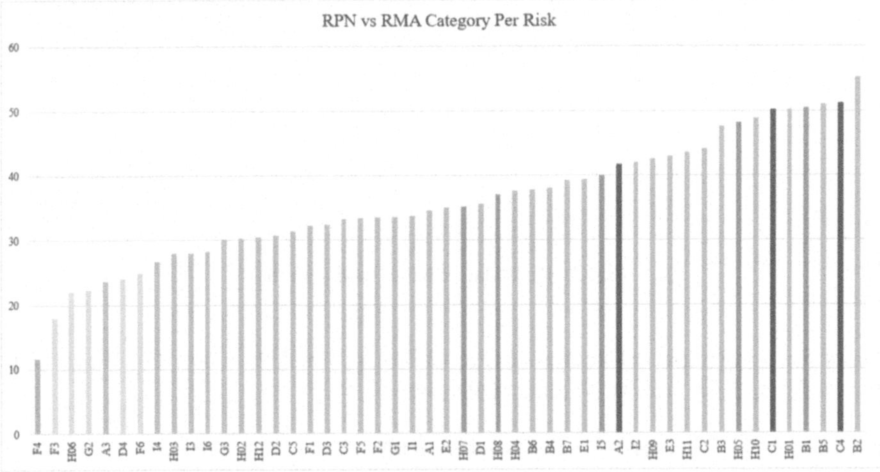

Fig. 5. RPN vs Matrix Category

5.2 Risk Mitigation and Recommendations

To address the identified risks, a set of actionable recommendations was developed, categorized into process improvements, strategic planning, resource allocation, and customer engagement.

To mitigate operational bottlenecks arising from quarterly updates, organizations should adopt centralized and automated solutions. Establishing a dedicated support team to manage Jira ticket inflows ensures efficient resolution of customer issues and prevents resource overload during peak periods. Complementing this effort with automated monitoring tools enables the tracking of service levels and response times, allowing for the early detection and management of delays or backlogs. These measures collectively reduce operational disruptions and enhance overall efficiency.

Planning and strategy play a pivotal role in fostering cross-departmental coordination and setting clear priorities. Documenting exception-handling policies provides teams with clear guidelines to manage resource constraints, reducing ambiguity and ensuring consistent responses. Furthermore, assigning a dedicated project manager to oversee updates facilitates seamless communication across departments, tracks progress, and maintains accountability. Sharing a comprehensive upgrade schedule with customers further strengthens this strategy by allowing them to prepare adequately for updates, thereby minimizing disruptions.

To address the increased workload and technical demands associated with quarterly updates, organizations must invest in resource and competency development. Targeted training programs can enhance employee skills in areas such as continuous deployment, customer support, and risk management. Additionally, hiring additional personnel in critical areas, such as customer support and technical operations, ensures that peak workloads are managed effectively. These measures not only bolster operational capacity but also prepare the workforce to adapt to evolving demands.

Building customer trust is integral to the successful implementation of any update policy. Proactive communication, such as conducting webinars, workshops, and email campaigns, helps articulate the benefits of quarterly updates while addressing customer concerns. Providing comprehensive training resources, including tutorials and live support, empowers customers to navigate the changes with confidence. To maintain a feedback-driven approach, organizations should implement mechanisms such as surveys and post-update feedback loops to identify areas of improvement and refine future releases.

Certain risks require tailored strategies to ensure mitigation. For instance, the anticipated surge in Jira tickets during update cycles poses a challenge that can be addressed by increasing staff availability during peak periods and creating FAQs or self-help resources to resolve common issues. Resistance from customers, another potential risk, can be mitigated through proactive engagement. Co-creating update schedules and offering tailored support fosters collaboration and reduces dissatisfaction. Additionally, the lack of centralized monitoring across departments can lead to inefficiencies. Appointing a project manager ensures clear communication, progress tracking, and centralized accountability, addressing this issue effectively.

The broader implications of these recommendations extend beyond immediate operational improvements. By aligning with best practices in SaaS and continuous deployment, organizations can enhance their efficiency during updates, strengthen customer trust, and foster a culture of adaptability and collaboration. These strategies not only serve as a blueprint for organizations like Stibo Systems but also provide a roadmap for other companies transitioning to SaaS models. Emphasizing risk prioritization and stakeholder engagement, the proposed solutions underscore the importance of integrating operational resilience into long-term strategic planning.

6 Discussion and Conclusion

6.1 Implementation

To validate the hybrid-RMA framework, future application within live or simulated deployment scenarios is essential. This would involve pilot testing the risk prioritization model across selected projects, followed by stakeholder feedback to evaluate its accuracy and usability. Implementation at Stibo Systems can begin with training internal teams on using the framework, integrating it into quarterly planning cycles, and establishing ownership for updating risk scores. The use of familiar tools like risk matrices and RPN ensures easy adoption, while Fault Tree Analysis enhances root-cause focus. Benefits of the framework include structured risk visibility, optimized resource allocation, and quicker decision-making in dynamic environments. It supports continuous improvement by offering a standardized method for evaluating and responding to risks across departments. Over time, consistent use will foster a proactive risk culture, enabling Stibo to strengthen its SaaS operations and maintain customer trust during continuous deployment cycles.

6.2 Discussion

The findings revealed that internal risks—such as increased Jira tickets, service delays, and unclear cross-departmental strategies—were more prevalent and immediate compared to external risks, like customer resistance and market perceptions. These internal challenges highlight gaps in organizational readiness for continuous deployment, especially in resource allocation and process coordination.

From an academic standpoint, the reliance on frameworks like the Risk Matrix Approach [17] and Gioia methodology [10] provided a structured way to assess and prioritize risks. However, these methods have limitations in addressing complex, interconnected risks. For instance, while RMA simplifies risk visualization, it struggles with multi-dimensional relationships. Similarly, qualitative approaches, though insightful, can be subjective and prone to bias.

The proposed strategies—centralized support management, proactive customer engagement, and cross-departmental monitoring—align with best practices in SaaS literature [11, 14]. However, implementing these measures requires significant organizational change, cultural shifts, and investment in employee training. The findings underscore the tension between theoretical ideals of continuous deployment and the practical challenges of execution, particularly for companies transitioning from legacy systems.

6.3 Conclusions to Research Questions

Key risks are seen in the form of Stibo facing internal risks (e.g., resource overload and unclear strategies) and external risks (e.g., customer resistance). Process inefficiencies and service-level impacts are the most critical.

Risk Prioritization: Using quantitative surveys, risks were ranked based on likelihood, impact, and recoverability. Internal risks dominated due to their operational immediacy.

Risk Mitigation Strategies: Recommendations include centralized support, automated tools for monitoring, enhanced employee training, and proactive customer communication.

The study concludes that while continuous updates promise operational and customer benefits, their success depends on addressing risks through structured management, cross-departmental coordination, and robust stakeholder engagement. These findings provide actionable insights for Stibo Systems and similar SaaS organizations.

References

1. Adams, J.: Conducting semi-structured interviews. In: Handbook of Practical Research Methods, pp. 321–335. Sage Publications, Thousand Oaks (2015)
2. Ali, M.A., Khan, S.U., Vasilakos, A.V.: Security in cloud computing: opportunities and challenges. Inf. Sci. **305**, 357–383 (2017)
3. Ard, C., Tomasi, S., Li, J.: Overcoming architectural challenges in continuous deployment. IEEE Softw. **31**(6), 50–56 (2014)
4. Boehm, B.W.: Software risk management: principles and practices. IEEE Softw. **8**(1), 32–41 (1991)
5. Chandrasekara, K., Herath, H.M.T.: Agile methodologies in cloud computing: a continuous delivery perspective. ACM Comput. Surv. **52**(5), 1–38 (2020)
6. Chou, S., Chiang, P.: Understanding the role of trust in cloud computing. Comput. Hum. Behav. **33**, 251–259 (2013)
7. Creswell, J.W.: Research Design: Qualitative, Quantitative, and Mixed Methods Approaches, 3rd edn. Sage Publications, Thousand Oaks (2009)
8. Flyvbjerg, B.: Five misunderstandings about case-study research. Qual. Inq. **12**(2), 219–245 (2006)
9. Fowler, M.: Continuous Integration. ThoughtWorks (2006). https://martinfowler.com/articles/continuousIntegration.html
10. Gioia, D.A., Corley, K.G., Hamilton, A.L.: Seeking qualitative rigor in inductive research: notes on the Gioia methodology. Organ. Res. Methods **16**(1), 15–31 (2013)
11. Humble, J., Farley, D.: Continuous Delivery: Reliable Software Releases through Build, Test, and Deployment Automation. Addison-Wesley, Upper Saddle River (2010)
12. Kaltenecker, N., Hess, T., Huesig, S.: Managing the transition to cloud business models: what should companies do? IEEE Trans. Eng. Manage. **62**(3), 351–362 (2015)
13. Keil, M., Cule, P.E., Lyytinen, K., Schmidt, R.C.: A framework for identifying software project risks. Commun. ACM **41**(11), 76–83 (1998)
14. Lal, R., Bharadwaj, S.: Managing risks in cloud-based SaaS solutions. MIS Q. Exec. **14**(1), 45–59 (2015)
15. Marston, S., Li, Z., Bandyopadhyay, S., Zhang, J., Ghalsasi, A.: Cloud computing: the business perspective. Decis. Support Syst. **51**(1), 176–189 (2011)
16. Saunders, M., Lewis, P., Thornhill, A.: Research Methods for Business Students, 8th edn. Pearson Education, Harlow (2019)
17. Szaciłło, A., Wójcik, M., Piotrowski, J.: Advanced risk matrix approach for managing complex risks in IT projects. J. Risk Res. **25**(2), 178–193 (2022)
18. Tan, C.-W., Kim, J.: Cloud computing and SaaS models: current trends and future directions. MIS Q. Executive **14**(3), pp. 73–90 (2015). Author, F.: Article title. Journal 2(5), 99–110 (2016)
19. Demirkan, H., Cheng, H.K., Bandyopadhyay, S.: Coordination strategies in a SaaS supply chain. J. Manage. Inf. Syst. **26**(4), 119-143 (2010)

20. Neely, S., Stolt, S.: Continuous delivery? Easy! Just change everything. (Well, maybe it is not that easy.). Nashville, TN, USA, IEEE, pp. 121–128 (2013)
21. Cusumano, M.: Technology strategy and management. cloud computing and SaaS as new computing platforms. Commun. ACM **53**(4), 27–29 (2010)
22. Dehmlow, M.: Editorial board thoughts: the importance of staff change management in the face of the growing "cloud." Inf. Technol. Libr. **35**(1), 3–6 (2016)
23. Agarwal, P.: Continuous scrum: agile management of SaaS products, pp. 51–60. ACM, Thiruvananthapuram Kerala India (2011)
24. Rodríguez, P., Haghighatkhah, A., Lwakatere, L. Teppola, S.: Continuous deployment of software intensive products and services: a systematic mapping study. J. Syst. Softw. **123**(2) (2016)
25. Kutlu, A., Ekmekçioglu, M.: Fuzzy failure modes and effects analysis by using fuzzy TOPSIS-based fuzzy AHP. Expert Syst. Appl. **39**, 61–67 (2012)
26. Pillay, A., Wang, J.: Modified failure mode and effects analysis using approximate reasoning. Reliab. Eng. Syst. Saf. **79**, 69–85 (2003)
27. Ruan, X., Yin, Z., Frangopol, D.: Risk matrix integrating risk attitudes based on utility theory. Risk Anal. **35**(8), 1437–1447 (2015)
28. Khomh, F., Dhaliwal, T., Zou, T., Adams, B.: Do faster releases improve software quality? An empirical case study of Mozilla Firefox, pp. 179-188. IEEE, Zurich (2012)
29. Kritzinger, D.: Fault tree analysis. In: Aircraft System Safety. Assessments for Initial Airworthiness Certification, pp. 59–99. Woodhead Publishing, s.l. (2017)
30. Kabir, S.: An overview of fault tree analysis and its application in model based dependability analysis. Expert Syst. Appl. **77**, 114–135 (2017)
31. Kavis, M.: Leveraging a DevOps culture to deliver software faster and more reliably. In: Architecting the Cloud: Design Decisions for Cloud Computing Service, pp. 163–170. John Wiley & Sons, Inc., Hoboken, New Jersey (2014)
32. McGaughey, R., Snyder, C., Carr, H.: Implementing information technology for competitive advantage: Risk management issues. Inf. Manage. **26**, 273–280 (1994)
33. Mell, P., Grance, T.: Effectively and securely using the cloud computing paradigm. NIST, Inf. Technol. Lab. **2**(8), 304–311 (2009)

The Influence of Agile IT Departments: Appreciation as a Catalyst for Organizational Transformation

Christopher Reichstein[1(✉)], Ralf-Christian Härting[2], Han Phan[2], and Nicole Luk[2]

[1] Cooperative State University, Baden-Württemberg, Marienstr. 20, 89518 Heidenheim an der Brenz, Germany
christopher.reichstein@dhbw-heidenheim.de

[2] Aalen University of Applied Sciences, Beethovenstr. 1, 73430 Aalen, Germany
rh@kmu-aalen.de

Abstract. Future-relevant research areas such as agility or the role of information technology in the disruptive world are gaining more and more range and presence. Whether it is the transformation to an agile IT or the evolution of the IT department to a strategic role, companies are under growing pressure to adapt to the challenges of digitization. To merge individual elements together to form a holistic picture, the following two questions are addressed in this paper: *To what extend does an IT department contribute to the entire company, by being agile and influencing other departments? Is the opportunity for influence mainly dependent on the appreciation of the IT department by these departments?* Starting with a structured literature analysis (SLA) followed by a structural equation model based on valuable data from a quantitative study by Reichstein (2019), the results are as follows: The IT departments ability to influence has a significant impact on its level of agility. When IT is receiving increasing appreciation, it tends to have a higher ability to influence.

Keywords: Agile IT · Strategic IT Management · Importance of IT Departments · Appreciation towards IT

1 Introduction

In recent years, companies have increasingly seen the urgency of adaptation due to the speed of digitization. Nevertheless, the cautious, small-step approach is not enough to survive the wave of digitization. Above all, the role of IT in the company moves into the center of interest. Even here, classical IT is no longer sufficient to withstand the uncertain and volatile circumstances. Driven by agility, it is supposed to equip companies for turbulence and contribute to strategic corporate planning. In particular, the organizational agility attracted great attention not only of researchers but also of companies. It describes companies that can achieve velocity and accuracy as well as cost efficiency in their business processes through exploitation of potentials for innovation [1]. Especially in times of uncertainty it is necessary to be able to react to changes and adapt. A company

K. Węcel (Ed.): BIS 2025, LNBIP 554, pp. 269–280, 2025.
https://doi.org/10.1007/978-3-031-94193-1_20

might be capable to implement their strategy with rapidity and ease, if the IT department can scale and adapt the IT infrastructure flexibly. In respect of a company's performance, agility therefore can have a protecting role and in the long run it also can improve [2].

However, an agile IT department on its own is not enough to ensure a company's success. It is much more of an enabler and therefore it exerts a certain influence on other departments in its cross-sectional function in some way or another. According to Chen's study in 2007 for example, the crux of the matter is that the IT only delivers its full potential benefits, through stronger communication and coordination between cross-functional teams, since IT facilitates people to discuss and share information [3].

Abstain from this, it is particularly interesting to find out what determining factor must be adjusted to increase the degree of impact of the IT on other components in a company. Regarding the development of the perception towards the IT for instance, one may say that it has changed tremendously over time. Sambamurthy and Zmud (2000) picture the change of the IT in the digital era as a move from an operational to strategical operator [4]. Due to obsolete areas of responsibilities and outdated corporate structure, one may have the perception that the IT department is simply taking care of maintenance issues from soft- and hardware [5]. Therefore, it can be assumed that the IT department is not valued enough yet by other departments as an equally important unit in terms of strategic alignment but more of a supporting role.

Motivated by valuable studies mentioned above, our intent is therefore to combine the two grand fields of research of agility and the position of the IT department in companies. Since little is known about the connection between these two fields, this paper will focus on the extent to which an agile IT department can influence others in the company and whether the degree of influence is dependent on the level of appreciation towards the IT department. The aim of the study is to answer the following research questions:

- To what extend does an IT department contribute to the entire company, by being agile and by influencing other departments?
- Is the opportunity for influence mainly dependent on the appreciation of the IT department given by other departments?

2 Conceptual Model and Methodology

The purpose of this study is to evaluate the value contribution of an IT department for the entire company when it is agile and being able to generate a significant impact on other departments. This study investigates if there is a positive relationship between the agility of an IT department and its ability to influence. In addition, it will be examined if the ability to influence depends on the appreciation of the IT department through other units. The conceptual model (see Fig. 1) is based on Reichstein (2019) and Tallon/Pinsonneault (2011) and examines the IT department in the context of the three factors agility, influence and appreciation. The basis of the investigation are theories and findings from the existing literature (see Table 1) and empirical research. The data of the research originates from a survey from 2019 which investigated the strategic influence of the IT department on the business performance. Given the high degree of data-topicality and the thematic overlap, parts of the survey results, regarding the factors agility, ability to influence and appreciation, are included in the designed model. Further information on the execution of the survey can be found in Reichstein (2019).

Table 1. Literature overview of the three research factors.

Factor	Reference
IT department's appreciation	Bhattacherjee/Sanford (2006), Lazic et. al. (2011), Reichstein (2019)
IT department's ability to influence	Croteau/Bergeron (2001), Andersen (2001), Bassellier/Benbasat (2004), Reichstein (2019)
IT department's agility	Sambamurthy et al. (2003), Lu/Ramamurthy (2011), Tallon/Pinsonneault (2011), van Oosterhout et al. (2006), Kaim et al. (2019)

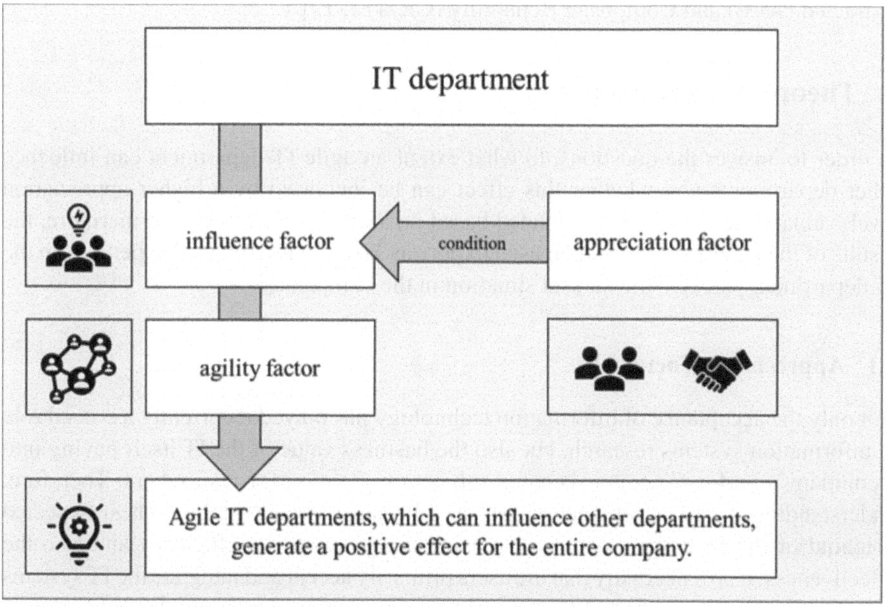

Fig. 1. Conceptual model based on Reichstein (2019) and Tallon/Pinsonneault (2011).

Based on the empirical investigation, a correlation analysis of the construct's *agility*, *influence* and *appreciation* is conducted. This serves as an upstream step to check if there could be a significant relationship between these constructs. After having obtained successful results, the next step is a multiple regression analysis, based on a structural equation modeling (SEM) approach, to test following assumptions:

- Hypothesis 1: The ability to influence has a positive impact on the agility of the IT department.
- Hypothesis 2: The appreciation of the IT department by other departments has a positive impact on the ability to influence.

To be able to analyze the given data, the so-called structural equation model (short for SEM) was applied. The aim is to analyze the theoretically causal model with empirical data and to visualize the connections between different variables. For the evaluation SmartPLS was used to calculate the structural path coefficients and its significance via Bootstrapping, which, in contrast to the two methods AMOS or LISREL for example, would not have been possible [6]. The SEM can be subdivided into two parts: the measurement model and the structural equation model. The measurement model checks the latent variables, whereas the structural equation model analyzes the accordance between the given variables and the hypotheses, assumed for the study model [7]. The statistical software was used to test the measurement and structural model due to its robustness and low data requirements [8–10]. The causal model uses multi-item measurement scales, considering common quality criteria such as Cronbach's Alpha (CA), Average Variance Extracted (AVE) and Composite Reliability (CR) [11, 12].

3 Theoretical Framework

In order to answer the questions, to what extent an agile IT department can influence other departments and whether this effect can be increased by a higher appreciation level, valuable content is first collected based on a literature analysis. Furthermore, the results of the expert survey will be used to portray how the interviewed experts from the IT department perceive the current situation in the companies.

3.1 Appreciation Factor

Not only the acceptance of information technology has played recurrently a crucial role in information systems research, but also the business value of the IT itself paying into a company's performance has become subject of attention for researchers. Therefore, understanding the IT acceptance is essential because to exploit the highest expected potential of the IT usage, such as increasing the productivity, efficiency and also the effectiveness, it is of necessity that the users primarily accept and integrate the IT systems into their task performance [13]. According to a study by Lazic et. al. (2011) the IT department has become decisive for the support, sustainability and growth of companies. Furthermore, the IT governance has taken on a key role in the success of the company [14]. Reichstein (2019) mentions in his study, besides the size of the IT department, the IT experience of the management and the degree of digitization, the value assessment of the IT as a crucial role in growing the appreciation of the IT in the company. Furthermore, he adds that to achieve greater appreciation, it is necessary to boost the acceptance of the company's employees towards IT-related topics [5].

Nevertheless, managers usually do not recognize the added value of cost-intensive IT infrastructure's properties, which provides the company the possibility to implement required application in due time. The same applies to the cost and value relation of technological innovations [15].

Not only the research above has shown the appreciation towards IT as an essential prerequisite in order to create impact, but also the experts from the field express the same opinion. Taken from the survey, the respondents were asked to assess their IT department based on various aspects of appreciation. Speaking of the general recognition of the IT department, the experts were asked whether the IT department is important for the company's success, most respondents (83.06%) agreed on the statement. Additionally, it is to mention that one third of the total sample (N = 124) strongly agreed. In contrast, only 8.87% were neutral or disagreed (8.06%). Similarly, the answers of the experts towards the statement that the IT department is seen as an important component are mostly affirmative (73.39%). On the contrary, only 12.10% disagreed and nearly 15% of the participants were undecided.

Despite the general approval of the interviewees that the IT department, as an important component, is also essential for the success of the company, the assessment of the statements regarding the aspects of respect and appreciation itself is not necessarily as expected. Being asked whether the IT department is respected compared to other divisions, only more than half of the experts (65.32%) agreed, whereas 19.35% of them disagreed. Around 15% remained neutral towards this statement. Directly asking for an assessment of appreciation, 62.10% of the experts agreed that the IT department is appreciated compared to other departments. One fifth (21.77%) did not agree, as well as the remaining 16.12% neither agreed nor disagreed to this statement. The results show that the awareness of the importance of IT is given, however, the respect and the necessary appreciation can be improved in the practical work.

3.2 Influence Factor

An important characteristic that enables the IT department to positively contribute to company performance could be its ability to influence other departments and their work processes. Bassellier and Benbasat (2004) note that the skills of IT experts are increasingly changing and interaction with employees from other functional areas is becoming more and more crucial [16]. Reichstein (2019) also points out that the IT competence area is no longer limited to software and hardware maintenance but is developing from operational to strategic responsibilities [4, 5]. Similarly, Croteau and Bergeron (2001) have investigated the influence of IT on corporate performance in a study. They observed that companies which react quickly to changing market conditions and are highly innovative in new product or service development often incorporate their IT department into strategic meetings. The IT department thereby takes on the role of a technological mediator and innovation driver. In this context, it is of importance that the technological architecture is flexible and can cope with changes required for new projects [17]. Andersen (2001) also proves in his study that the involvement of the IT department in the strategic orientation and decision making of different business units is evidently also beneficially related to the company's performance [18].

Hence, the theory implies that an IT department with a high degree of influence can contribute positively to business performance by supporting other departments in strategic decision-making, by influencing and adapting work processes, and by driving digitization projects or further projects to encourage innovation.

The survey of IT experts on this topic, however, revealed no more than mediocre results. To be influential, the IT section requires, among other things, the right to have a say. However, only 41.13% of the 124 respondents stated that their IT department has a lot of say in internal work processes of other divisions. It is consequently not surprising that only 46.77% agreed to have an IT department with significant influence. When asked whether the IT division has influenced other organizational units in the last six months, again only two thirds (66.13%) agreed. The fact that it has moreover also altered various work processes in other departments in the last six months was also only approved by about two thirds (67.74%). Less than half of the respondents (46.77%) stated that the IT department can influence the strategic orientation of other divisions. When asked whether other units accept and execute the recommendations of the IT department, only 59.68% said yes. This indicates that in practice the IT department is not yet able to sufficiently influence other organizational units. As a result, the question can be asked whether the reason for this an undervaluing of the IT department by other departments may be.

3.3 Agility Factor

Nowadays, companies are faced with very uncertain, dynamic and volatile market conditions in which they must adapt rapidly to changing requirements in order to remain competitive and exist in the market [19]. Many scientific studies have shown that agility enables companies to identify and react quickly to potential threats and opportunities while having a positive impact on company performance [20]. Tallon und Pinsonneault (2011) also observed that agility in a highly volatile business environment has a more positive effect on the company's performance than it has in a stable market [2].

The notion of agility is multifaceted and has been defined differently in the literature. Basically, agility is the ability of an organization to respond quickly to uncertain changes and use them as an opportunity for innovation [21, 22]. Sambamurthy et al. (2003) also highlight the innovative power of agile operations: 'Agility underlies firms' success in continually enhancing and redefining their value creation, capture, and competitive performance through innovations in products, services, channels, and market segmentation. Moreover, they classify agility into three essential dimensions: customer agility, partnering agility and operational agility [1]. The latter dimension is particularly crucial for this study, as it enables to redesign existing business processes quickly and establish new processes to benefit from dynamic market conditions. In this respect, IT also has a decisive role, as it drives the modularization and automatization of business processes and allows their combination and recombination to create new business processes [23].

Other research has also shown the impact of IT on business agility by accelerating decision making, facilitating communication and enabling companies to respond quickly to changing environments [20, 24]. However, it was also found that IT cannot only be beneficial but can also hinder the agility of companies. Van Oosterhout et al. (2006) found that inflexible IT systems can lead to a rigid IT architecture and, due to their limited ability to change, can impair business agility. In contrast, an agile process and information system architecture can promote corporate agility [25]. Tallon and Pinsonneualt (2011) also concluded that a flexible IT infrastructure has a positive significant influence on agility [2]. A flexible IT infrastructure can scale and adjust essential IT resources for different purposes [26].

In theory an agile IT department can have an important positive impact on business performance by providing flexible IT infrastructures and a flexible, agile organizational structure that encourages business innovation in a company.

Likewise, the survey of IT experts on the topic of agility shows that there is a consensus on the research results listed above. Most of the experts (84.68% of a total of 124 respondents) stated that an agile IT organization supports the implementation of digitization projects in companies. Only 9.68% disagreed with the opinion and 5.65% neither agreed nor disagreed. A high level of agreement was also found in the statement that a relatively high degree of agility of the IT department has a positive effect on the entire company. 83.06% of the IT experts asked agreed with the opinion, while 8.87% were neutral about the statement and only 8.06% disagreed. However, in response to the question whether the experts believe their IT department is agile, barely more than half (56.45%) agreed. 24.19% said they did not have an agile IT organization and 19.35% were undecided. Although only about half of the experts surveyed have an agile IT department, they are aware of the advantages of an agile IT organization and have confirmed this fact.

4 Research Results

Following the data analysis through the structural equation modelling (SEM) to validate the hypotheses by, the results are shown in Fig. 2.

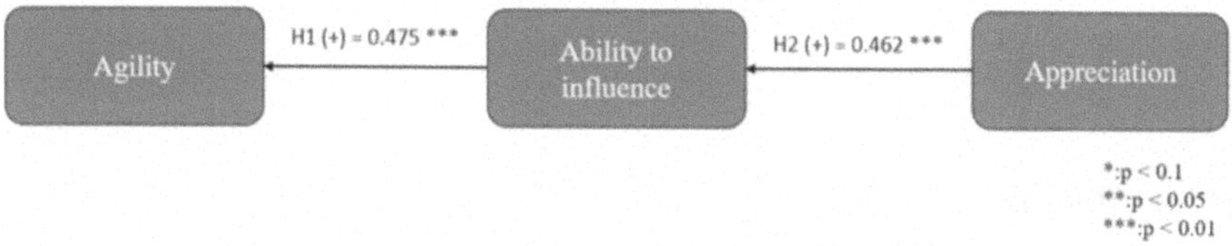

Fig. 2. Structural equation model with coefficients.

Hypothesis 1 (The ability to influence has a positive impact on the agility of the IT department) explores the direct impact of the ability to influence on the agility. As shown in the regression analysis, the ability to influence has a positive impact (+0.475), at the significance level of 0.000, on the agility. Thus, the hypothesis can be confirmed. Once the IT department can influence the processes or even the strategical direction of other business units, its level of agility also increases. Additionally, the indirect effect of appreciation on the agility has been explored. The result (+0.22) shows a positive effect of appreciation on the IT department's agility.

Similarly, positive results can be observed for *Hypothesis 2 (The appreciation of the IT department by other departments has a positive impact on the ability to influence)*. It examines the direct effect of appreciation on the ability to influence. Due to a positive (+0.462) and highly significant impact ($p < 0.01$), this hypothesis can be confirmed. As a result, the level of appreciation does have a positive impact on the IT department's ability to influence. According to these results, the more the IT department is appreciated, the greater is their ability to influence.

Furthermore, the coefficient of the determination (R2) is in a satisfying range 0.214 (ability to influence) and 0.226 (agility) > 0.19 [10]. In respect of the common quality criteria Cronbach's Alpha (CA), Average Variance Extracted (AVE) and Composite Reliability (CR), the measures indicate good to moderate psychometric properties. To ensure a high quality of measurement model and its constructs, the reliability and validity has been examined as shown in Table 2. All three constructs exceed the recommended threshold with values > .50 for Cronbach's Alpha, > .60 for Composite Reliability and > .70 for Average Variance Extracted [8].

Table 2. Cronbach's Alpha, Average Variance Extracted, Composite Reliability.

	Cronbach's Alpha (CA)	Average Variance Extracted (AVE)	Composite Reliability (CR)
Ability to influence (6 items)	0.889	0.642	0.915
Agility (3 items)	0.759	0.681	0.864
Appreciation (4 items)	0.871	0.722	0.912

5 Conclusion

In summary, uncertain and dynamic market conditions require companies to react and adapt quickly to remain viable and competitive. The IT can enable companies to gain these skills by providing an agile IT infrastructure to ensure the agile operation of companies. Thus, companies can scale and adapt their IT resources to different requirements. An agile IT organization can moreover support the implementation of digitization projects and drive innovation processes. In this way, an agile IT organization can generate a positive effect for the company. In this respect it needs a certain ability to influence other departments to be able to adjust these processes. This was also confirmed by the results of the regression analysis, thus confirming *Hypothesis 1 (The ability to influence has a positive impact on the agility of the IT department)*. The more the IT department can influence the processes or even the strategic orientation of other departments, the higher is its level of agility.

Hypothesis 2 (The appreciation of the IT department by other departments has a positive impact on the ability to influence) was also confirmed. The higher the appreciation of the IT department, the higher the degree of decision influence on other departments. This also means that an IT department that is not sufficiently appreciated is hardly able to influence other departments and thus cannot generate an agility effect.

Overall, this means that an IT department can make a positive value contribution to the entire company if it is agile and able to exert a significant influence on other departments. However, this is only possible if other departments value IT accordingly and accept and allow suggestions and changes.

Looking at the limitations of the work, there are good opportunities for further research. It has to be noted that the research model is based on the work of Reichstein (2019) and Tallon/Pinsonneault (2011). In the context of this paper, the IT department was examined only in the context of the three factors agility, influence and appreciation, since a very strong correlation was assumed. Other factors, such as the size of the IT department or the degree of digitization, were not considered, but could also have an impact on the agility of the IT department. A reason for different results could also be the company size. This could have an influence on the possibilities to react agile and get connections to other departments. The investigation concept can be continued in further research. For example, it could be examined what contribution the IT department made to companies during the Corona crisis and if agility was also an important success factor here.

In view of praxeological implications, the study concludes that an agile and influential IT department can beneficially contribute to a company. Therefore, the recommendations of the IT experts must be more widely accepted and applied. Although other departments are aware of the high importance of the IT for the company's success, the value of the IT department is still not yet sufficiently appreciated. This is urgently needed to change, as only with an appropriate level of acceptance the IT can influence other departments and drive digitization processes and other innovations in the company. In this respect, companies are recommended to rely more on the expertise of the IT, as this can be very valuable for their strategic business planning in times of digitization. Starting with the optimization of the current IT infrastructure, such as providing a smooth VPN connection for the company's employees to enable the possibility of working remotely from home to the early identification of relevant digital innovations or even completely new digital business models [27, 28].

Furthermore, companies must ensure that they no longer consider their IT department primarily as an internal service provider responsible for hardware and software maintenance. Basic IT issues and demands of the employees should be dealt with as far as possible by themselves to avoid losing valuable IT capacity. This, of course requires effort from the management. They should guide and lead the employees of their company towards IT-relevant topics. The lack of appreciation that we observed in this study may have changed because of the Corona crisis, so that IT experts are now more respected and acknowledged for their work.

Appendix

Excerpt of the survey results (mean and standard deviation)		
Agility factor		
Question	**mean**	**std**
1. An agile IT organisation helps your company with the implementation of digital projects.	5.65	1.38
2. A relative high degree of agility of the IT department has a positive effect on the entire country.	5.68	1.30
3. Our IT department is agile.	4.55	1.55
Influence factor		
Question The IT department...	**mean**	**std**
1. has a great say in internal work processes of other departments.	4.12	1.41
2. has influenced other departments in the last six months.	5.10	1.52
3. has changed various work processes in other departments in the last six months.	5.02	1.61
4. can influence the strategic direction of other departments.	4.26	1.59
5. proposes solutions, which are accepted and implemented by other departments.	4.65	1.44
6. has a considerable impact.	4.35	1.54
Appreciation factor		
Question The IT department…	**mean**	**std**
1. is respected.	5.06	1.47
2. is important for company's success.	5.69	1.31
3. is appreciated.	4.83	1.48
4. is a company's important component.	5.25	1.33

References

1. Sambamurthy, V., Bharadwaj, A., Grover, V.: Shaping agility through digital options. reconceptualizing the role of information technology in contemporary firms. MIS Q. **27**(2), 237–263 (2003)
2. Tallon, P., Pinsonneault, A.: Competing perspectives on the link between strategic information technology alignment and organizational agility. insights from a mediation model. MIS Q. **35**(2), 463–486 (2011)
3. Chen, C.: Information technology, organizational structure, and new product development. the mediating effect of cross-functional team interaction. IEEE Trans. Eng. Manage. **54**(4), 678–698 (2007)

4. Sambamurthy, V., Zmud, R.W.: Research commentary. the organizing logic for an enterprise's IT activities in the digital era – a prognosis of practice and a call for research. Inf. Syst. Res. **11**(2), 105–114 (2000)
5. Reichstein, C.: Strategic IT management. How companies can benefit from an increasing IT influence. J. Enterp. Inf. Manage. **32**(2), 251–273 (2019)
6. Wong, K.K.-K.: Partial least squares structural equation modeling (PLS-SEM) techniques using SmartPLS. Mark. Bull. **24**(1), 1–32 (2013)
7. Markus, K.: Principles and practice of structural equation modeling. by Rex B. Kline. Structural Equation Modeling: A Multidisciplinary Journal **19**(3), 509–512 (2012)
8. Hair Jr, J.F., Hult, G.T.M., Ringle, C., Sarstedt, M.: A Primer on Partial Least Squares Structural Equation Modeling (PLS-SEM). Sage publications (2016)
9. Ringle, C.M., Wende, S., Becker, J.M.: SmartPLS 3. Boenningstedt: Smart-PLS GmbH (2015).https://www.smartpls.com/. Accessed 10 Apr 2018
10. Chin, W.W.: The partial least squares approach to structural equation modeling. Mod. Methods Bus. Res. **295**(2), 295–336 (1998)
11. Ringle, C.M., Sarstedt, M., Straub, D.: A critical look at the use of PLS-SEM. MIS Q. (2012)
12. Coreynen, W., Matthyssens, P., Van Bockhaven, W.: Boosting servitization through digitization: pathways and dynamic resource configurations for manufacturers. Ind. Mark. Manage. **60**, 42–53 (2017)
13. Bhattacherjee, A., Sanford, C.: Influence processes for information technology acceptance. an elaboration likelihood model. MIS Q. **30**(4), 805–825 (2006)
14. Lazic, M., Groth, M., Schillinger, C., Heinzl, A.: The Impact of IT governance on business performance. In: Americas Conference on Information Systems, Detroit, MI (2011)
15. Bharadwaj, A.S.: A resource-based perspective on information technology capability and firm performance. An empirical investigation. MIS Q. **24**(1), 169–196 (2000)
16. Bassellier, G., Benbasat, I.: Business competence of information technology professionals. conceptual development and influence on IT-business partnerships. MIS Q. **28**(4), 673–694 (2004)
17. Croteau, A.-M., Bergeron, F.: An information technology trilogy. business strategy, technological deployment and organizational performance. J. Strateg. Inf. Syst. **10**(2), 77–99 (2001)
18. Andersen, T.J.: Information technology, strategic decision-making approaches and organizational performance in different industrial settings. J. Strat. Inf. Syst. **10**(2), 101–119 (2001)
19. Kaim, R., Härting, R., Reichstein, C.: Benefits of agile project management in an environment of increasing complexity. a transaction cost analysis. In Czarnowski, I.; Howlett, R.; Jain, L. (eds.): Smart Innovation, Systems and Technologies. Proceedings of the 11th KES International Conference on Intelligent Decision Technologies (KES-IDT 2019), vol. 143, no. 2, pp. 195–204. Springer (2019)
20. Lu, Y., Ramamurthy, K.: Understanding the link between information technology capability and organizational agility. an empirical examination. . MIS Q. **35**(4), 931–954 (2011)
21. Dove, R.: Response Ability. The Language, Structure, and Culture of the Agile Enterprise. New York, John Wiley & Sons (2001)
22. Goldman, S.L., Nagel, R.N., Preiss, K.: Agile Competitors and Virtual Organizations. Strategies for Enriching the Customer, New York, Van Nostrand Reinhold (1995)
23. Malone, T.W., Crowston, K., Lee, J., Pentland, B.: Tools for inventing organizations. toward a handbook of organizational processes. Manage. Sci. **45**(3), 425–443 (1999)
24. Lucas, H.C., Jr., Olson, M.: The impact of information technology on organizational flexibility. J. Organ. Comput. Electron. Commer. **4**(2), 155–176 (1994)
25. van Oosterhout, M., Waarts, E., van Hillegersberg, J.: Change Factors Requiring Agility and Implications for IT. Eur. J. Inf. Syst. **15**(2), 132–145 (2006)

26. Byrd, TA, Turner, D.E.: Measuring the flexibility of information technology infrastructure: exploratory analysis of a construct. J. Manag. Inf. Syst. **17**(1), 167–220 (2000)
27. Härting, R., Reichstein, C., Jozinovic, P.: The potential value of digitization for business. insights from German-speaking experts. In: Eibl M.; Gaedke, M. (eds.): Informatik 2017, Vol. 47. Jahrestagung der Gesellschaft für Informatik, Lecture Notes in Informatics (LNI), Gesellschaft für Informatik, Bonn 2017 (2017)
28. Härting, R., Reichstein, C., Schad, M.: Potentials of digital business models – empirical investigation of data driven impacts in industry. In: Robert J. Howlett, Carlos Toro, Yulia Hicks, Lakhmi C. Jain (eds.): Knowledge-Based and Intelligent Information & Engineering Systems: Proceedings of the 22nd International Conference (KES-2018), vol. 126, pp. 1495–1506. Elsevier B.V. (2018)

Sustainable Operations and Smart Automation

Activity-Based Cost & GHG Accounting: Conceptualization, Design and Implementation

Walter S. A. Schwaiger(✉), Josef Baumüller, and Lalla Hasnae Alaoui

Institute of Management Science, TU-Wien, Theresianumgasse 27, 1040 Vienna, Austria
{walter.schwaiger,josef.baumueller,lalla.alaoui}@tuwien.ac.at

Abstract. The Greenhouse Gas (GHG) Protocol initiative established an activity-based standard for accounting and management of GHG emissions that is used by regulators all over the world for setting up climate-change legislations. In the 'Activity-Based Cost & GHG Accounting' the GHG Protocol's accounting metrics are integrated with the cost accounting metrics that underly the flexible standard-cost accounting. In the 'Activity-Based Life-Cycle Accounting' the GHG Protocol's accounting metrics and corresponding informational and computational requirements are defined and aligned within an ERP-Control information system. The alignment with the ERP-control system gives the providers of such systems precise guidance for integrating the activity-based GHG accounting system into their existing cost accounting systems. On the other hand, the alignment also allows the implementation of the integrated cost & GHG accounting system as an add-on solution to any ERP-Control system. For being able to apply the integrated system in the generic context of a multi-level product structure and a multi-level production structure, i.e. the multi-level 'Bill-of-Material' (BOM) and the multi-level 'Routing' an adequate integration of the two multi-level structures is needed. Solving this 'multi-level integration problem' for ensuring the system's generic applicability is this paper's primary research objective.

Keywords: Activity-based cost & GHG accounting system · GHG Protocol standards · domain engineering methodology

1 Introduction

'We set the standards to measure and manage emissions' [1] is the slogan of the Greenhouse Gas (GHG) Protocol initiative. The GHG Protocol started its GHG accounting initiative in 2004 by publishing the 'Corporate Accounting and Reporting Standard' [2]. In 2011 the standards on 'Product Life Cycle Accounting and Reporting Standard' [3] and 'Corporate Value Chain (Scope 3) Accounting and Reporting Standard' [4] were published. In all three GHG Protocol's standards an 'activity-based approach' is considered as the preferred way to assess the GHG emissions of corporates and products and in the value chain. The reasoning for this preference is the given cause-and-effect relationship that prevails when the GHG emissions are linked to activities. This relationship is a generic one as it holds for any activity. Furthermore, it captures the core of managing

K. Węcel (Ed.): BIS 2025, LNBIP 554, pp. 283–296, 2025.
https://doi.org/10.1007/978-3-031-94193-1_21

GHG emissions as the only way to reduce and mitigate GHG emissions is by changing activities. Finally, it can be modeled with Leontief's input-output production theory [5] so that it also can be linked with activity-based cost accounting as well.

The cause-and-effect relationship of the activity-based GHG accounting and its connectivity to activity-based cost accounting provides a common grounding of the cost and GHG accounting domains. Furthermore, the GHG Protocol's standards received international recognition by climate-change legislation regulators. Accordingly, the GHG Protocol's activity-based approach moves the earlier developed ISO's 'Life Cycle Assessment' (ISO-LCA) methodology [6] into the background. Emblemsvåg/Bras [7, 8] were the first ones who addressed the activity-based integration of cost and environmental accounting. They accounted for the company's environmental impact in form of energy and waste by the resources that are applied in all of its activities. "For determining the cost, energy and waste of an activity, multiply each resource driver in the activity with its respective consumption intensity and sum the totals for the activity." [7, p. 8]. Accordingly, energy and waste are conceptually calculated in the same manner, i.e. by considering drivers that measure the quantity of consumed resource and that are multiplied with the intensity of the resource consumption to measure the energy consumed as well as the waste generated within the activity.

The 'activity-based cost & GHG accounting' was established by Baumüller/Schwaiger [9] by extending the cost accounting metrics that underly the 'flexible standard-cost accounting' with the GHG accounting metrics from the GHG Protocol product standard [10]. Specifically, activity-based energy consumption metrics in form of time-driven (TD) metrics and resource consumption driven (RCD) metrics are defined for covering all kinds of production, assembly, logistics and service activities as well as activity-based emission-liability ('E-Liability') allocation metrics according to the 'E-Liability' concepts of Kaplan/Ramanna's 'Accounting for Climate Change' [11] for the materials consumed and the equipment used in the activities.

For integrating the accounting (business) domain with the IT domain, Alaoui/Schwaiger specified cost and GHG accounting in the light of the domain engineering methodology [12]. The resulting 'activity-based life cycle accounting methodology' [13] closes the gap that normally exists in software development projects between experts from the business domain and the IT domain by applying the domain-driven design (DDD) methodology [14]. Bridging the gap is based upon the three pillars, i.e. (1) the conceptualization of the integrated accounting domain in order to explicitly specify the domain language in a formal modeling language and then using the language of the specified domain system in order (2) to operationalize the design of the intended software artifact and (3) to implement the artifact in an executable language capable of deriving results for application cases.

By explicitly specifying the domain language and its subsequent use for the design and implementation of the software artifact, the domain language serves as 'ubiquitous language' in the sense of the Domain Driven Design (DDD) methodology [14]. Furthermore, the domain language is based on the definition of ISO's 'Enterprise Control System Integration' standard ISO/IEC-ECSI [15], so that the language underlies all ERP-systems and Production-Control-systems that are aligned with that language.

As the specified language obeys the ISO/IEC-ECSI standards it is called 'ERP-Control domain' language.

The accounting engineering methodology of Alaoui/Schwaiger [13] starts with the conceptualization of the activity-based life cycle accounting by considering the concepts of 'Bill-of-Materials' (BOM) that specifies the product structure and 'Routing' that specifies the production structure. As they only focus on the costs and GHG emissions related to the materials consumed in manufacturing activities, the activities to be performed and their arrangement in the routing are not investigated. Only a multi-level structure of the BOM is considered. As such, the applicability scope of the methodology in real cases is limited. For extending the methodology's application range, also for the routing a multi-level structure has to be specified and the integration of the two multi-level structures has to be addressed.

Solving this 'multi-level integration problem' constitutes the primary research objective of this paper. For solving this problem the concept of 'Generic Node Tree' is introduced that relates to the concepts of Yeh [16] and Jiao et al. [17] for the integration of BOM and routing. The Generic Node Tree-based solution is valid as it enables all accounting calculations needed. Finally, its practical applicability is demonstrated in the use case 'Slot Car Manufacturing'. By covering the problem specification, the solution design, the validation and the demonstration, the primary research objective is investigated with the 'design science research methodology' [18].

The paper is structured according to the three pillars of the accounting engineering methodology: After this introduction, the activity-based cost & GHG accounting domain is conceptualized in the UML language with the Generic Node Tree concept to solve the multi-level integration problem. Based on the accounting domain's UML conceptualization the thereof derived accounting system is presented in the next section. This system is the 'activity-based cost & GHG accounting system'. It also is defined in the UML language what has the advantage that the system's operationalization can be directly seen via the additional information items (attributes of the classes) and accounting metrics (methods of the classes). In the following implementation section, the accounting system is implemented in a practical use case with the R-Tidyverse data science language. Finally, the paper is concluded.

2 Cost and GHG Accounting Domain: Conceptualization

Conceptualization is the first pillar of the accounting engineering methodology. In the conceptual modeling of the domain the key concepts of domain are specified. The key concepts' terminology defines the language that is spoken in the domain.

The language spoken in the activity-based cost & GHG accounting domain is based upon Leontief's input-output production theory that translate into the language spoken not only in the accounting (business) domain [19, 20] and but also in the ERP-Control (IS) domain [15]. According to its production theoretical foundation, activity-based cost & GHG accounting relates to an integrated management accounting model where cost and GHG emissions are assessed via input-output ratios concerning the resources applied in the activities. This cost & GHG accounting 'model' can be considered from different abstractions levels. At the conceptual level the model specifies the key accounting concepts in form of entity types and relationship types. At the operational level the model gets extended by considering entity classes and relationship classes that have in contrast to the types also specific attributes and methods. At the implementation level the classes are translated into entity tables and relationship tables and corresponding computational functions.

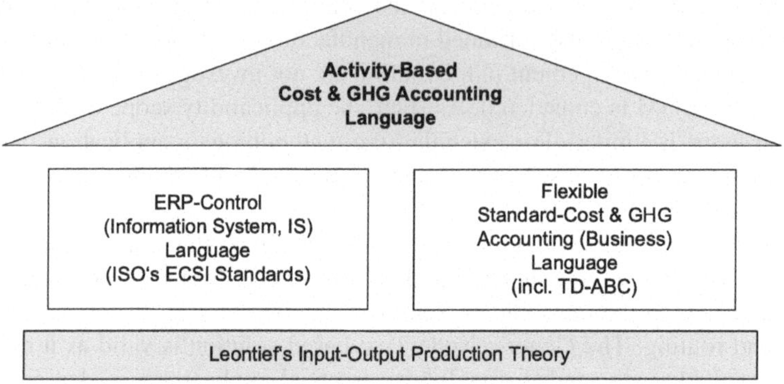

Fig. 1. Language spoken in the Cost & GHG Accounting Domain

2.1 Cost and GHG Accounting Domain: Theoretical Foundation

Activity-based cost & GHG accounting has its roots in the 'flexible standard-cost accounting' that was developed in the last century in Germany. More recently, its production theoretical foundation also was introduced into the Activity-Based Cost (ABC) accounting when switching from the traditional ABC to the Time Driven (TD)-ABC accounting.

Ad flexible standard cost accounting) "One of the main features of 'flexible standard costing' [19, p. 306, fn 6] lies in its production theoretic foundation which reflects a company as a complex input-output system. This theory is based on Leontief's work and has been generalized for quite arbitrary production functions… This foundation is widely used both for tracing direct costs to products and for allocating indirect costs." [19, p. 306]. "In Germany flexible standard costing finally found a better theoretical basis in the production theory… The production theory… Was developed to study the input-output relations of a business enterprise in great detail… The main characteristic of this theory is that there are two different input-output relations in a firm – direct relations between output and input of direct materials and direct labour, and indirect relations. In cases of indirect relations the input depends on the number of elementary work units of cost centres wanted, the intensity or speed of production and the technical features of the assets used. Thus flexible standard costing has a theoretical basis for direct and indirect costs which, by referring to work units, is similar to activity-based costing." [20, p. 265].

Ad TD-ABC accounting) By switched from the original activity-based cost (ABC) accounting [21] that is based upon transactional cost drivers to the Time Driven (TD)-ABC accounting [22] that is based upon effort cost drivers, i.e. input-output relating cost drivers, the input-output production theoretical foundation was taken over by

Kaplan/Anderson. "The basis for the new approach (TD-ABC) is highlighted in an early cost management article, where Robin Cooper articulated the difference between transactional and "effort" cost drivers. Transactional cost drivers count the number of times an activity is performed." [23, p. 6]. "An alternative approach for estimating an ABC model, which we call 'time-driven activity-based costing', addresses all the above limitations. It is simpler, less costly, and faster to implement, and allows cost driver rates to be based on the practical capacity of the resources supplied." [23, p. 5]. "The essence of activity-based costing and activity-based management is the measurement and management of the organization's capacity. For this purpose, ABC systems require two estimates:

1. The unit cost of supplying capacity, and
2. The consumption of capacity (unit times) by the activities the organization performs for products, services, and customers." [23, p. 6].

"In retrospect, we wish that the evolution of ABC in the 1980s had taken a different path so that this method could have been implemented at the outset. But the underlying theory for ABC had not been developed when it was first introduced in the mid-1980s so the elegance and conceptual clarity of this new approach were not obvious at the time." [23, p. 5].

Focusing upon the consumptions of capacity resources in the activities shows two important properties of TD-ABC accounting, firstly, it actually is a Resource Consumption Driven (RCD)-ABC and secondly, it can be extended to also include materials and other directly consumed resources. Seeing and extending TD-ABC in that way gives exactly the scope covered by the flexible standard-cost accounting's costing metrics.

2.2 Cost and GHG Accounting Domain: Conceptualization

Modeling cost & GHG accounting at the conceptual level specifies the accounting's key concepts in form of entity types and relationship types. Figure 2 shows the conceptualization of the cost & GHG accounting domain which is the application domain of the accounting engineering methodology from Alaoui/Schwaiger [13, p. 268]. The conceptualization is expressed in the formal modeling language UML so that it can be operationalized consistently into an accounting system design. This conceptual model of the Cost & GHG accounting domain contains the concepts Activity, Equipment, Material and Finished Good which are modeled as entity types as well as the concepts Routing and BOM modeled as relationship types as they are specified in the ERP-Control domain language [15]. The entity types are associated with the corresponding relationships types and the cardinalities in the associations are specified. The BOM has a circular association to itself, so that it constitutes a multi-level structured type. As the Routing does not have such an association, it is a single-level type.

Yeh [16] distinguishes two possibilities for linking BOM and routing, i.e. the 'product oriented' and the 'process oriented' approach. "A better method is to specify each component material in the BOM as required by the relevant operation of the routeing for making its parent product. This approach of linking the BOM and reouteing data can be classified as product oriented, as the linking is derived from the BOM. The product-oriented approach cannot handle situations where a component material is required by more than one operation in a routeing." [16, p. 53]. "The integration of the BOM and routeing is based on a process-oriented approach emphasizing the production process. The process-oriented approach specifies the sequence of operations needed for making a product, together with the material and resources required at each operation." [16, p. 53].

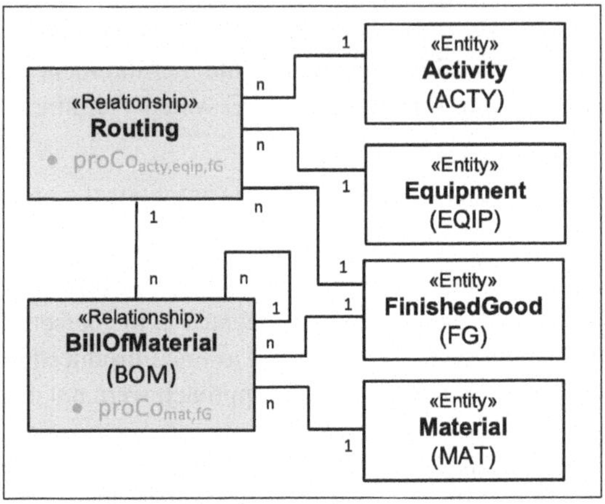

Fig. 2. Cost & GHG ACC Domain – UML-Conceptualization (1/2)

In Fig. 2 Routing and BOM are linked by a 1:n association, so that each element of the Routing can have assigned multiple elements from the BOM. This restriction, which corresponds to the Yeh's product oriented linking approach, is fine as long the Routing is of single-level nature and each component material is required by only one activity in the routing. For extending the domain model to the more general case of linking the BOM with a multi-level Routing the process-oriented approach is taken. But instead of directly linking BOM and Routing, the idea of 'shared common product and process structures' from Jiao et al. [17, p. 297] is used for intermediating them via a generic construct that contains the common multi-level structure of the BOM and the Routing.

In Fig. 3 the generic concept called 'Generic Node Tree' contains the common multilevel structure of the 'Production Structure' (Routing) and the 'Product Structure' (BOM). The Generic Node Tree is modeled as an abstract relationship type with a circular association to itself that incorporates the common multi-level structure of the Routing and the BOM. Furthermore, it is a generalization association to the Routing and the BOM so that both inherit the common multi-level structure as well.

Next to their names and stereotyped entity or relationship nature, the key concepts defined in the conceptual model of the Cost & GHG accounting domain in Fig. 3 also include the production coefficients in the production and product structures. The reason for this 'semantic sugar' is to hint to the accounting domain's production theoretical foundation where the production coefficients express the input-output ratios in form of the activity's unit-input of capacity, materials and other (direct) resources. (also see [16, p. 56, 62]).

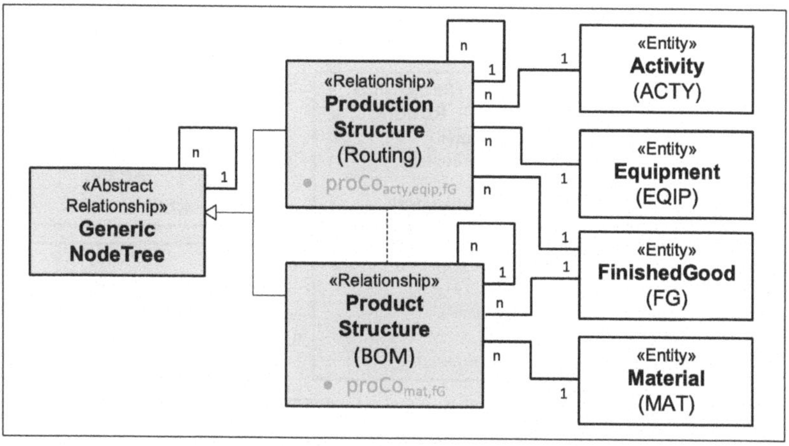

Fig. 3. Cost & GHG ACC Domain – UML-Conceptualization (2/2)

3 Cost and GHG Accounting System: Design

Due to the conceptual modeling of the cost & GHG accounting domain in the formal modeling language UML, the model can be directly operationalized in form of entity classes and relationship classes by switching from types to classes and adding attributes and methods to the classes. The transformation from the accounting domain's conceptual to its operational model specifies the design of the accounting information system.

3.1 Cost and GHG Accounting System: Design

Figure 4 shows the cost & GHG accounting system in form of entity classes and relationship classes. The class Acc(ounting)NodeTree translates the abstract concept Generic Node Tree into a relationship class. It incorporates the common structure of the BOM and the Routing which is expressed in its attribute chilNodeID that identifies the children nodes of each node in the AccNodeTree. By incorporating the common structure, the BOM and the Routing classes can be anchored directly into the generic AccNodeTree. This also has computational advantages as the calculations for accumulating cost and GHG emissions over the multi-level structure can directly be performed by recursive programming in the AccNodeTree.

The attributes and methods of the accounting systems classes are arranged as follow. On top the attributes for identifying the elements of the class are shown. In the multi-level classes the goes-into relationships among the different levels follow. In the AccNodeTree class it relates to the nodes' children, whereas in the Routing and BOM classes it relates to the elements' parent, i.e. parRouteID and parMatID. In the Routing and BOM class the production coefficients follow next. Afterwards follow the cost and then the GHG emission related attributes. At the end the methods for the calculating cost and GHG emission information are shown. Finally, it can be seen that the AccNodeTree contains all methods needed for accumulating costs and GHG emissions over the multi-level structures of the BOM and the Routing.

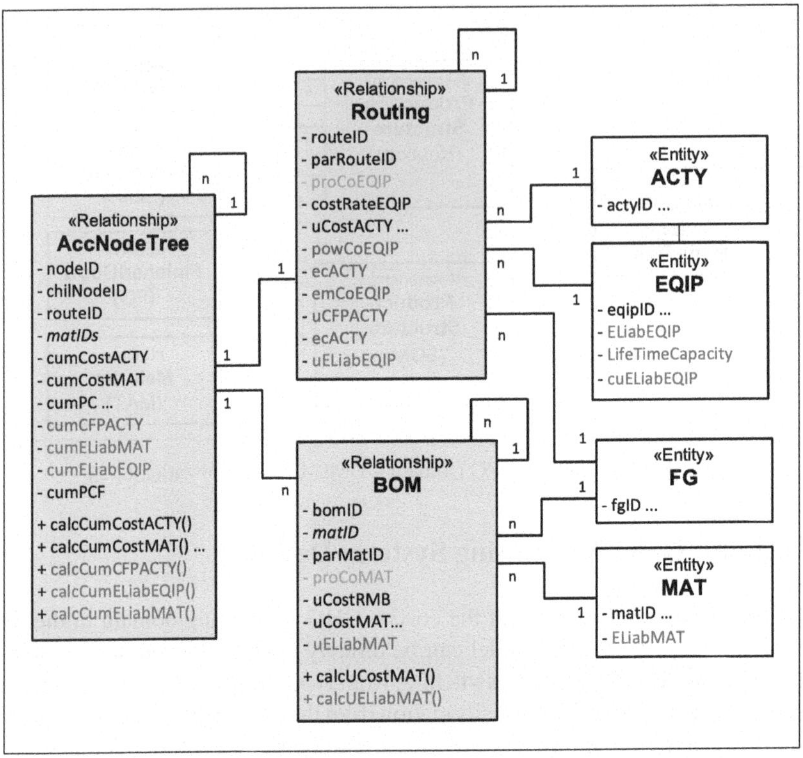

Fig. 4. Cost & GHG Accounting System – UML-Design

3.2 Cost and GHG Accounting System: Metrics

For completely designing the accounting system, next to its UML design also its computational metrics have to be specified. All metrics of cost & GHG accounting are based upon production coefficients due to their grounding in Leontief's input-output production theory. In the cost metrics the activities' production coefficients are evaluated with their corresponding cost rates to give the activities' unit-costs for the finished product [19, 20]. Due to the coverage of not only indirect costs but also direct costs in the activity-based cost & GHG accounting, all components of the finished product's total costs, i.e. product cost (PC), are calculated with metrics that rest on the same production theoretic foundation. The same holds true for the finished product's total GHG emission as also they are grounded on the same foundation. The product carbon footprint (PCF) consists of the emissions from the company's activities that are calculated with the 3-Levers-of-Control (3-LoEC) metrics and the emissions that are calculated with the environmental liabilities (E-Liability) metrics for the materials consumed and the equipment used in the activities [10].

The inclusion of the AccNodeTree class in the cost & GHG accounting system shows its benefits in the accumulation of costs and GHG emissions over the common multi-level structure it contains. All accumulation calculations can be implemented by recursive algorithms in the AccNodeTree class by using its incorporated common node structure (nodePath). For the cost accounting the costs of the activities (cumCostACTY) and the costs of the materials (cumCostMAT) are accumulated. For the GHG accounting the emissions (Carbon FootPrint) of the activities (cumCFPACTY) and the emissions (ELiab) for the used equipment (ELiabMAT) and the consumed materials (ELiabEQIP).

Ad cumCostACTY, cumCostMAT) The accumulated costs of the activities as well as materials are calculated in Eq. (1) as well as Eq. (2) by successively adding the costs of the activities as well as materials – all of them are derived by multiplying their production coefficient with the corresponding cost rate – over the generic multi-level structure (nodePath up to the fG, i.e. finished product) specified in the AccNodeTree's chilNodeID.

$$cumCostACTY_{fG:nodePath} = \sum_{\substack{node\,\in \\ nodePath}} proCoEQIP_{acty,eqip,fG} \cdot costRateEQIP_{eqip} \tag{1}$$

$$cumCostMAT_{fG:nodePath} = \sum_{\substack{node\,\in \\ nodePath}} uCostMAT_{fG:parMat} \tag{2}$$

Ad cumCFPACTY, cumELiabEQIP, cumELiabMAT) The accumulated GHG emissions of the activities as well as materials consumed and equipment used in the activities are calculated in Eq. (3) as well as (4) and (5) by successively adding the GHG emissions of the activities as well as materials consumed and equipment used in the activities – all of them are derived via their corresponding production coefficients – over the generic multi-level structure specified in the AccNodeTree's chilNodeID.

$$cumCFPACTY_{fG:nodePath} = \sum_{\substack{node\,\in \\ nodePath}} uCFPACTY_{acty,res,fG} \tag{3}$$

$$cumELiabEQIP_{fG:nodePath} = \sum_{\substack{node\,\in \\ nodePath}} uELiabEQIP_{eqip,3,fG} \tag{4}$$

$$cumELiabMAT_{fG:nodePath} = \sum_{\substack{node\,\in \\ nodePath}} uELiabMAT_{fG:parMat} \tag{5}$$

Finally, for calculating the finished product's cost (PC) the accumulated activity and material costs are added, and for calculating the finished product's carbon footprint (PCF) the accumulated emission of the activities as well as the materials consumed and the equipment used in the activities are added.

4 Cost and GHG Accounting System: Implementation

The 'Slot Car Manufacturing' (SCM) use case applied at TU-Wien for study activities in the 'Pilot Factory' [24] and the 'Mechanical and Industrial Engineering' study program is used for demonstration purposes. Figure 5 shows an excerpt from the BOM.tbl table containing the finished slot car (fgID = 120) in the first line and subsequently the material components needed for its manufacturing.

A tibble: 10 x 8

	bomID	fgID	matID	matName	parMatID	proCoMAT	uOM	uCostMAT
	<int>	<dbl>	<dbl>	<chr>	<dbl>	<dbl>	<chr>	<dbl>
1	1	120	NA	<NA>	NA	1	pc	74.4
2	2	120	2	Cover	120	1	pc	28
3	3	120	7	CScrew_M2x6	120	1	pc	0.02
4	4	120	3	DriveUnit	120	1	pc	10.1
5	5	120	4	Motor	3	1	pc	10
6	6	120	5	MotorMountingBracket	3	1	pc	0.12
7	7	120	7	CScrew_M2x6	3	1	pc	0.02
8	8	120	25	AluminiumL_Shape_Cut	5	1	pc	0.12
9	9	120	6	AluminiumL_Shape	25	0.015	m	0.12
10	10	120	1	Chassis	120	1	pc	36.2

Fig. 5. BOM.tbl with parMatID (goes-into relationship) and proCoMAT

Figure 6 shows all cutting, machining and assembly activities needed for manufacturing the slot car. The 'Assembly of Slot Car' (routeID = 16) is the 'marriage' in the slot car manufacturing where the 'Chassis' (matID = 1) with 'DriveUnit' (matID = 3) is combined with the 'Cover' (matID = 2) by using a 'CScrew' (matID = 7).

A tibble: 16 x 8

	routeID	routeName	fgID	actyID	eqipID	parRouteID	proCoEQIP	uOM
	<dbl>	<chr>	<dbl>	<dbl>	<dbl>	<dbl>	<dbl>	<chr>
1	1	Cutting of Aluminium Shape L	120	1	1	5	1	min
2	2	Cutting of Aluminium Shape U	120	1	1	6	2	min
3	3	Cutting of Aluminium Cylinder (front)	120	1	1	7	1.5	min
4	4	Cutting of Aluminium Cylinder (back)	120	1	1	8	1.5	min
5	5	Machining of Motor Bracket Shaping	120	2	2	14	6	min
6	6	Machining of U-Body Shaping	120	2	2	15	21	min
7	7	Machining of Rim Shaping (front)	120	2	2	9	9	min
8	8	Machining of Rim Shaping (back)	120	2	2	10	9	min
9	9	Assembly of Wheel (front)	120	3	3	12	0.8	min
10	10	Assembly of Wheel (back)	120	3	3	13	0.8	min
11	11	Assembly of Power system	120	3	3	15	1	min
12	12	Assembly of Wheel front	120	3	3	15	1.1	min
13	13	Assembly of Wheel (back with gear)	120	3	3	15	1.3	min
14	14	Assembly of Unit Drive	120	3	3	16	0.5	min
15	15	Assembly of Chassis	120	3	3	16	2	min
16	16	Assembly of Slot Car	120	3	3	NA	1.5	min

Fig. 6. Routing.tbl with activities, parRouteID and proCoEQIP

The combination of this Routing and BOM information is contained in the AccNodeTree.tbl table in Fig. 7 in the first line (nodeID = 1). Furthermore, this table contains the common multi-level structure of the Routing and the BOM what is specified in the second column that contains the nodes of the children (chilNodeID). E.g., the child node 2 of the first node contains the activity 'Assembly of Unit Drive' (routeID = 14) and the materials 'Motor' (matID = 4), 'Motor Mounting Bracket' (matID = 5) and 'CSrew' (matID = 7). Hence, the AccNodeTree.tbl table in Fig. 7 contains the common structure of the Routing.tbl and BOM.tbl tables together with their routeID's and matIDs'. This tree is 'rooted' with root node (#1) and 'unordered' [25, p. 3] as the temporal arrangement of the activities in the production structure is not completely taken into account. In the right panel of Fig. 7 the columns for the chilNodeID and matIDs are 'list columns' from R-Tidyverse that allow the implementation of the accumulation functions in Eq. (1) to (5) for the different cost and GHG emission components via recursive programming.

```
# A tibble: 16 x 4
   nodeID chilNodeID    routeID matIDs
    <dbl> <chr>           <dbl> <chr>
1       1 "2, 3"             16 1, 2, 3, 7
2       2 "4"                14 4, 5, 7
3       3 "5, 6, 7, 8"       15 9, 10, 14, 22
4       4 "9"                 5 25
5       5 "10"                6 26
6       6 ""                 11 11, 12, 13
7       7 "11"               12 15, 16, 17
8       8 "12"               13 15, 16, 17, 20, 23
9       9 ""                  1 6
10     10 ""                  2 8
11     11 "13"                9 18, 19, 20
12     12 "14"               10 18, 19, 20
13     13 "15"                7 27
14     14 "16"                8 27
15     15 ""                  3 21
16     16 ""                  4 21
```

```
# A tibble: 16 x 4
   nodeID chilNodeID routeID matIDs
    <dbl> <list>       <dbl> <list>
1       1 <dbl [2]>       16 <dbl [4]>
2       2 <dbl [1]>       14 <dbl [3]>
3       3 <dbl [4]>       15 <dbl [4]>
4       4 <dbl [1]>        5 <dbl [1]>
5       5 <dbl [1]>        6 <dbl [1]>
6       6 <NULL>          11 <dbl [3]>
7       7 <dbl [1]>       12 <dbl [3]>
8       8 <dbl [1]>       13 <dbl [5]>
9       9 <NULL>           1 <dbl [1]>
10     10 <NULL>           2 <dbl [1]>
11     11 <dbl [1]>        9 <dbl [3]>
12     12 <dbl [1]>       10 <dbl [3]>
13     13 <dbl [1]>        7 <dbl [1]>
14     14 <dbl [1]>        8 <dbl [1]>
15     15 <NULL>           3 <dbl [1]>
16     16 <NULL>           4 <dbl [1]>
```

Fig. 7. Common generic structure for BOM and Routing in AccNodeTree.tbl

The routeID's and matIDs' from the Routing.tbl and BOM.tbl tables come from their structures as expressed via the 'goes-into' relationship [17, p. 299] of each element in the table to its parent, i.e. to 'parRouteID' and to 'parMatID'. Furthermore, each table contains the corresponding production coefficients, i.e. proCoEQIP und proCoMAT.

```
# A tibble: 16 x 7
   cumCostACTY cumCostMAT  cumPC cumCFPACTY cumELiabMAT cumELiabEQIP cumPCF
         <dbl>      <dbl>  <dbl>      <dbl>       <dbl>        <dbl>  <dbl>
1        138         74.4  212.   0.0517      2.27          1.32   3.64
2         17.8       10.2   27.9  0.00713     0.301         0.178  0.485
3        118         36.2  154.   0.0445      1.82          1.14   3.00
4         17          0.12  17.1  0.00709     0.0006        0.175  0.183
5         56.5        1.44  57.9  0.0227      0.0072        0.575  0.605
6          1.5        3.26   4.76 0.0000633   0.09          0.005  0.0951
7         28.4       26.9   55.2  0.0108      1.54          0.272  1.82
8         28.6       31.9   60.6  0.0108      2.32          0.273  2.60
9          2          0.12   2.12 0.00139     0.0006        0.025  0.0270
10         4          1.44   5.44 0.00279     0.0072        0.05   0.0600
11        26.7        8.1   34.8  0.0107      0.873         0.266  1.15
12        26.7        8.1   34.8  0.0107      0.873         0.266  1.15
13        25.5        1.2   26.7  0.0106      0.00144       0.262  0.275
14        25.5        1.2   26.7  0.0106      0.00144       0.262  0.275
15         3          0.6    3.6  0.00209     0.00072       0.0375 0.0403
16         3          0.6    3.6  0.00209     0.00072       0.0375 0.0403
```

Fig. 8. Cost & GHG Accounting – cumPC and cumPCF in AccNodeTree.tbl

Figure 8 shows the AccNodeTree.tbl table that contains the results for the costs as well as the GHG emissions. The results for the manufactured slot car are shown in the table's first row. The slot car's product cost (PC) amounts to EUR 212,4 which is the sum of the activity cost (cumCostACTY) of EUR 138 and the material cost (cumCostMAT) of EUR 74,4. The slot car's product carbon footprint (PCF) amounts to 3.64 $kgCO_2e$ which is the sum of the product's manufacturing activities (cumCFPACTY) of 0.052 $kgCO_2e$, the E-Liability from the consumed material resources (cumELiabMAT) of 2.27 $kgCO_2e$ and the E-Liability from the used equipment resources (cumELiabEQIP) of 1.32 $kgCO_2e$. Furthermore, the table also contains the costs and GHG emissions for all intermediate material components that were produced in the cutting, machining and assembling activities contained in the Routing.tbl table in Fig. 6.

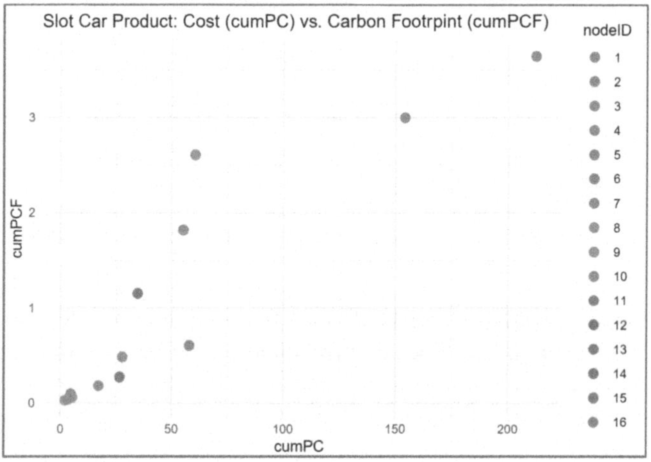

Fig. 9. Slot Car Manufacturing: Product Cost vs. Product Carbon Footprint

All results are visualized in Fig. 9. This impressively reveals the main practical contribution of the activity-based cost & GHG accounting, i.e. the now 2-dimensional view that contains in the second dimension next to the costs also the GHG emissions for the finished good and all intermediate material component needed for its manufacturing.

5 Conclusion

The primary research objective of this article is solving the problem of integrating multi-level structures in both the Routing and the BOM. This objective was achieved by using a 'process-oriented approach' [16] for identifying the 'shared common product and production structure' [17] from the BOM and the Routing and extracting it into the Generic Node Tree. Providing the solution started with the definition of the language spoken in the cost & GHG accounting domain (Fig. 1). Equipped with that language the cost and GHG accounting domain was conceptualized (Fig. 3). Thereof the design of the cost & GHG accounting system (Fig. 4) was operationalized by specifying the attributes and methods of the systems classes and the metrics in Eq. (1) to (5) for accumulating costs and GHG emissions in the class AccNodeTree. This tree represents in the accounting system the Generic Node Tree from the conceptual model. Finally, the cost & GHG accounting system was implemented in the Slot Car Manufacturing use case with the R-Tidyverse data science language.

Applying the three pillars of the accounting domain engineering methodology [13] combines the domain-driven design methodology [14] with the design science research methodology [18]. In this 'domain-driven-design (DDD) science research methodology' the ubiquitous language gets consistently incorporated in all three pillars. The validity of the conceptualized cost & GHG accounting domain shows by its consistent reflection in the operationalized accounting systems. The applicability of the domain conceptualization and the system design shows by implementing the designed system in the practical Slot Car Manufacturing use case. There, all calculations perform well concerning costs and GHG emissions of the finished goods and all intermediate material components that were produced in the different activities.

There are different benefits from the conceptualization and operationalization of the activity-based cost & GHG accounting domain. The applicability scope of the accounting engineering methodology [13] is greatly enhanced for covering many more constellations. The activity-based cost & GHG accounting domain [9] gets a solid foundational representation specified in the formal modeling language UML. The alignment of the integrated cost & GHG accounting with ERP-Control systems gives providers of such systems precise guidance for consistently integrating the GHG accounting into their existing cost accounting systems. And, on the other hand, the alignment also allows the implementation of the integrated cost & GHG accounting system as an add-on solution to any ERP-Control system.

Finally, it has to be noted that by solving the multi-level integration problem form the Routing and the BOM does not restrict the universality of the activity-based cost & GHG accounting to the production domain. Due to the different possibilities of specifying the production coefficients for the unit-inputs of the resources in the activities, the cost & GHG accounting can be applied also e.g. to logistics and services. In the logistics domain time is not the key driver for the resource inputs, but weight and distance of transported materials are more important. In the service domain also capacity resources are applied, but the materials play only a minor or no role at all.

References

1. GHG-Protocol Initiative: GHG Protocol Homepage. https://ghgprotocol.org/
2. GHG-Protocol's Corporate Standard: A Corporate Accounting and Reporting Standard (2004)
3. GHG-Protocol's Product Standard: Product Life Cycle Accounting and Reporting Standard (2011)
4. GHG-Protocol's Scope 3 Standard: Corporate Value Chain (Scope 3) Accounting and Reporting Standard (2011)
5. Leontief, W.: Input-Output Economics. Oxford University Press, Oxford (1986)
6. ISO-LCA (14040): Environmental management - Life Cycle Assessment - Principles and framework, ISO 14040:2006 (2006). https://www.iso.org/standard/76121.html
7. Emblemsvåg, J., Bras, B.: An activity-based life-cycle assessment method. In: DETC'97: 1997 ASME Design Engineering Technical Conference - Sacramento, California, pp. 1–13 (1997)

8. Emblemsvåg, J., Bras, B.: Activity-Based Cost and Environmental Management - A Different Appraoch to the ISO 14000 Compliance. Springer Science & Business Media, New York (2001)
9. Baumüller, J., Schwaiger, W.S.A.: Aktivitätsbasierte Standard-Kosten- & Treihausgas-Rechnung - Integration der CSRD-Treibhausgas-Rechnung in die Standard-Kostenrechnung. Austrian Manag. Rev. **14**, 106–115 (2024)
10. Alaoui, L.H., Baumüller, J., Schwaiger, W.S.A.: 3-Levers of Emission Control-Modeling Framework: Modeling GHG emissions when direct measurement is not possible. In: ISEC 2024 – 3rd International Sustainable Energy Conference - Solutions for Climate Neutral Industrial Production, pp. 1–15 (2024)
11. Kaplan, R.S., Ramanna, K.: Accounting for Climate Change. Harv. Bus. Rev. **99**, 120–131 (2021)
12. Arango, G.: Domain analysis - from art form to engineering discipline., 152–159 (1989)
13. Alaoui, L.H., Schwaiger, W.S.A.: Activity-based life-cycle accounting (AB-LC-ACC) methodology: calculating finished Good' s material cost and material carbon footprint within a hierarchical bill-of-material (BOM). In: 2024 26th International Conference on Business Informatics (CBI), pp. 264–273. IEEE (2024)
14. Evans, E.: Domain-Driven Design - Tackling Complexity in the Heart of Software, Boston et al. (2004)
15. ISO/IEC-ECSI: Enterprise-control system integration - Part 1: Models and terminology, IEC 62264-1:2013 (2013). https://www.iso.org/standard/57308.html
16. Yeh, C.H.: Production data modelling: an integrated approach. Int. J. Oper. Prod. Manag. **15**, 52–62 (1995)
17. Jiao, J., Tseng, M.M., Ma, Q., Zou, Y.: Generic bill-of-materials-and-operations for high-variety production management. Concurr. Eng. Res. Appl. **8**, 297–321 (2000)
18. Geerts, G.L.: A design science research methodology and its application to accounting information systems research. Int. J. Account. Inf. Syst. **12**, 142–151 (2011)
19. Kloock, J., Schiller, U.: Marginal costing: cost budgeting and cost variance analysis. Manag. Account. Res. **8**, 299–323 (1997)
20. Schildbach, T.: Cost accounting in Germany. Manag. Account. Res. **8**, 261–276 (1997)
21. Cooper, R.: Five steps to ABC system design. Accountancy., 78–81 (1990)
22. Kaplan, R.S., Anderson, S.R.: Time-Driven Activity-Based Costing: A Simpler and More Powerful Path to Higher Profits. Harvard Business School Press, Boston, MA (2007)
23. Kaplan, R.S., Anderson, S.R.: Time-driven activity-based costing. Harvard Work, pp. 1–18 (2004)
24. TU Wien: Pilotfabrik. https://www.pilotfabrik.at/
25. Kashkoush, M., ElMaraghy, H.: Product family formation by matching Bill-of-Materials trees. CIRP J. Manuf. Sci. Technol. **12**, 1–13 (2016)

Call Detail Record Generator for Modeling Real-World Telecommunication Scenarios

Mariusz Rafało[(✉)]

Warsaw School of Economics, Warsaw, Poland
mrafalo@sgh.waw.pl

Abstract. The paper introduces a novel Call Detail Record generator designed to produce massive synthetic datasets for telecommunications research and operational testing. The generator employs a multi-distribution approach and Markov chains to simulate realistic telecommunications data, encompassing both standard user behaviors and complex scenarios such as SIM box fraud or telecommunication probes. The generator serves multiple practical applications, including network planning and optimization, fraud detection and the training and testing of AI and machine learning models for predictive analytics. Moreover, the generator is characterized by very high speed of operation, allowing the generation of millions of events in seconds.

Keywords: simulation · call detal record · markov chains · statistical distributions · synthetic data · fraud detection

1 Introduction

In the landscape of telecommunications, the analysis and simulation of network behaviors is important for optimizing performance, ensuring reliability, and enhancing security [22]. However, access to telecommunications data is often limited, posing challenges for researchers and data analysts. To circumvent these limitations, we introduce a method for generating synthetic telecommunications traffic data. Synthetic data can simulate less common occurrences, such as fraud, anomalies, or probing devices. This paper primarily explores the generation of Call Detail Record (CDR) data and the simulation of typical more and less common situations within in telecommunication network.

The CDR data are essential for various operational and business purposes, including billing, network operations, and performance analysis. Insights derived from CDR analysis can help in enhancing customer service, optimizing network operations, and improving network security against fraudulent activities. CDR dataset contains detailed information about each individual communication transaction handled by the telecommunication system. The specifics can vary among different service providers, but typical CDR include the elements

K. Węcel (Ed.): BIS 2025, LNBIP 554, pp. 297–309, 2025.
https://doi.org/10.1007/978-3-031-94193-1_22

like: date and time of a call, phone numbers of both the caller (originating) and the receiver (terminating), call duration, Mobile Station International Subscriber Directory Number (MSISDN), International Mobile Equipment Identity (IMEI), contact type (e.g. voice, SMS, MMS), Base Transceiver Stations (BTS) used during the call, and other information about services used during the contact [24]. MSISDN is connected with single Subscriber Identity Module (SIM card).

CDR records are not only a foundational element for the operational and financial aspects of a telecommunications provider but also a rich source of data for analytics. Insights derived from CDR analysis can help in enhancing customer service, optimizing network operations, and improving security measures against fraudulent activities. The analysis of CDR focuses on:

- billing and accounting: CDRs are primarily used to bill customers for the services they have used. They provide all the necessary details to calculate charges based on the type, duration, and nature of the service consumed.
- network management and planning: analyzing CDRs helps network operators understand traffic patterns, peak usage times, and load distribution across the network. This information is crucial for capacity planning and ensuring optimal network performance.
- fraud detection: CDRs are analyzed to detect unusual patterns that might indicate fraudulent activities, such as unusually long call durations, frequent calls to premium-rate numbers, or unexpected international calls which could suggest cloning or theft of service [21].
- regulatory compliance: telecommunications companies are often required to maintain and provide CDRs to regulatory bodies to ensure compliance with local and international laws regarding data retention and consumer protection.
- customer service: CDR data helps customer service representatives resolve disputes by providing them with precise details of customer usage. Furthermore, analyzing CDRs can help identify customer preferences and behavior, aiding in customer relationship management and tailored marketing initiatives [20].
- legal enforcement: CDRs can be requested by law enforcement agencies for investigative purposes under legal provisions, as they contain detailed information about communications between individuals.

The goal of this paper is to present a system that simulates realistic telecommunications data, encompassing user behavior, as well as rare events, that can occur in the telecommunication network. The generator can capture key attributes of interactions, such as:

1. family dynamics: frequent calls and SMS exchanges during specific times (e.g., mornings or evenings), shared geographical proximity, and common service providers,
2. business relationships: structured and repetitive communication patterns, such as regular conference calls, peak-hour activities, and inter-departmental communications,

3. social interactions: diverse communication behaviors, including group chats, periodic check-ins, and varied engagement levels over time.

The paper presents a high speed, multi-stage CDR generator that leverages Markov chains and statistical methods to produce realistic and comprehensive telecommunication datasets. Markov chains provide a robust framework for modeling the stochastic processes inherent in telecommunications systems. By capturing the probabilistic transitions between different states of a system, Markov chains facilitate the generation of traffic dynamics [16,19].

2 Related Work and Motivation

While it is acknowledged that the telecommunications industry inherently produces large volumes of data, access to such real-world datasets is often restricted due to privacy, regulatory and proprietary concerns. Consequently, despite the data abundance in operational environments, researchers and developers face a scarcity of accessible and representative telecommunications datasets. The motivation for generating synthetic data is rooted in the lack of accessibility and variability - especially concerning rare or anomalous events such as SIM box fraud or network probing, which are often underrepresented in real datasets [3]. Furthermore, the use of preselected features and synthetic generation enables controlled experimentation under configurable parameters and distributions, which would not be feasible using full, real-world contextual data. While a detailed quantification of accuracy trade-offs between full-context real data and feature-selected synthetic data is beyond the current scope, the generator's design emphasizes structural and statistical fidelity, with calibration against known distributions and empirical patterns.

Research on the use of CDR-based analysis includes, among others, trends in system dynamics [4], multi-agent approach [25], classic machine learning models [9,12] and multi-stage simulations based on Markov chains [1,14]. Current research streams in this area mainly investigate telecommunication frauds, employing various clustering and machine learning techniques. Due to the lack of public repositories of reliable CDR data (due to sensitivity of these data), most studies are based on small samples or synthetic data [6,11].

For example, a study [15] uses synthetic CDR data to examine deep learning techniques for classifying traffic patterns to distinguish machine-to-machine (M2M) communication from human-to-human (H2H) interactions. The primary goal of this analysis is to enhance network resource management and support critical applications, such as detecting anomalous behavior and ensuring the efficient operation of Internet of Things (IoT) devices. By leveraging deep learning algorithms and generated CDR data, the study aims to improve the overall quality of service in telecommunications networks. Other study [13] investigates the implementation of Bearer Independent Call Control (BICC) protocols to optimize the synthesis and monitoring of CDRs. Using hashing techniques, the research proposes an efficient method for generating lightweight and accurate CDRs. This approach reduces computational overhead while maintaining data integrity, ensuring that CDR analysis remains reliable and scalable. The primary goal is to enable real-time monitoring and effective traffic management, with a focus on supporting network operations and detecting irregularities, such as fraud or network misuse. Both studies underscore the importance of CDR generation and analysis in advancing telecommunication technologies and ensuring secure, efficient network management.

There are few CDR data generators and even fewer that are capable of generating large amounts of data in bulk. For example, the authors in [5] present a web based data generator. The generator can produce synthetic datasets encompassing normal behavior and fraudulent network attack scenarios. Another approach to generating CDR data [18] utilizes graph database to produce synthetic data. However, this generator bases on static rules. When working with synthetic data to model telecommunications behaviors, the use of a single static data distribution is often insufficient to capture the complex dynamics of customer interactions. Instead, employing multiple data distributions ensures that the synthetic data generator can adequately represent the variability and heterogeneity inherent in real-world data. Another study [17] explores the generation of data by replicating user behaviors through social network modeling. This approach allows for representing social relationships, but generates relatively small data sets. In fact, most systems suffer from performance problems, especially when generating larger datasets and using Java [7] or Python [23] programming languages. A workaround to performance issues is to use distributed computing in CDR generation. It is therefore possible to generate data in parallel on multiple servers, using Apache Spark technology [2]. However, the limitation of this type of generators is the need to have an extensive hardware infrastructure and appropriate software.

By integrating multiple parameters and scenarios into the CDR generator, researchers and analysts can create more nuanced and realistic models that are capable of adapting to and predicting changes in customer behavior over time. For instance, a synthetic CDR generator should have the capacity to adjust its parameters to reflect seasonal variations, special events, or the introduction of new products and services. This involves varying the statistical distributions used for call frequencies, calls duration, and network traffic patterns to align with expected changes in user behavior during different times or under different conditions.

Our approach addresses limitations of current CDR generators. The proposed system is characterized by the inclusion of complex scenarios and the use of multiple statistical distributions. Furthermore, our generator allows for very efficient generation of hundreds of millions of events. The system was created in the Rust programming language and can be run on both Windows and Linux computers. The system does not require a specific hardware infrastructure, and there are no dependencies on other modules or libraries. The source codes are published in the author's public GitHub repository[1].

[1] https://github.com/mrafalo/fastcdrgen.

3 System Overview

The core of the data generation process is based on Markov chains. A Markov chain is a statistical model that predicts future states based on the current state, without considering the sequence of events that preceded it [8]. In the context of CDR generation, Markov chains are utilized to model the sequence of calling behaviors, including the initiation, duration, and termination of calls. Each state in the Markov chain represents a group of network events, like business calls, family calls, MMS messages, etc. Probabilities are assigned to transition groups between states base on configuration parameters. These probabilities indicate how likely a user is to perform specific action, ensuring that the generated data mimics the stochastic nature of actual telecommunications activity.

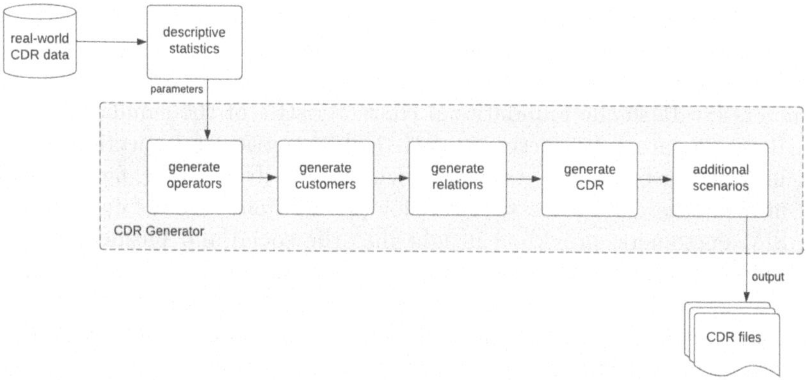

Fig. 1. CDR generator framework

The system's configuration is defined in a file. This configuration file specifies multiple parameters for the generation process, including parameters controlling the telecommunications market (e.g.: number of operators, number of customers, market shares, etc.), parameters related to the distribution of variables (e.g.: average call duration for specific customer types), parameters controlling scenarios regarding SIM Box or telecommunication probe, parameters controlling the infrastructure (e.g.: number of base transceiver stations). The generator operates through distinct stages, each designed to replicate specific characteristics and interactions within the network (see 1):

1. The creation of a diverse and realistic customer base. The customers are generated, divided into the private customers, business customers (and other types). Each client is assigned to an operator. A customer can have multiple SIM cards (identified as MSISDN) and multiple devices (identified as IMEI). For each client, the individual behavior profile is randomly generated.
2. The generation of relationships graph. The connections between customers are based on customer profiles. The relations are categorized into private

(family and friends), business, and random relationships. This involves setting probability distributions for interactions based on relationship types.
3. The generation of CDR events. CDR data is generated based on customer relationships, control parameters (Table 1) and configured data distributions. The stochastic nature of this module reflects missed calls, dropped calls, coverage issues, etc.
4. The incorporation of specific scenarios. The generator implements telecommunication probes, multi-SIM devices, and SIM Box fraud by embedding predefined patterns and irregularities within the CDRs.

The parameters used in the system can be grouped into categories that capture different aspects of its design and functionality: general, graph of relations, call patterns, distribution used during generation and technical aspect of data processing.

General parameters define the overall scope and structure, such as the number of local and international customers, operators, and BTS, along with proportions for various customer types like SIM box, probe, multi SIM devices. These parameters establish the foundational characteristics of the simulated environment. Relation parameters describe statistical properties of connections between users, including average and standard deviations for friendships, family ties and other interactions. These are specified for private and business customers and multi SIM customers, providing insight into the social and business dynamics modeled in the system. Call parameters focus on communication behaviors, specifying average and standard deviations for the counts of calls, SMS, and MMS per relation. These parameters help to simulate realistic communication patterns across different user groups. Duration parameters further refine the model by defining statistical properties of call durations for private, business, probe, SIM box, and multi SIM users, reflecting the time spent on interactions. Distributional parameters define the statistical distributions used for modeling key behaviors such as call durations, number of calls, number of text messages, etc. For instance, by default, SIM box durations are modeled with a Weibull distribution, while probe durations use a Poisson distribution and other groups typically follow a normal distribution. These choices influence the variability and characteristics of generated data. Finally, technical parameters address the operational aspects of the system, such as the start date of the simulation, batch size for processing and the output file name. These settings ensure efficient execution and proper management of results.

The default values parameters from the general, distributional, and technical groups are presented in detail in Table 1, which provides an organized view of their roles, values, and application within the system. The table shows sample parameters out of over 60 possible configurations.

Data is generated in batch (batch size is configurable); each batch of generated data is saved to a file. This is a unique approach, because the generated CDRs are not stored in memory, but are recorded directly to file.

We use a number of random generators that draw data from a specific distribution. The generators are prepared in such a way that sampling is based

Table 1. Default values for selected parameters

Group	Parameter	Value
general	number of local customers	10000
general	number of international customers	100
general	number of local operators	4
general	number of international operators	2
relations	average relation friends count	5
relations	std. dev. of relation friends count	10
relations	average relation family count	5
relations	std. dev. of relation family count	3
calls	average calls count per relation (probe)	20
calls	std. dev. of calls count per relation (probe)	10
calls	average calls per relation (SIM box)	20
calls	std. dev. of calls per relation (SIM box)	10
durations	average call duration (priv)	413
durations	std. dev. of call duration (priv)	60
durations	average call duration (business)	600
durations	std. dev. of call duration (business)	70
distributions	SIM box duration distribution	WEIBULL
distributions	probe duration distribution	POISON
distributions	business duration distribution	NORMAL
distributions	priv duration distribution	NORMAL
technical	start date	2024-05-01
technical	batch size	100000
technical	results filename	results/cdr.csv

on specific distribution parameters. To ensure that the synthetic data is realistic and useful for analytical purposes, the parameters of these distributions are often calibrated based on historical data from real networks. This calibration process involves statistical analysis and machine learning techniques to fit the distribution parameters such that they closely replicate observed behaviors. Validation is typically conducted by comparing statistical summaries and key metrics of the synthetic data against actual data, ensuring that the generated records accurately represent the target environments in terms of behavior and variability. Sample distributions of selected variables are presented in Fig. 2 (for various customer types) and 3 (for various scenarios).

We utilize the Cullen and Frey graph to visualize the characteristics of various theoretical and empirical distributions used in generating data for different scenarios. The graph is a tool commonly used in statistical analysis to plot the kurtosis versus the skewness of a dataset, helping to understand the type of distribution the data follows. For our analysis, we consider three theoretical distri-

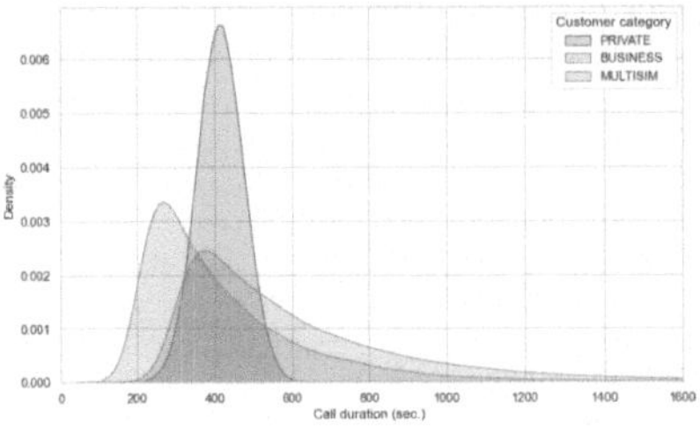

Fig. 2. Generated data distributions for different customer types

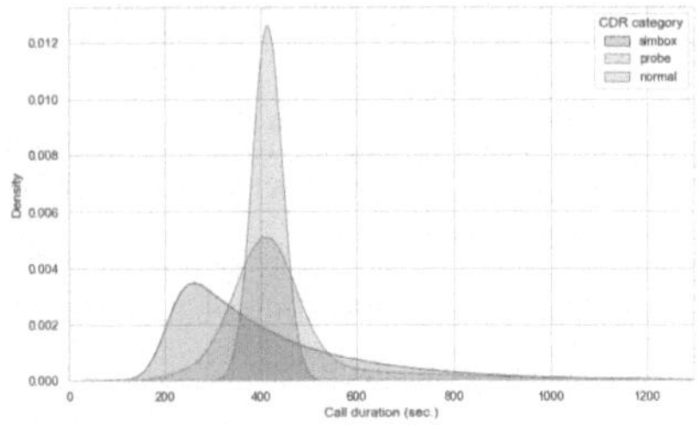

Fig. 3. Generated data distributions for different activity scenarios

butions: normal, Weibull and Poisson. These distributions were chosen based on their relevance to different patterns observed in telecommunications data (however the system can also use other distributions like uniform or Cauchy). By default normal distribution represents general scenarios where customer behaviors are randomly distributed around a mean. In our approach, we have a multimodal distribution, due to the varied nature of customer usage (different means and variances for different customer types). Weibull distribution is used for modeling SIM box scenario, because it reflects the irregular, burst-like patterns of calls associated with fraudulent activities. The Poisson distribution is used to model events occurring independently over time, which matches the behavior of probe traffic.

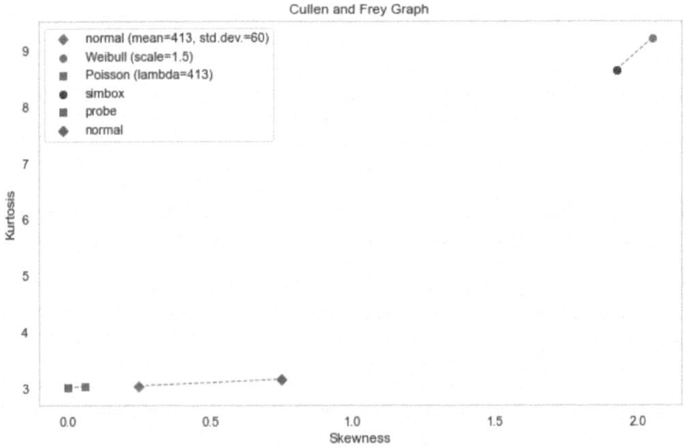

Fig. 4. Cullen and Frey graph for sample generated dataset

In Fig. 4, we present the Cullen and Frey graph for theoretical distributions along with the empirical data generated for the scenarios. It presents how closely generated data align with the chosen theoretical distribution. Lines were drawn connecting the empirical data points to the corresponding theoretical distribution, indicating the degree of similarity between the actual patterns observed in the data and the idealized statistical models.

4 Special Scenarios

We explore several scenarios, including telecommunication SIM box fraud, probe behavior and multi-SIM usage. Each scenario addresses specific aspects of telecommunications operations and customer behavior.

4.1 SIM Box Fraud

SIM box fraud involves using a device that has multiple SIM cards to illegally reroute calls through alternative, cheaper routes, bypassing the official telecommunication network. This can lead to significant revenue losses for operators. The typical scenario of SIM Box fraud process involves purchasing a wholesale number of minutes on international telecommunications platforms, then using the Internet (IP networks), bypassing official contact gateways, introducing this traffic to the network of a given operator [10]. SIM box fraud exhibits features like exclusive or large share of outgoing voice calls with no or almost no SMS and MMS messages, connections mainly within the network of a given operator, no movement of SIM cards. By simulating SIM box fraud scenarios, telecommunication companies can train their fraud detection systems to identify and mitigate such activities effectively. This scenario is vital for testing the capability of AI systems to detect unusual patterns that may indicate fraud, ensuring that losses are minimized and network integrity is maintained.

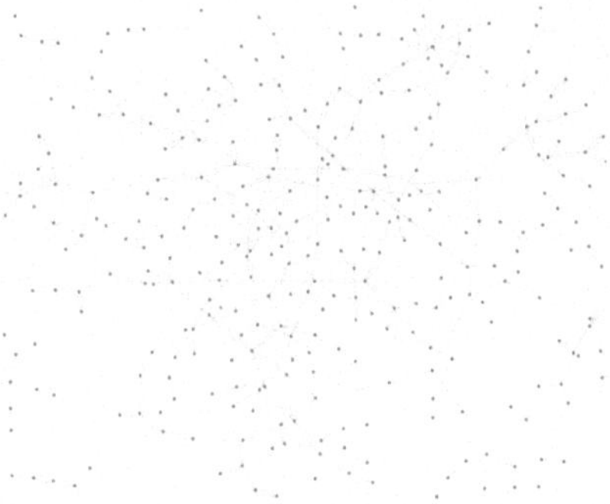

Fig. 5. Sample graph presenting normal customer behavior

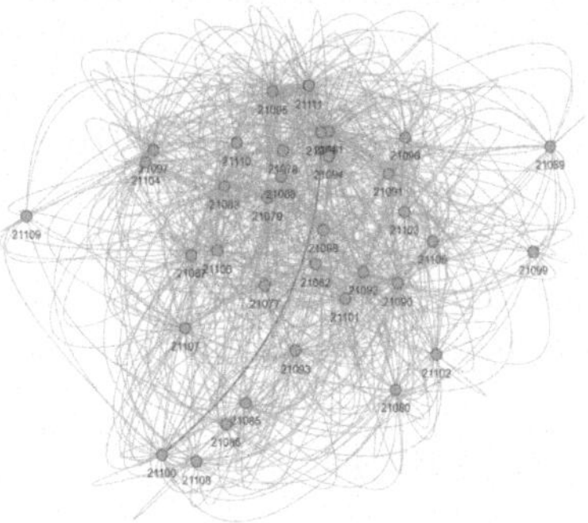

Fig. 6. Sample graph presenting SIM box activity

The Figs. 5 and 6 illustrate a contrast between the network activities of regular customers and fraudulent SIM box activities. In the first chart, representing normal customer behavior, the network exhibits a high number of nodes with relatively sparse connections. This reflects the typical interaction pattern of individual SIM card, where connections are distributed among diverse entities, leading to a web-like graph with a wide reach but low-density links. Conversely, the second chart, depicting SIM box fraud, shows a significantly different structure. Here, the network has a small number of nodes with disproportionately high connectivity. This configuration is characteristic of fraudulent activity, where a few SIM cards handle a large volume of traffic, mimicking multiple customers while bypassing legitimate network operations. The dense connections among a limited set of nodes make SIM box activity visually and analytically distinct.

4.2 Probe Scenario

Probe scenario involves simulating the actions of network testing equipment or external entities attempting to evaluate network performance and security. Probes might mimic regular user behavior or malicious activities to test the robustness of network defenses. Including this scenario helps in developing and testing systems that can differentiate between benign probe activities meant for network testing and potentially malicious activities that resemble probing but aim to exploit network vulnerabilities. This is crucial for maintaining network security and performance. Probes can simulate normal user behavior to test how well the network is functioning under typical and peak load conditions. This helps in planning network expansions or adjustments before actual demand arises. This includes testing for latency, bandwidth availability, throughput and packet losses, etc. In CDR data, probes often exhibit similar behavior to the SIM Box scenario, generating false positive alarms.

However, it is essential to note that a telecommunication probes and other devices with multiple SIM cards (e.g. dual SIM phones or IoT devices) can sometimes produce graph structures that superficially resemble SIM box behavior. For example, probes might create dense connections in testing environments, and multi SIM users-legitimate customers with multiple SIM cards-may show similar network density. The CDR generator software provides the capability to emulate these scenarios with nuanced variations in activity characteristics. Our generator allows to define subtle differences in different scenarios by configuring parameters related to the distribution parameters. In this way, the generator ensures that synthetic datasets accurately reflect real-world complexities.

4.3 Multi SIM

Multi SIM behavior refers to the use of multiple SIM cards within a single device. This behavior can be legitimate, as seen in dual-SIM phones used by many consumers for convenience, but it can also signal fraudulent activity, because it can resemble SIM box pattern. The second area of application of multiple SIM cards is IoT communications. IoT devices in machine-to-machine communication can exhibit behavioral patterns similar to multi SIM scenarios, particularly when a single device or gateway manages multiple connections or SIM cards. In M2M communication, a gateway device often aggregates data from multiple IoT endpoints, transmitting this information to central servers or other nodes within the network. This results in a dense connectivity pattern, akin to what is observed in multi SIM activities, where a legitimate user holds and uses multiple SIM cards for varied purposes. However M2M communication is done using data protocols, rather than voice or SMS.

Multi SIM scenario complicates network monitoring and requires advanced detection techniques to differentiate between legitimate use and fraud. Moreover, while multi SIM scenarios typically involve human-driven activity, IoT-based M2M communication is primarily automated, with distinct temporal and transactional characteristics. For instance, IoT devices may generate periodic, uniform traffic or follow predefined communication protocols, distinguishing them from the irregular, human-like patterns often observed in multi SIM usage. Analyzing the structural differences in social network graphs for these scenarios allows fraud detection systems to better identify and separate fraudulent activities from legitimate multi SIM usage, ensuring the accuracy and reliability of detection mechanisms.

5 Summary

The presented CDR generator allows for efficient generation of CDR data in a way that allows for adapting the data to actual activities in the telecommunications network. The practical applications of the generator include network planning and optimization and simulation of fraud scenarios that can provide a valuable resource for developing fraud detection systems. Additionally, the realistic and complex datasets produced are ideal for training and testing AI and machine learning models. The generator is characterized by very high work efficiency. It allows to generate millions of CDR events in a few seconds.

Despite its robust capabilities, the generator has limitations that include the reliance on random distributions. Markov chains, while effective for modeling certain types of dependencies, may not capture long-term dependencies or complex patterns involving more extensive historical data. Furthermore, the scenarios currently supported are limited to predefined types of behaviors, which could be expanded to enhance the generator's applicability.

Future research could potentially address these limitations by expanding the range of scenarios to make the generator more versatile. Additionally, integrating advanced machine learning techniques to capture longer-term dependencies and more complex behavioral patterns could further improve the realism and applicability of the generated data.

References

1. Al-Habashna, A., Al-Dala'in, S., Al-Qudah, F.: A Markov chain model for predicting the evolution of network traffic states. J. Netw. Comput. Appl. **157**, 102628 (2020)

2. Analytics, R.I.: CDR generator (2024). https://github.com/RealImpactAnalytics/cdr-generator. Accessed 09 June 2024
3. Azimi, A., et al.: zGAN: an outlier-focused generative adversarial network for realistic synthetic data generation (2024). arXiv preprint arXiv:2410.20808, https://arxiv.org/abs/2410.20808
4. Banda, P.K., Tembo, S.: Application of system dynamics to mobile telecommunication customer churn management. J. Telecommun. Electron. Comput. Eng. **9**(3) (2018)
5. Boggs, N., et al.: Synthetic data generation and defense in depth measurement of web applications. Tech. rep, Unknown (2014)
6. Chouiekh, A., Haj, E.: Convnets for fraud detection analysis. Procedia Comput. Sci. **127**, 133–138 (2018). https://doi.org/10.1016/j.procs.2018.01.107
7. Deshpande, T.: CDR data generator (2024). https://github.com/deshpandetanmay/cdr-data-generator. Accessed 09 June 2024
8. Gagniuc, P.A.: Markov Chains: From Theory to Implementation and Experimentation. Wiley, USA, NJ (2017). this book includes practical applications and case studies, illustrating how Markov chains are implemented in real-world business analytics
9. Hastie, T., Tibshirani, R., Friedman, J.: The Elements of Statistical Learning, 2nd edn. Springer, Cham (2011). https://doi.org/10.1109/SITIS.2013.106
10. Karunathilaka, A.V.V.S.: Fraud Detection on International Direct Dial Calls. Ph.D. thesis, University of Colombo School of Computing, Colombo (2020)
11. Kashir, M., Bashir, S.: Machine learning techniques for sim box fraud detection. In: 2019 International Conference on Communication Technologies, ComTech 2019. Institute of Electrical and Electronics Engineers (2019)
12. Kouam, A.J., Viana, A.C., Tchana, A.: Simbox bypass frauds in cellular networks: strategies, evolution, detection, and future directions. IEEE Commun. Surv. Tutor. **23**(4), 2295–2323 (2021). https://doi.org/10.1109/COMST.2021.3100916
13. Li, D., Zhang, Z.: The protocol research and application of bearer independent call control in the NGN network monitoring system. In: Proceedings - 2012 9th International Conference on Fuzzy Systems and Knowledge Discovery, FSKD 2012, pp. 2261–2264 (2012). https://doi.org/10.1109/FSKD.2012.6234005
14. Li, X., Wang, Y., Zhang, M.: Stochastic modeling and analysis of cellular network traffic using Markov chains. IEEE Trans. Commun. **69**(4), 2501–2514 (2021)
15. Lu, Y., Ma, Y., Shi, L., Chen, L.: A deep learning approach for m2m traffic classification using call detail records. In: 2021 6th International Conference on Signal and Image Processing, ICSIP 2021, pp. 964–968 (2021). https://doi.org/10.1109/ICSIP52628.2021.9688785
16. MDPI: Markov and semi-markov chains, processes, systems and emerging related fields. Mathematics **2023**, Special Issue (2023). https://www.mdpi.com/journal/mathematics/special-issues/Markov-SemiMarkov-Processes-Systems/. Accessed 09 June 2024
17. Murtic, A., et al.: SNA-based artificial call detail records generator. In: 2018 41st International Convention on Information and Communication Technology, Electronics and Microelectronics, MIPRO 2018 - Proceedings, pp. 1226–1230. Institute of Electrical and Electronics Engineers Inc. (June 2018). https://doi.org/10.23919/MIPRO.2018.8400222
18. Neo4j: Call detail records (cdr) analytics (2023). https://neo4j.com/graphgists/call-detail-records-cdr-analytics/. Accessed 02 June 2024

19. Osorio-Lird, A., Chamorro, A., Videla, C., Tighe, S., Torres-Machi, C.: Application of Markov chains and Monte Carlo simulations for developing pavement performance models for urban network management. Struct. Infrastruct. Eng. 1–13 (2017). https://doi.org/10.1080/15732479.2017.1402064
20. Park, S., Huh, S., Oh, W., Han, S.: A social network-based inference model for validating customer profile data. MIS Q. **36**(4), 1217–1237 (2012)
21. Rzepakowski, P.: Supporting telecommunication product sales by conjoint analysis. J. Telecommun. Inform. Technol **2008**(3), 28–34 (2008)
22. Shafiq, G., Liu, S., Jiang, F., Gifford, W.M., Zhang, Z., Liu, A.X.: Characterizing and modeling internet traffic dynamics of cellular devices. In: Proceedings of the ACM SIGMETRICS International Conference on Measurement and Modeling of Computer Systems, pp. 305–316. ACM (2011)
23. Songailaite, M.T.K.: Synthetic call detail records generator. In: IVUS 2021, Proceedings of the 26th International Conference on Information Society and University Studies, vol. 2915. Aachen: CEUR-WS (2021). 9
24. Sultan, K., Ali, H., Zhang, Z.: Call detail records driven anomaly detection and traffic prediction in mobile cellular networks. IEEE Access **6**, 41728–41737 (2018)
25. Wojewnik, P., Kaminski, B., Zawisza, M., Antosiewicz, M.: Social-network influence on telecommunication customer attrition. In: KES-AMSTA11 Proceedings of the 5th KES International Conference on Agent and Multi-agent Systems: Technologies and Applications (2011)

Automating Water Management Using an Online Planning Approach

José G. Quenum(✉)

Namibia University of Science and Technology, Windhoek, Namibia
jquenum@nust.na
http://www.nust.na

Abstract. This paper proposes an online planning approach to automate complex task execution by autonomous agents. We model the complex task as a Partially Observable Markov Decision Process (POMDP) to capture the inherent uncertainty. Then, using an adaptation of the Monte Carlo Tree Search (MCTS) algorithm, we generate a *policy* that governs how gradually the complex task should be accomplished. Our approach demonstrates a reliable and scalable procedure for automating a complex task using autonomous entities.

Keywords: Online Planning · Monte Carlo Tree Search · POMDP

1 Introduction

This work is developed in the context of complex task *full* automation integrating *autonomous* entities (e.g., autonomous agents [1]) in the decision-making during task automation. We define a complex task as one that requires the composition of smaller-scale actions. This composition might involve several operators, including *sequential*, *parallel*, *conditional*, etc. We note $\tau_\imath \equiv (\sigma(\mathrm{a}_{0\imath}, \ldots, \mathrm{a}_{\ell\imath}))$, where σ represents the composition[1] operator. Furthermore, we assume that each action $\mathrm{a}_{k\imath}$ can be represented *abstractlly* and can thus be fulfilled by several possible *concrete* functions $\mathrm{f}_{\jmath\imath}$. For example, consider a wastewater treatment task encompassing several water filtration steps/actions to remove different types of particles and organic matter. Each step can be refined using specific concrete functions based on the particles or organic matter in a water sample.

A naive solution would be to construct an offline plan $\pi_\jmath$ encompassing *abstract* actions $\hat{\mathrm{a}}_k$ and then perform a match-making between $\hat{\mathrm{a}}_k$ and concrete functions f_{km}. However, this approach is not suitable for a dynamic environment[2]. Moreover, given a problem with a large state space, an offline plan has to explore the entire state space to find an optimal sequence of actions. Instead, an online planning approach focuses on the reachable state space, a subset of the

[1] The reader should note that we do not exclude multiple composition layers.

[2] A dynamic environment is one where its configuration, i.e., the attributes defining it, might change dynamically.

K. Węcel (Ed.): BIS 2025, LNBIP 554, pp. 310–318, 2025.
https://doi.org/10.1007/978-3-031-94193-1_23

state space. Furthermore, the history[3] explored during these algorithms grows exponentially, rendering the offline techniques inefficient (see [14]).

Existing research contributions to sequential decision-making propose various techniques, including POMCP [10] and Partially Observable Markov Decision Process (POMDP) variants [17], Large Language Model (LLM) [5], a reinforcement learning integrated approach [12]. While these approaches help model the problem adequately, they prove generally inefficient, computationally intractable or not fully autonomous. This research aims to optimally select functions required to fulfil a complex task following an online planning method. We assume the complex tasks discussed in this work imply *uncertainty* in their modelling[4]. As such, we represent the tasks as a POMDP [7,13]. Roughly, a Markov Decision Process (MDP) models a decision problem where the transition is *stochastic* (i.e., transitioning from a given state after performing an action is no longer deterministic), and the next state only depends on[5] the current one and not on all the states visited in the past. A POMDP is a generalisation of an MDP where the states cannot be perceived directly. Rather, an observation is stochastically generated from the new state. The generated observation provides a hint to the agent about the current state. To solve the POMDP, we propose an adaptation of the Monte Carlo Tree Search (MCTS) [2,15,18] algorithm to approximate the best action(s) to fulfil a complex task. In this paper we present our planning algorithm and illustrate it using water management application.

The remainder of the paper is organised as follows. Section 2 elaborates on our online planning approach. Section 3 discusses the evaluation of our approach, and concludes the paper and sheds light on future work.

2 Proposed Approach

Definition 1. *Formally, a POMDP can be defined as a tuple* $< \mathcal{S}, \mathcal{A}, \mathcal{O}, \mathbf{T}, \mathbf{R}, \mathbf{O}, \gamma >$ *where* $\mathcal{S}$, $\mathcal{A}$ *and* $\mathcal{O}$ *represent the state, action and observation spaces.* $\mathbf{T} = \mathbb{P}(\mathsf{S}' \mid \mathsf{S}, \mathsf{a})$ *is the transition function. It represents the probability for landing in state* S' *after executing action* a *in state* S. $\mathbf{R} = \mathbb{E}[\mathsf{R} \mid \mathsf{S}, \mathsf{a}]$ *is the reward function, i.e., the expected reward* R *after executing action* a *in state* S. $\mathbf{O}$ *is the observation function. It provides observation probabilities for each state. Finally,* γ *represents the discount factor, which is used in the value function estimate for each state.*

To determine the initial state of a POMDP, we use a belief state materialised by a probability distribution over a subset of the state space. As well, a *history* is a sequence of action-observation pairs in. A typical solution to a POMDP is a *policy*, a rule that prescribes which action to take given the current state. The selected action maximizes the value function estimate (or utility) for the state.

[3] A history should be understood as a set of action-observation pairs in this context.
[4] In this work, we focus on state uncertainty.
[5] This property is referred to as the Markov property.

Several algorithms have been proposed to solve POMDPs. These include offline methods such as *exact value iteration* [8] and *policy iteration* [11]. The existing range of solutions also include online planning techniques [16]. In this section, we present an overview of online planning to solve POMDPs. Specifically, we discuss the MCTS algorithm and an adaptation we propose for this work. We refer the reader to [15] for a detailed discussion on online planning. Online planners [5,12,20] for POMDPs use a *forward search* on the current history to compute a (local) approximation of the value function. A common technique to sample the next state from the current belief state is by using a simulation algorithm.

MCTS [4,6,9,19] is one such algorithm which uses Upper Confidence Bound (UCB) as its simulation mechanism. It is a tree search technique that balances exploration and exploitation, using four iterative steps: (1) selection, (2) expansion, (3) roll-out, and backpropagation. The selection step uses a strategy (or tree policy) to explore the tree, looking for an expandable node. Once identified, a new child node is added to the expandable node. The roll-out step runs a simulation until a terminal node to compute the action value function, which is later propagated up to the root to help identify which branch is worth exploring.

$$Q(h, a) + c\sqrt{\frac{\log \sum_a N(h, a)}{N(h, a)}} \tag{1}$$

Furthermore, we assume that during the exploration, the agent does not fully observe the environment. As such, we adapt the MCTS algorithm following a POMDP [3]. This in turn implies that we cannot determine the actual states of the system. Rather, we observe the environment, and through each observation map a *belief* state. Consequently, the explored history does not involve the states, but combines actions and observations. Algorithm 1 presents our adaptation of the MCTS algorithm which selects the action that maximizes the value estimate (defined in Eq. (1)). In Eq. (1), $N(h, a)$ represents the visit count for a history-action pair, c represents the exploration parameter, and $Q(h, a)$ represents the value estimate for a history-action pair. Algorithm 1 inputs a search tree (π) encompassing a POMDP. It also takes an exploration depth (d), several iterations (m) and a state inferred from a probability distribution representing a belief state. It returns an action, the approximately optimal action for the state. On line 1, the algorithm calls the main search function, $search\_MCTS(|)$, which returns the optimal action. $search\_MCTS()$ (lines $2-5$) simulates the tree exploration for a given number of iterations and returns the approximately optimal action. The algorithm approximates the optimal action using the $simulate()$ function (lines $7-23$). The function expands the tree and rolls out several nodes until the depth is exhausted to determine which action yields the best reward. The exploration (see function $explore()$ (lines $25-27$)) factor also determines the evaluation outcome.

Algorithm 1: Partially Observable MCTS.

```
Input: S                                        // Current state
Input: π                                        // Search tree
Input: d                                        // Exploration depth
Input: m                                        // Number of iterations
Input: h                                        // History
Output: a                                       // Optimal action
return search_MCTS(π, S, h, m, d)
Function search_MCTS(π, S, h, m, d)
    for ı = 1 to m do
        simulate(π, S, h, d)
    return argmax_a f(a) = {a ∈ P.A | π.Q(h, a)}

Function simulate(π, S, h, d)
    if d ≤ 0 then
        return value_fn(S)

    if ∄y · ((h, first(a)), y) ∈ π.N then
        for a ∈ π.P.A do
            π.N(h, a) ← 0
            π.Q(h, a) ← 0

        return value_fn(S)

    a ← explore(π, h)
    (S', r, o) ← TRO(S, a)
    q ← r + γ(simulate(π, S', h ⋃{(a, o)}, d − 1))
    π.N(h, a) += 1
    π.Q() += (q − π.Q(h, a)) / π.N(h, a)
    return q

Function explore(π, h)
    Nh ← Σ_{a∈π.P.A} π.N(h, a)
    return argmax_a g(a) = {a ∈ P.A | π.Q(h, a) + π.c sqrt(log π.N(h, a) / Nh)}
```

We employ a water management application as an example to illustrate our approach. The latter is a multi-faceted application with various components, including *water source* management, *demand* and *supply*, *wastewater* management and *usage* and *governance*.

The wastewater treatment applies various treatment procedures to remove contaminants from domestic, industrial or agricultural wastewater. We define a six-step workflow (see Table 1): $W^{\omega} = \sigma_1(a_0^{\omega}, a_1^{\omega}, a_2^{\omega}, a_3^{\omega}, a_4^{\omega}, a_5^{\omega})$ which captures all actions executed during the treatment. Note that σ_1 is the sequential composition operator.

Table 1. Wastewater Treatment Workflow

Action	Step	Designation
a_0^ω	Step 1	Water filtration to remove solid particles
a_1^ω	Step 2	Screening to remove large floating objects
a_2^ω	Step 3	Remove cinder, sand and small stones
a_3^ω	Step 4	Remove bio-degradable organic matter
a_4^ω	Step 5	Remove dissolved substances
a_5^ω	Step 6	Reduce amount of organic matter

Figure 1 presents a simplified[6] model corresponding to the wastewater treatment example. The complete POMDP is presented in Table 2. The model encompasses six states ($\{S_0, \ldots, S_5\}$). Each node in Fig. 1 represents a state, while the edges represent the transitions, labelled by action name and probability. When several actions can be executed from a state, we add the list and their probabilities to the transition label. As discussed earlier, we use observations to determine the current state of the system. Both states and observations are defined in Table 3. The state representation in our system should be understood as a pendulum between unclean and clean water. We use levels of impurities (from L_5 to L_1). In the table, BDO matter stands for bio-degradable organic matter.

We use the steps in Algorithm 2 to simulate Algorithm 1. Algorithm 2 inputs belief probabilities, a POMDP, an exploration depth, a number of iterations and returns the history along the simulation. The algorithm initialises the history on line 1. It performs the search on lines 2 – 11 and returns the history. During the search, when the depth is negative or equal to zero, the returned history maps the first action in the action space and the corresponding observation. On the contrary, it iterates over each reachable state and approximately determines its optimal action. It invokes the *get_next_reachable_state*() function to generate the next state. Note that the function *construct_MC_Tree*() constructs a tree structure, which the MCTS algorithm uses for its simulation and exploration. The tree encompasses a POMDP and additional parameters required for the exploration. Moreover, the function *get_obs*() returns an observation once an action has been executed. Algorithm 2 yields the following result: $\{: S_0 \rightarrow a_0 \dashrightarrow o_1 \rightarrow a_2 \dashrightarrow o_4 \rightarrow a_3 \dashrightarrow o_5 \rightarrow a_5 \dashrightarrow o_7\}$. In short, it starts by removing solid particles from an unclean water sample, then removes sand and stones and bio-degradable organic matter and finally reduces organic matter.

[6] For simplicity we omitted the observations and rewards. Moreover, the probability distributions as well as the rewards used in this example are for illustrative purposes. The real values should emanate from experiments in each specific context.

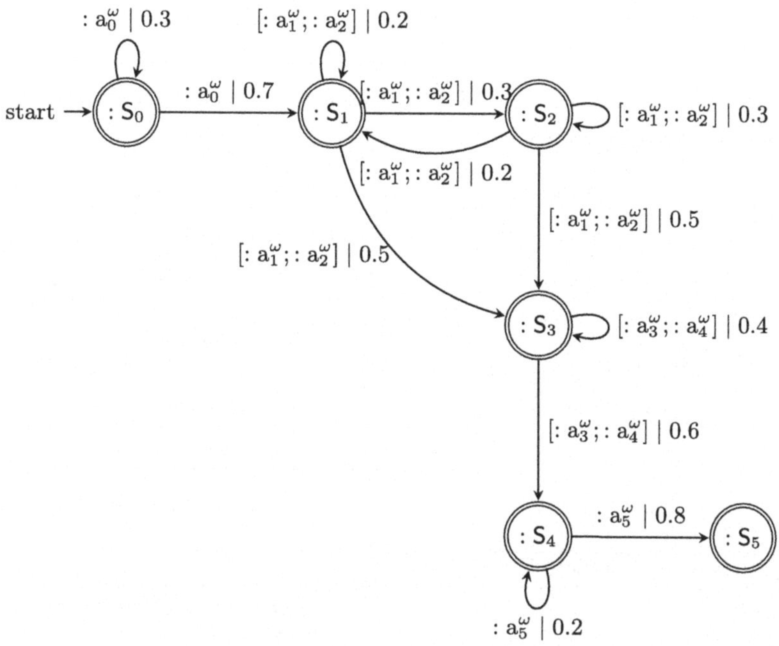

Fig. 1. Simplified Wastewater POMDP

Table 2. Complete POMDP Specification

State	Observation	Action	Reward
: S_0	$(O_0 \mid 1.0)$	: a_0^ω	+1
	$\mathcal{O} \setminus \{O_1\}$		−20
: S_1	$(O_1 \mid 0.5)$	: a_1^ω |: a_2^ω	+2
: S_2	$(O_2 \mid 0.5)$	: a_1^ω |: a_2^ω	+2
	$\mathcal{O} \setminus \{O_1, O_2\}$		−20
: S_3	$(O_3 \mid 0.5)$	: a_3^ω |: a_4^ω	+3
	$(O_4 \mid 0.5)$	: a_3^ω |: a_4^ω	+3
	$\mathcal{O} \setminus \{O_3, O_4\}$		−20
: S_4	$(O_5 \mid 0.5)$	: a_5^ω	+10
	$(O_6 \mid 0.5)$	: a_5^ω	+10
	$\mathcal{O} \setminus \{O_5, O_6\}$		−5

Table 3. State and Observation Descriptions

State	Designation
: S_0	Unclean water L_5
: S_1	Unclean water L_4
: S_2	Unclean water L_3
: S_3	Unclean water L_2
: S_4	Unclean water L_1
: S_5	Clean Water

Observation	Designation
O_0	Has solid particles
O_1	Has floating objects
O_2	Has stones
O_3	Contains BDO matter
O_4	Contains organic matter
O_5	Has floating objects
O_6	Has floating objects
O_7	Appears clean

Algorithm 2: Complete Execution.

```
Input: b                                              // Belief probabilities
Input: P                                              // POMDP
Input: d                                              // Exploration depth
Input: m                                              // Number of iterations
Output: h                                             // History
h ← {}
if d ≤ 0 then
    h ← {(first(P.A), get_obs(first(P.A)))}
else
    π ← construct_MC_tree(P)
    cur_st ← get_next_reachable_state(P, b)
    while cur_st ≠ NIL do
        a* ← search_MCTS(π, cur_st, h, m, d)
        o ← get_obs(a*)
        h ← h ∪ {(a*, o)}
        cur_st ← get_next_reachable_state(P, o, a*)
return h
```

3 Evaluation and Conclusion

The time complexity of MCTS is $\mathcal{O}(\text{mbd})$, with m the number of iterations for simulation, b the tree branching factor, and **d**the search depth. Practically, Fig. 2 presents the performance obtained from our implementation. It shows that the example discussed in Sect. 2 runs on average in 65 ms ($59.7\,\text{ms} \leq \mathsf{T} \leq 72.2\,\text{ms}$) based on 78 samples. Furthermore, the memory used for this execution is less than 200 MB. This indicates that, realistically, the MCTS algorithm can be used to support the automation process of complex tasks and be executed in a timely fashion.

```
BenchmarkTools.Trial: 78 samples with 1 evaluation.
 Range (min … max):  59.680 ms … 74.645 ms  ┊ GC (min … max):  7.99% … 12.01%
 Time  (median):     64.290 ms              ┊ GC (median):    10.47%
 Time  (mean ± σ):   64.667 ms ±  3.006 ms  ┊ GC (mean ± σ):  10.65% ±  2.24%
```

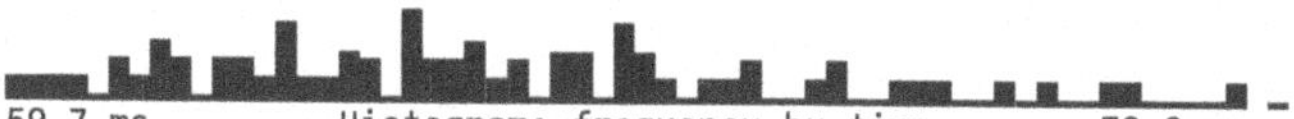


```
Memory estimate: 171.89 MiB, allocs estimate: 102733.
```

Fig. 2. MCTS Benchmark

In the future, we wish to involve multiple agents in the task execution, thus modelling the problem as a decentralised POMDP. Particularly, we wish to study the cooperation strategies among the agents during the execution. Furthermore, we wish to introduce a parallel and distributed version of the MCTS algorithm to enable the agents to execute partial simulations and explorations in parallel.

References

1. Albrecht, S.V., Stone, P.: Autonomous agents modelling other agents: a comprehensive survey and open problems. Artif. Intell. **258**, 66–95 (2018). https://doi.org/10.1016/j.artint.2018.01.002
2. Auger, D.: Multiple tree for partially observable Monte-Carlo tree search. In: Di Chio, C., et al. (eds.) EvoApplications 2011. LNCS, vol. 6624, pp. 53–62. Springer, Heidelberg (2011). https://doi.org/10.1007/978-3-642-20525-5_6
3. Bernstein, D.S., Zilberstein, S., Immerman, N.: The complexity of decentralized control of Markov decision processes. In: Proceedings of the Sixteenth Conference on Uncertainty in Artificial Intelligence, pp. 32–37. UAI'00, Morgan Kaufmann Publishers Inc., San Francisco, CA, USA (2000)
4. Browne, C.B., et al.: A survey of Monte Carlo tree search methods. IEEE Trans. Comput. Intell. AI Games **4**(1), 1–43 (2012). https://doi.org/10.1109/TCIAIG.2012.2186810
5. Chen, D., Zhang, Q., Zhu, Y.: Efficient sequential decision making with large language models (2024). https://arxiv.org/abs/2406.12125
6. Coulom, R.: Efficient selectivity and backup operators in Monte-Carlo tree search. In: van den Herik, H.J., Ciancarini, P., Donkers, H.H.L.M. (eds.) Computers and Games. LNCS, vol. 4630, pp. 72–83. Springer, Berlin, Heidelberg (2006). https://doi.org/10.1007/978-3-540-75538-8_7. http://dblp.uni-trier.de/db/conf/cg/cg2006.html#Coulom06
7. Hauskrecht, M.: Value-function approximations for partially observable Markov decision processes. J. Artif. Int. Res. **13**(1), 33–94 (2000)
8. Kaelbling, L.P., Littman, M.L., Cassandra, A.R.: Planning and acting in partially observable stochastic domains. Artif. Intell. **101**(1–2), 99–134 (1998)
9. Kocsis, L., Szepesvári, C.: Bandit based Monte-Carlo planning. In: Proceedings of the 17th European Conference on Machine Learning, LNAI, pp. 282–293. ECML'06, Springer-Verlag, Berlin, Heidelberg (2006). https://doi.org/10.1007/11871842_29

10. Lee, J., Kim, G.H., Poupart, P., Kim, K.E.: Monte-Carlo tree search for constrained pomdps. In: Proceedings of the 32nd International Conference on Neural Information Processing Systems, pp. 7934–7943. NIPS'18, Curran Associates Inc., Red Hook, NY, USA (2018)
11. Ni, Y., Liu, Z.Q.: Policy iteration for bounded-parameter pomdps. Soft. Comput. **17**(4), 537–548 (2013). https://doi.org/10.1007/s00500-012-0932-3
12. Núñez-Molina, C., Mesejo, P., Fernández-Olivares, J.: Towards a unified framework for sequential decision making (2023). https://arxiv.org/abs/2310.02167
13. A Concise Introduction to Decentralized POMDPs. SIS, Springer, Cham (2016). https://doi.org/10.1007/978-3-319-28929-8_9
14. Pineau, J., Gordon, G., Thrun, S.: Anytime point-based approximations for large pomdps. J. Artif. Int. Res. **27**(1), 335–380 (2006)
15. Ross, S., Pineau, J., Paquet, S., Chaib-draa, B.: Online planning algorithms for pomdps. J. Artif. Int. Res. **32**(1), 663–704 (2008)
16. Silver, D., Veness, J.: Monte-Carlo planning in large pomdps. In: Proceedings of the 24th International Conference on Neural Information Processing Systems, vol. 2, pp. 2164–2172. NIPS'10, Curran Associates Inc., Red Hook, NY, USA (2010)
17. Sunberg, Z., Kochenderfer, M.J.: POMCPOW: an online algorithm for pomdps with continuous state, action, and observation spaces. CoRR **abs/1709.06196** (2017). http://arxiv.org/abs/1709.06196
18. Swiechowski, M., Godlewski, K., Sawicki, B., Mańdziuk, J.: Monte Carlo tree search: a review of recent modifications and applications. Artif. Intell. Rev. **56**(3), 2497–2562 (2022). https://doi.org/10.1007/s10462-022-10228-y
19. Vodopivec, T., Samothrakis, S., Šter, B.: On Monte Carlo tree search and reinforcement learning. J. Artif. Int. Res. **60**(1), 881–936 (2017)
20. Wu, P., Huang, H., Liu, Z.: Online sequential decision-making with unknown delays. In: Proceedings of the ACM Web Conference 2024, pp. 4028–4036. WWW '24, Association for Computing Machinery, New York, NY, USA (2024). https://doi.org/10.1145/3589334.3645388

Author Index

K. Węcel (Ed.): BIS 2025, LNBIP 554, pp. 319–320, 2025.
https://doi.org/10.1007/978-3-031-94193-1